4-11-96

GROUNDWATER RECHARGE and WELLS

A Guide to Aquifer Storage Recovery

R. David G. Pyne

LEWIS PUBLISHERS
Boca Raton   Ann Arbor   London   Tokyo

Library of Congress Cataloging-in-Publication Data

Pyne, R. David G.
    Groundwater recharge and wells : a guide to aquifer storage
recovery / R. David G. Pyne.
        p.   cm.
    Includes bibliographical references and index.
    ISBN 1-56670-097-3
    1. Aquifer storage recovery.   I. Title.
TD404.5.P95   1994
627′.56—dc20
                                                                                    94-27575
                                                                                    CIP

Fight for the Waterhole

*by Frederic Remington symbolizes man's primal need
for water. A sustainable water supply for people of all
nations can defuse competition for this limited resource, the
fight for which is taking on global proportions. Aquifer
storage recovery technology is directed toward this goal.*

LIST OF FIGURES

CONTENTS

Albert Muniz, Deerfield Beach, Florida, has pushed back the technical and regulatory frontiers through his energetic and capable direction of several challenging ASR projects in Florida. Modelling of ASR systems has benefitted from the strong capabilities of John Glass, Peter Kwiatkowski, Al Aikens, Margaret O'Hare and Sumant Gupta. Their valuable work is reflected in these pages.

Through their efforts on several ASR projects, many others within CH2M HILL have helped to move this technology forward, among whom are Mark Lucas, Clifford Bell, Doug Dronfield, Tom Buchanan, Greg Tate, Bryan McDonald, Ross Sproul, Robert Peterson, Dr. Jose Ignacio Garcia-Bengochea, Bart Zeigler, Sean Skehan, Tim Sharp, Phil Waller, Mark McNeal, Mike Micheau, Paul Wallace, Rene Brewster, John McLeod, Paul Thornhill, Larry Amans, Courtney Hemenway, Andrea Aikin, Peter Livingston, Derrik Williams, Fritz Carlson, Jeff Barry, Stuart Brown, Pamela Brody, David Livise and Ken Trotman. Many others are now sharing in the ASR vision and will build upon the foundations laid by those who have gone before.

Outside CH2M HILL, many have contributed through their suggestions, support and encouragement. Through his early grasp of the ASR vision and his continuing attention to details as our shared understanding of the ASR technology evolved, Sam Stone at the Peace River Water Treatment Plant, Florida, has made a valuable and enduring contribution to the field. The American Water Works Association Research Foundation (AWWARF) deserves much appreciation for funding the research into reduction of disinfection byproducts that has been observed to occur during ASR storage, as presented in Chapter 4. The Las Vegas Valley Water District, Nevada, and Thames Water Utilities, England, also shared in funding this important work. The Southwest Florida Water Management District deserves great credit for funding the initial ASR test program at Lake Manatee in 1979, at a time when conventional wisdom was that the project would undoubtedly fail. Without their early financial support, ASR might still be a concept instead of a reality. Similarly, the South Florida Water Management District has grasped the potential value of ASR technology to a region with seasonally available supplies and many water resource challenges. Their early financial support for the Marathon ASR program in the Florida Keys has contributed greatly to definition of potential ASR applications in brackish and seawater aquifers.

Writing this book has required a lot of work, made easier through the efforts of Tammy Nelson, Mark and Joyce Bradley, Michelle Colletti and Gary Mardock, who assisted with the word processing, editing and graphics. Mr Hang Soo-Hoo assisted with library support, unearthing sources from many strange places. I am also grateful to Brian Lewis, whose early and continuing encouragement regarding the writing and publishing of this book provided a timely stimulus. And for her skill in editing the manuscript and preparing it for printing, and also for her humor, Julie Spadaro has earned my gratitude for a job well done.

Most importantly, my family has waited patiently for too long for this book to be finished so that we can get on with our lives. To Emily, thank you for your support and encouragement. To Christopher, yes, it is finally, really, 100% finished.

ACKNOWLEDGMENTS

The author's experience with aquifer storage recovery has evolved from the glimpse of an idea in 1976, to a grand vision during the mid-1980s, and to the widespread and rapidly growing implementation of this vision as of 1994. The experiences and understanding of many people have guided the development of ASR technology during this period. Some have assisted directly with the writing of this book, while many others have played important roles during the development and refinement of ASR technology at many locations.

CH2M HILL Inc., consulting engineers, has provided a strong technical platform, without which it is unlikely that ASR would have developed as successfully and as rapidly as it has. By providing ready accessibility to technical experts in a wide variety of fields, it has been possible to address and rapidly resolve many issues that have arisen along the way, both technical and otherwise. By providing a broad network of office locations and client needs, throughout the United States and also overseas, we have been able to continually redefine ASR technology to meet a rapidly growing number and variety of applications, thereby deepening our understanding of the technology and how it may best be applied to meet these needs.

Through their hard work and creative ideas, many of my friends at CH2M HILL have earned recognition for their valuable contributions to the development of ASR technology. Richard P. Glanzman, Denver, Colorado, has provided continuous enthusiastic encouragement and sage guidance on geochemistry for virtually all ASR projects. Chapter 5, Geochemistry, is his contribution, for which he has achieved widespread recognition. Some of the words are mine but many are his, and the deep understanding upon which they are all based is his.

Kevin Bral, Denver, Colorado, has similarly contributed greatly to the technical success of many ASR systems and has earned recognition for his efforts in many areas, particularly including well and wellhead design, test procedures and the development of the column testing equipment and downhole control valve. His contributions to Chapter 3, Design of ASR Systems, are greatly appreciated. Richard Randall, Phoenix, Arizona, has worked for many years to gain improved understanding of well clogging issues. The current status of his work is included in Chapter 4, Selected ASR Technical Issues. Margaret Ibison, Reston, Virginia, has contributed her special knowledge of coring, core handling and analysis procedures along with her enthusiasm and boundless perseverance. Ken McGill, Philadelphia, Pennsylvania, has persevered through a daunting sequence of technical challenges on a most complex project at Swimming River, New Jersey, emerging successfully and contributing valuable experience to guide many other projects with similar challenges. Dan Wendell, Los Angeles, has contributed his understanding that comes from having helped to develop and operate the early ASR system at Goleta, California. His insightful questions, suggestions and observations have added much. Also in Los Angeles, Terry Foreman has guided the local development of ASR to meet the needs of water users in southern California, contributing greatly to the broadening applications of this technology.

ABOUT THE AUTHOR

David Pyne was born and raised in England, emigrating to the United States in 1957. He received a BS degree in Civil Engineering from Duke University in 1966 and an MSE from the University of Florida in 1967. He is employed by CH2M HILL, consulting engineers, where he directs groundwater technologies, including artificial recharge of groundwater. Since 1976, he has pioneered development of the aquifer storage recovery (ASR) technology for seasonal, long-term and emergency storage of large volumes of water underground in suitable aquifers through wells. The success of these projects and the rapid implementation of ASR technology by water users throughout the United States in recent years has provided the foundation of experience for the writing of this book. During the past 24 years, Mr. Pyne has provided consultant assistance to utility, agricultural and industrial clients and government agencies in the United States and Canada. He has also assisted the United Nations and several countries in Europe, the Middle East, Asia, Africa and the Caribbean with the development and management of water supply systems. He is the author of many papers and has presented short courses as well as expert-witness testimony on water resources and supply issues. He resides in Gainesville, Florida (CH2M HILL, PO Box 147009, Gainesville, FL 32614-7009; Phone: 904-331-2442; Fax: 904-331-5320; CompuServ 74404.2640).

plugging. These dual-purpose wells are called "aquifer storage recovery," or ASR wells. They are designed and operated differently than normal production wells or injection wells. This book presents ASR technology, as it has evolved over the past 25 years in the United States.

During the past six years, many interested in the field of groundwater recharge have urged the writing of this book, so that they and others can more easily grasp the ASR vision and implement this technology to meet their various needs. As time has gone by and new technical issues have been met and resolved, the body of knowledge that comprises ASR technology and experience has evolved and matured. With 20 ASR systems now operational in the United States and about 40 more in various stages of development, it is appropriate to distill the variety of technical and other experiences into this book, a guide to aquifer storage recovery. Following the procedures suggested in these chapters should enhance the likelihood of achieving success with groundwater recharge through wells.

Although ASR is not "high tech," neither is it "low tech."Understanding the issues that have been encountered at other sites, and the steps that have lead to successful resolution of these issues, can provide great help to those considering, planning or implementing new groundwater recharge projects.

ASR is a new, efficient and cost-effective tool for water resources management. Although developed primarily within the United States, it builds upon prior experience, primarily in The Netherlands and Israel, relating to artificial recharge of groundwater through wells. ASR is therefore equally applicable in other countries, many of which have severe water supply challenges.

It is hoped that, by presenting the ASR technology and demonstrating its feasibility and cost-effectiveness to meet global needs for sustainable water development, this book will help to defuse political tensions, improve human welfare and enhance the reliable supply of good quality water at reasonable cost to people around the world.

R. David G. Pyne

PREFACE

Water is power. Control of water is, therefore, the currency of personal, regional and national ambitions. Existing dams, surface reservoirs, major pipelines and pumping stations continue to serve us well, and appeal to that side of us which needs a monument to our achievements. The days are numbered, however, for many major water resources development projects that are in the planning stages. Changing priorities are causing us to reconsider the wisdom of traditional engineering solutions that were quite acceptable just a few years ago. More sophisticated, but less visible monuments are needed. The time has come to take a harder look at storing water below ground in reservoirs provided by nature.

Groundwater levels continue to decline around the world in response to increasing withdrawals to meet the needs of an expanding population. At the same time, surface waters are proving increasingly unreliable to meet these growing needs, despite numerous major and costly programs to store and divert water from a diminishing number of uncontrolled rivers. Intensive urban and agricultural development is draining our lands during times of rainfall, and depleting our groundwater resources at an alarming rate during times of drought. As the global nature of this challenge becomes more clear, we are slowly coming to grips with the need for sustainable water management.

Adequate storage is the key to sustainable water management. We have sufficient water in most cases, however we have difficulty storing it when it is available so that it will also be available when and where we need it. Storage in surface reservoirs is expensive and increasingly perceived as an unacceptable exchange for valued ecosystems. Storage below ground has been limited for a variety of reasons. Among these, the principal constraints have been technical, although political, legal and other constraints have proved significant.

In recent years, development of technology for artificial recharge of aquifers has accelerated. Most of the technical constraints have been addressed and resolved through research and experience at many sites. Systems to recharge aquifers through surface methods such as basins and in-channel structures are functioning reasonably well, however their widespread application is frequently limited by hydrogeologic constraints and the availability of land at reasonable cost. Many areas would benefit from aquifer recharge but can do so only through wells. While many test programs have been conducted, there are few operational well recharge systems around the world.

Recent technical advances and operational experience have demonstrated that well recharge is feasible and cost-effective. While many papers have been written regarding specific recharge projects, the author is aware of only a few papers and books that assemble the technology. This book provides a guide to those who would endeavor to successfully recharge aquifers through wells.

A key element of this technology is the need to control well plugging due to suspended solids in the recharge water, bacterial activity, gases and other causes. Experience has demonstrated that dual-purpose wells, equipped for both recharge and recovery, are best able to achieve recharge objectives while controlling

*To Emily
and
Christopher*

Fight for the Waterhole, by Frederic Remington; The Museum of Fine Arts, Houston; The Hogg Brothers Collection, Gift of Miss Ima Hogg

- Groundwater supplies 97% of the water used in Saudi Arabia, yet water levels are declining at such a rate that groundwater reserves will soon be exhausted, probably within 20 to 50 years.
- About two-thirds of India is underlain by basalt aquifers that supply water to hand dug wells for domestic and agricultural use. Increasing withdrawals are causing these wells to dry up in many areas, creating the need for deep well pumping equipment and accelerating the rate of water level decline.
- Groundwater withdrawals in the Denver Basin of Colorado in the U.S. have caused water levels to decline from near land surface to a depth of almost 275 m (900 ft).
- Wells supplying growing populations in many coastal areas have been lost to saltwater intrusion directly attributable to increasing groundwater withdrawals.
- Loss of wetlands and natural areas to urban development and agricultural production is accelerating the rates of storm runoff and erosion in many countries, creating pronounced increases in the intensity of flooding and droughts in downstream areas, and reducing natural recharge to aquifers.
- In Beijing, China, water levels have been dropping about 1 to 2 m per year and about one-third of the wells have reportedly gone dry [1].
- With a 1992 population of 56 million, Egypt has a renewable water supply averaging only about 82 LPC/day (22 GPC/day). The population of Egypt is doubling about every 28 years [2].

At the time of Christ, world population has been estimated to have been approximately 300 million people [3]. It required 1700 years for the population to double to slightly over 600 million people. By about 1860, a span of 160 years, it had doubled again to 1.2 billion. The population reached 2.4 billion by about 1945, doubling within 85 years. It had doubled again to 4.8 billion by about 1984, 39 years, and is expected to double again by the end of the century, another 16 years. Annual population growth rates for the period 1980 to 1988 averaged about 1.7% worldwide, within a range of 0.3% for Europe to 2.8% for Africa.

Per capita water demands have tended to rise, associated with standards of living that have improved in many parts of the world. However, per capita water supplies have fallen rapidly, associated with increasing population growth. The widening difference between per capita demand and supply represents a growing potential for problems and a growing challenge for water managers. **Figure 1.1** shows the estimated annual world water use between 1900 and 2000. During this period, water use has increased almost ten times to over 5000 km^3/year [4].

Global water supplies are generally believed to be constant. About 40,000 km^3/year constitutes the world's renewable freshwater supply [5].

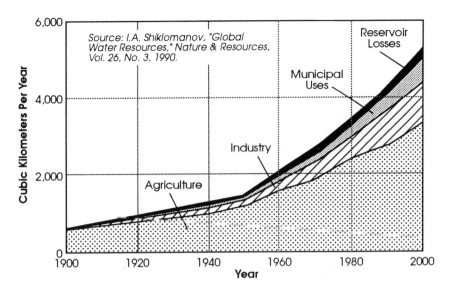

Figure 1.1 Estimated annual world water use, total and by sector, 1900–2000. From Shiklomanov, I. A., Global Water Resources, *Nature and Resources,* Vol. 26, No. 3, 1990. With permission.

It is apparent that water is being utilized more intensively to meet the growing needs of the global population. Examples include flood control and channelization projects, levees and pumping stations, dams, reservoirs, regional irrigation systems, and water transmission pipelines.

These types of major regional water supply facilities were first constructed over 2500 years ago to irrigate the region of Mesopotamia around the Tigris and Euphrates rivers, including such cities as Babylon, Nineveh, and Ur. They have become quite common throughout the world during this century.

Even more intensive water management measures have been implemented relatively recently in a few areas to meet local needs. Some of these include pumped storage projects to meet peak power requirements, deep injection wells to dispose of wastewater and to form salinity intrusion barriers, desalination of brackish water and seawater, reclaimed water irrigation systems, increasingly sophisticated treatment plants to treat water and wastewater to potable standards, and artificial recharge facilities to replenish aquifers.

Artificial recharge is therefore but one of many tools available to achieve more efficient utilization of limited available water supplies. Improvements in artificial recharge technology in recent years have reduced the cost of water supply facilities expansion substantially. As a result, future use of this technology is expected to accelerate.

1.2 ASR: A NEW WATER MANAGEMENT TOOL

Interest in artificial recharge has strengthened in recent years, in response to declining groundwater levels, increased vulnerability of surface water supplies to contamination, environmental opposition to increased reliance upon surface water supplies, and many other reasons. Conventional artificial recharge methods have included surface infiltration systems and injection wells, both of which have technical constraints that have tended to limit their widespread implementation.

Surface recharge systems work well in situations where soils are permeable from ground surface to the water table and where adequate land area is available at reasonable cost to accommodate the recharge facilities. Solids that accumulate at the surface are periodically removed following a series of wet-dry cycles that maintain the long-term infiltration rate. Where low permeability soils are present between ground surface and the water table, or where land availability at reasonable cost is limited, surface recharge may not be viable.

Injection wells tend to plug, requiring periodic redevelopment to maintain their capacity. Since they are usually not equipped with pumps, this is achieved by redeveloping the well using a temporary pump or an air line, assuming the degree of plugging is slight. However, if plugging has been allowed to deteriorate to the point that this is inadequate to clear the well, then it is necessary to use physical scrubbing, acidification, jetting, surging, pumping, disinfection, and other more intensive methods to restore capacity.

Both surface recharge and injection well systems have been utilized to achieve the single limited objective of getting water into the ground. Since the quality of water required for injection well systems to minimize plugging generally has to be much better than that required for surface recharge systems, injection wells have generally been perceived as a higher cost recharge alternative, to be considered only at such time as all possible alternatives for surface recharge have been proven non-feasible or too costly. As a result, there have been relatively few applications of injection well technology to achieve artificial recharge objectives.

The author proposes a broader vision of artificial recharge, one in which the objective is not only to get water into the ground, but also to recover it for a beneficial use **at the same location.** A key element of this broader vision is that the storage zone may contain native water of poor or brackish quality, in addition to freshwater zones previously considered for recharge. With this broader vision and dual-purpose approach, recharge is accomplished with one or more wells, and the same wells are used for recovery of the stored water. Pumps provided in the wells to enable recovery are also used periodically to redevelop the wells, thereby maintaining their injection capacity. Such dual-purpose wells are called **aquifer storage recovery** or **ASR** wells.

This slight shift in approach radically alters the economics of artificial recharge, and is therefore altering the direction of its future development. In particular, operating experience has shown that if water is treated to a level that will avoid rapid plugging of the well, such as meeting potable drinking water standards, the water may be stored and recovered from the well, generally without the need for retreatment other than disinfection. This is true for freshwater storage zones and also for brackish water zones. For potable and other higher quality water uses, the option exists to use the same facilities for both recharge and recovery, without the need for retreatment other than disinfection. For lower quality water uses such as irrigation, the same advantage may possibly apply; however, greater care will be needed to avoid well plugging and also to avoid aquifer contamination (see Chapter 8, Future Directions).

With surface recharge systems, recovery of the water at the same site would require additional cost for construction of wells, for piping and pumps, and perhaps for construction of the associated treatment facilities to meet water quality requirements prior to the ultimate use. Where both surface and well recharge are feasible, well recharge may therefore tend to be more cost-effective in situations where a need exists for the recovered water at the recharge site and where treatment of the water would be required anyway. As discussed subsequently in this chapter, there are many applications that meet this description.

In situations where surface recharge is not feasible, ASR wells will tend to be cost-effective relative to systems that rely upon separate injection and recovery wells, due to the lesser costs of construction and operation for dual-purpose ASR wells. Probably the only applications where single-purpose injection wells are preferable are those where it is desired to maximize blending between stored water and native groundwater, and where aquifer hydraulics or recharge water quality are such that plugging is not an operating constraint. Even in situations where it is desired to use the aquifer to convey water from the point of injection to a distant point of recovery, providing a pump in the well will usually be less costly as a means of maintaining injection capacity than periodically removing all injection piping and redeveloping the well.

The ASR concept therefore represents a significant new development in how we manage water. First and foremost it is an idea, or change in thinking, about how to approach artificial recharge. However, it is also a new technology. The technology is usually not complicated; however, experience suggests that there are several technical and other elements

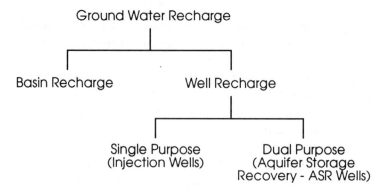

Figure 1.2 Groundwater recharge.

unique to ASR that, if understood, can lead to a successful and cost-effective operation. Similarly, neglecting these elements may contribute to system failure due to plugging, improper well design or operation, poor location, geochemical problems, inappropriate regulatory actions, and other consequences.

1.3 WHAT IS ASR?

Aquifer Storage Recovery may be defined as the storage of water in a suitable aquifer through a well during times when water is available, and recovery of the water from the same well during times when it is needed. The concept is simple; that this ASR technology has only been implemented fairly recently by the water industry reflects changing needs and also successful technical resolution of several issues that previously hindered recharge well performance.

In the wide range of activities that comprise artificial recharge, ASR is a thriving newcomer, as shown in **Figure 1.2.** While most recharge still occurs through surface methods such as basins and river channels, an increasing amount of recharge occurs through wells.

Historically, recharge or injection wells have tended to plug, requiring periodic redevelopment. The relatively high cost of well redevelopment and the frequency with which this is often required have combined to effectively preclude cost-effective well recharge activities at many sites. As a recent technological development, ASR resolves the inherent operational drawbacks of single-purpose injection wells by equipping each well with a pump and operating it in a dual-purpose mode for both recharge and recovery. The pump used for recovery of the stored water is also used periodically to redevelop the well and thereby maintain its capacity. No additional facilities are required for recovery of the stored water. The ASR approach to aquifer recharge overcomes the plugging associated with most injection wells, the hydraulic limitations of many surface recharge sites, and the large land area requirements of these sites. Because of these and other advantages, ASR is being implemented at an accelerating rate in the U.S. and overseas.

Typically, the same volume of water stored in an ASR well can be recovered. In some situations, it may be possible to recover a greater volume than the amount injected, relying upon mixing between the stored water and the surrounding native water in the aquifer to provide a blend of acceptable quality. In other situations, leaving a small percentage of the stored water in the ground may be desirable to restore depleted groundwater reserves; to address concerns regarding potential geochemical plugging; to form or maintain a buffer zone between stored water and surrounding brackish or poor quality native water; or to build up a reserve for future recovery during droughts, emergencies, or anticipated times of higher demand.

Where storage zone characteristics are more than usually challenging, it may not be possible to achieve 100% recovery efficiency. This may be true in storage zones with very high transmissivity and also poor water quality, zones with inadequate confinement separating them from adjacent zones with poor water quality, and zones with substantial groundwater velocity. In such situations, a careful evaluation of recovery efficiency and cost-effectiveness can lead to the right decisions regarding ASR feasibility and design.

In most instances, the water quality required for recharge must be rather good. It is no coincidence that all of the existing (1994) operational ASR systems store water that meets potable standards. However, it is important to distinguish between regulatory requirements for potable water quality and technical requirements for ASR well recharge. While meeting regulatory potable standards is probably suitable for ASR recharge, experience suggests that, in some cases, additional treatment may be necessary, such as reduction of total suspended solids or pH adjustment. Conversely, for some applications recharge water quality could probably fall short of meeting potable standards while still minimizing well plugging, assuming regulatory concerns regarding potential aquifer contamination can be addressed satisfactorily. Potable water standards provide a reasonable reference point against which to evaluate recharge and recovered water quality from ASR wells.

ASR recovered water usually requires no retreatment following recovery, other than disinfection for potable uses. In a few situations, pH adjustment of the recovered water may also be necessary.

Suitable storage zones for ASR may be confined, semi-confined, or unconfined (water table) aquifers; however, most experience to date is with semi-confined aquifers, some of which have been partially dewatered due to overdevelopment. Storage in unconfined aquifers can also be feasible; however, several factors adversely impact feasibility:

1. Groundwater velocity is usually higher in unconfined aquifers, a result of which is that the stored water bubble will tend to move away from the well, reducing recovery efficiency. Where the distance that the bubble moves between the time of recharge and the time of recovery exceeds the diameter of the bubble, it may not be possible to recover the stored water. Obviously this is of greater concern in situations where native water quality is not as high as recharge water quality.
2. The rate and duration of recharge may be limited by buildup of a mound in the water table that intersects either ground surface or the invert of local drainage systems, causing loss of the stored water.
3. Overlying land use in the vicinity of the ASR well may be inconsistent with the need for protecting the quality of stored water for its anticipated ultimate use following recovery.

These three considerations are of particular concern for unconfined aquifer ASR applications. Where the water table is deep and relatively flat, and overlying land use is not likely to contaminate the stored water, ASR can be viable and cost-effective. It is an alternative to surface recharge methods where land costs or availability preclude surface recharge, or where it is desired to recover the stored water at the site for irrigation purposes.

Volumes of water stored in ASR wells depend upon several factors, such as well yield, variability in water supply, and variability in water demand. Typical volumes for individual wells range between 0.04 and 2 Mm^3 (10 to 500 MG; 31 to 1535 acre ft). Where appropriate, multiple ASR wells are operated as a wellfield, the capacity of which meets system needs or opportunities during either recharge or recovery periods. As discussed in the next section of this chapter, there are potentially many different applications of ASR technology; however, all store sufficient volumes during times when water is available and recover it from the same well(s) when needed. The storage time is usually seasonal but may be diurnal, long-term, or for emergencies.

Aggregate storage volumes for ASR wellfields may be large. The largest three ASR wellfields currently in operation have design storage volumes in excess of 4 Mm^3 (1 BG; 3000 acre ft) and are able to augment seasonal peak water supplies with treated drinking water at rates of 30 to 380 ML/day (8 to 100 MG/day).

1.4 HISTORICAL DEVELOPMENT OF ASR

The old saying that "there is nothing new under the sun" is also true of ASR. The Kara Kum Plain is a 311,000 km^2 (120,000 miles^2) desert located on the southeast shore of the Caspian Sea in Turkmenistan, and is characterized by black sand, alluvial plains interspersed by sand dunes. Rainfall is usually less than 100 mm (4 inches). During the past several hundred years, the nomads of this area found that recharge during infrequent rainfall in the area occurred only beneath the sand dunes, while a shallow clay-silt layer and transpiration from desert vegetation prevented recharge in other areas. The groundwater occurs under water table conditions but is quite brackish except under the sand dunes. To ensure water supplies, the nomads dug long trenches, extending radially from the sand dunes and with lengths of up to 2 to 3 km/km^2 of tributary area. They were graded to convey intercepted surface runoff to a central pit excavated in the dunes. A system of hand-dug wells, cased with locally available brush wood woven with grass and camel wool, was then constructed surrounding the central pit, to augment recharge during rainfall events and also to supply water. Outer wells were utilized to provide poorer quality water for livestock, while central wells were utilized to meet potable water needs. With depletion of the stored water between rainfall events, outer wells would be abandoned as they became too salty, and central wells would be partially filled in to reduce upwelling of salty water from below, thereby skimming the residual freshwater in the stored water bubble. Some of these underground reservoirs were utilized routinely while others were reserved for use during severe droughts. Because of the highly turbid recharge water, the wells required cleaning every year [6].

It is perhaps ironic that the term "ASR," first utilized by the author in 1983 to describe the process of underground storage of water using dual-purpose wells, is also a word that means "capture" in Arabic. Certainly, for the nomads of the Kara Kum Plain, capturing the limited available rainfall was absolutely essential for their survival.

In western India, a tribal community has been practicing artificial recharge for several centuries to obtain drinking water for its members and livestock. The primary occupation of the tribal people is cattle breeding and selling milk products. The area is known as Banni and is located in the northern part of the Kutch District of Gujarat State in western India. The area is a raised tidal flat, about 5 m above sea level, and covers about 700 km^2. Annual rainfall is only 15–20 cm, most of which occurs during the monsoon season from July to September. During the rainy season, runoff from a large area is collected in a village tank through several collection channels. The tank is about one hectare in area and about 2 m deep, excavated in fine sand and clay. The quality of groundwater is brackish, but the recharge from the tank displaces the native groundwater below the tank bed and its vicinity. When the tank dries up by March, dug wells about 1 m in diameter and 2–3 m deep are excavated in the tank bed to recover the percolated runoff. During the summer months, it is necessary to dig several such wells as each well can only be used for about two weeks because of the less permeable strata. Dug wells that provide drinking water supply for people are located in the central portion of the tank while those for cattle are located near the margins [7].

Much has been written regarding artificial recharge experience in many countries, primarily pertaining to surface recharge projects. Recharge through wells has received less attention, but is more pertinent to this book in that it provides a basis for the subsequent development of ASR. Accordingly, a brief overview of well recharge operational experience in the U.S. and overseas is perhaps helpful. The term "well" is intended to be synonymous with "borehole" and "tubewell," as used in other English-speaking countries. The term "recharge well" covers all recharge through wells, whether by gravity or pumping, that is intended primarily to replenish groundwater supplies and meet drinking water needs. This includes ASR wells, injection wells for potable aquifer recharge and for salinity intrusion control, and reclaimed water injection wells that are used to replenish potable water supplies. It excludes those wells operated for disposal of wastewater or brine, drainage wells, air-conditioning return flows, and hazardous waste cleanup re-injection wells. It also excludes recharge through pits and shafts. Some specific well recharge projects are presented in greater detail in Chapter 9, Selected Case Studies.

United States

Many well recharge studies and field investigations have been conducted during the past few decades; however, few projects are currently operational. **Table 1.1** includes a list of 24 known recharge well projects, all of which are currently operational. Of these, 20 are ASR projects, all of which have become operational since 1968. **Figure 1.3** shows the location of ASR facilities in the U.S. as of 1994, including systems in operation and others in various stages of development.

Canada

The Mannheim aquifer recharge program at Kitchener, Ontario, is one of two known recharge well projects in Canada. Water from the Grand River is being diverted and treated to meet local water demands. The water treatment plant has a capacity of up to 16 migd (73 ML/day; 19 MG/day), exceeding current water requirements. During low demand winter and spring months, some of this water will be recharged through wells into the

TABLE 1.1 OPERATIONAL RECHARGE WELL SYSTEMS IN THE UNITED STATES (1994)

Location	Year Operation Began	Storage Zone	Application
Orange County, California	1950s	Sand	Salinity intrusion barrier
Los Angeles County, California	1950s	Sand	Salinity intrusion barrier
El Paso, Texas	1986	Sand	Reclaimed water injection
Gainesville, Florida	1978	Limestone	Reclaimed water injection
Wildwood, New Jersey	1968	Sand	ASR
Gordons Corner, New Jersey	1972	Clayey sand	ASR
Goleta, California	1978	Silty, clayey sand	ASR
Manatee, Florida	1983	Limestone	ASR
Peace River, Florida	1985	Limestone	ASR
Cocoa, Florida	1987	Limestone	ASR
Buell-Red Prairie, Oregon	1988	Basalt	Recharge
Las Vegas, Nevada	1988	Valley fill	ASR
Port Malabar, Florida	1989	Limestone	ASR
Oxnard, California	1989	Sand	ASR
Chesapeake, Virginia	1990	Sand	ASR
Kerrville, Texas	1991	Sandstone	ASR
N. Las Vegas, Nevada	1991	Valley fill	ASR
Seattle, Washington	1992	Glacial drift	ASR
Calleguas, California	1992	Sand	ASR
Pasadena, California	1992	Sand	ASR
Highlands Ranch, Colorado	1993	Sand	ASR
Swimming River, New Jersey	1993	Clayey sand	ASR
Boynton Beach, Florida	1993	Limestone	ASR
Murray Avenue, New Jersey	1994	Clayey sand	ASR
Marathon, Florida	1994	Sand	ASR

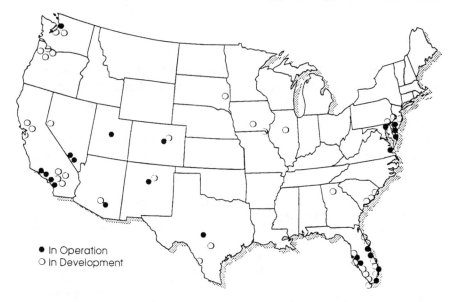

Figure 1.3 Location of ASR systems in the United States, 1994.

Mannheim glacial drift aquifer and possibly into an underlying brackish aquifer. During peak demand summer months, this water will be withdrawn to supplement the capacity of the regional water system. Major intake, pipeline, and treatment facilities have been constructed and are in operation. Construction and field testing of new ASR wells is expected to begin during 1994, and the recharge system should become operational within a few years to help meet peak water demands. Expected recovery capacity is about 20 migd (91 ML/day; 24 MG/day) to meet projected demands in 2006, increasing to 37 migd (168 ML/day; 44 MG/day) by the year 2036.

A second well recharge system in Canada is under development in Saskatchewan, to improve the quality of groundwater available for water supply to rural residents, using recharge of snowmelt or other high quality surface water through existing production wells.

Israel

Since about 1956, artificial recharge through wells has been an important element of the National Water System for Israel. Most of the long-term operating experience has been in the sandstone aquifer of the coastal plain and the limestone-dolomite aquifer of central Israel, although some investigations have also been conducted in the basalt aquifer of lower Galilee. Recharge has occurred primarily through dual-purpose injection/production wells (ASR wells), although single purpose recharge wells, recharge basins, and abandoned quarries are also utilized. The water source is primarily chlorinated water from Lake Kinneret, although storm runoff and groundwater from a limestone aquifer are also recharged. Radioactive tracers have been utilized to determine movement and mixing of the recharge water with the native groundwater. Annual recharge volumes in excess of 80 Mm³ (21 BG; 64,000 acre ft) have been achieved, utilizing over 100 wells and surface recharge facilities.

England

Thames Water Utilities is a private water company serving customers in London and the Thames valley. During winter months, a portion of the flow available from the River Thames and also the River Lee is treated and injected into several wells in north London. The storage zone is a partially dewatered confined aquifer composed of chalk overlain by sands. The recharge water is primarily intended for drought storage, to help meet demands during periods of low streamflow that are assumed to occur approximately every 8 years. The water is recovered when needed, and is retreated at the Coppermills Water Treatment Plant. Initial testing began during the 1950s, and the wellfield has recently been expanded. Recovery rates up to 90 ML/day (24 MG/day) are planned.

Well recharge has been tested at several other sites in England; however, none of these are believed to be currently operational.

The Netherlands

Extensive work relating to well recharge has been conducted in The Netherlands, particularly in the coastal dunes area near Amsterdam. For years, surface basins have been utilized to recharge treated water from the River Rhine and the River Meuse into the surficial sand aquifer. Wells are then utilized to recover the water for municipal water supply purposes. Environmental and land use constraints are forcing increased reliance upon wells (instead of basins) for recharge. Another important factor contributing to the development of well recharge technology in The Netherlands has been the need to reinject water pumped out of the ground during dewatering operations for building construction. As a result, considerable work has been conducted relating to well recharge, water quality effects, clogging mechanisms, and water facilities design.

As of 1994, two recharge well systems are operational and two more are under construction. Several other sites have been tested during the past few decades. In operation since 1990, the two operational recharge well systems each have recharge rates of about 4 MCM/year. Total annual recharge from both wells and surface systems in The Netherlands was about 180 MCM as of 1990 [8].

Australia

Farmers in the vicinity of Adelaide use irrigation supply wells to recharge their brackish aquifer with seasonally available fresh surface water. This water is then recovered to augment irrigation supplies during dry months. Recently investigations have been initiated to assess the feasibility of recharging urban runoff through wells in the Adelaide area, following pretreatment in natural swales and detention ponds. If technical feasibility and regulatory viability can be confirmed, this practice would offset increasing salinity of local aquifers, attributable to reduced natural recharge in this urban area.

Many other well recharge projects are in development or operation that do not appear on this list. A review of literature and other sources suggests well recharge activity in Italy, Spain, Kuwait, India, Japan, and Iran. However the list provides some idea of the global level of activity related to recharge wells and, in particular, the principal current locations of well recharge and ASR activity.

Prior to 1970 in the U.S., artificial recharge was little utilized and was accomplished mostly by surface recharge methods, using recharge basins or river channels in pervious soils to convey water into aquifer storage. Several experiments with injection wells were undertaken prior to 1968, among which those by the United States Geological Survey (USGS) are perhaps best documented. Cederstrom [9] conducted early injection and recovery experiments at Camp Peary, Virginia, in 1946, in which treated water was stored in a brackish aquifer and was recovered from the same well during several test cycles. Similar USGS investigations were conducted at several other sites.

A common denominator of these early investigations is that few of them were carried forward into continued operation by the utility systems and other local water management agencies. Reasons for this vary, reflecting site-specific challenges, needs and priorities. Several of the USGS programs were quite successful from a technical viewpoint. It may be inferred that, prior to 1970, water management needs in the U.S. were adequately met with conventional technology, including wellfields, dams, reservoirs, and river diversions. Environmental constraints had yet to achieve the impact upon water management policies that became evident during subsequent decades. Artificial recharge was in its infancy as a viable water management tool, and plugging problems with some recharge well projects tended to support the conclusion that recharge well cost-effectiveness was questionable when compared to other available alternatives.

That is not to imply that work prior to 1970 was of lesser value. The USGS has compiled two bibliographies of artificial recharge [10,11] that list numerous references on the subject. In 1973, the U.S. Department of the Interior, Office of Water Resources Research published a bibliography on the same subject [12]. The technical lessons learned in these investigations provided a foundation of experience, including both successes and problems, upon which later projects might build. A review of these bibliographies provides a rather complete perspective on artificial recharge in the U.S. prior to 1973. More recently the University of Oklahoma compiled an updated bibliography on artificial recharge in the U.S. through 1985 [13]. Some international projects are also referenced.

The widely acknowledged leading area for artificial recharge in the U.S. is southern California, which had a 1993 population of about 16 million and local average precipitation in the Los Angeles area of less than 500 mm (20 inches). The high cost of several imported water sources, environmental and other constraints upon their availability, and their vulnerability to emergency loss during earthquakes has provided a continuing incentive to make more efficient use of locally available water supplies. To control saltwater intrusion that became a serious problem during the 1940s, the Orange County Water District and Los Angeles County Department of Public Works began operation of a series of salinity intrusion barriers, a line of injection wells currently numbering over 200 wells and recharged with over 170 ML/day (45 MG/day; 50,000 acre ft/year) of reclaimed water from Water Factory 21 and from imported surface water. Portions of this barrier have remained in continuous operation since 1956, while new sections were added during the 1960s and 1970s. The barriers have proven effective in achieving the original objective of controlling saltwater intrusion. About 15% of the water flows seaward, while the balance flows landward and recharges the local aquifer system. Other recharge activities in southern California prior to 1970 were associated with surface recharge through basins and river channels.

About 1970, water management in the U.S. began taking a new direction in response to growing recognition that continued construction of dams and major water conveyance schemes were increasingly unacceptable for environmental reasons. Water management was also incapable, in some areas, of keeping up with rising water demands by agriculture, industry, and people. This opened the way for more serious and widespread consideration of artificial recharge as a means of augmenting the yield of limited available water supply sources. Several different areas within the U.S. then began considering artificial recharge as a water supply alternative deserving consideration. However, many had conditions unsuited to surface recharge methods. The occurrence of hardpan, clay, or other low permeability confining layers between the ground surface and the water table or potentiometric surface in the aquifer frequently provided a substantial constraint upon effective, long-term recharge rates and program cost-effectiveness. In other locations, including southern California, hydrogeologic characteristics were suitable but urban development encroached upon river channels and other potential surface recharge expansion areas, raising land values to the point that further acquisition for recharge purposes was not cost-effective. In such areas, recharge could only be achieved through injection wells. As discussed earlier, recharge experience with injection wells was quite limited, partly reflecting recurring technical problems with well plugging due to a variety of physical and geochemical causes.

Outside the water utility field, injection well technology advanced primarily through the needs of the petroleum industry, which increasingly relied upon brine reinjection and water injection for secondary recovery operations in order to produce oil. Wellhead filtration was required in order to avoid plugging of injection wells with particulates. In addition, increasingly sophisticated chemical pretreatment techniques were developed to avoid geochemical plugging of the aquifer receiving the injected water and to achieve other objectives related to secondary recovery operations. Thus, by 1970, alternatives had been developed and applied in the petroleum industry to address some of the reasons for well plugging that had previously hindered implementation of injection well technology for artificial recharge within the water utility field.

Wildwood, NJ was the first utility system (1968) to begin long-term operation of a wellfield for seasonal storage and recovery of treated drinking water. The purpose of this system was to meet short-term peak demands during summer months with water stored in the same wells during the remainder of the year. Four ASR wells have been constructed at Wildwood. This was followed in 1971 by a similar system at Gordons Corner, NJ. Both of these systems are still in operation as of 1994. These are the oldest known U.S. water systems operating in what we now refer to as an "aquifer storage recovery" operating mode.

During the early 1970s, the USGS continued with investigation of injection well recharge and recovery at two sites. One, at Hialeah, FL, obtained useful data on recovery efficiencies associated with cyclic storage and recovery of freshwater from a shallow aquifer in a deeper, brackish limestone aquifer [14]. At Norfolk, VA, Brown and Silvey [15] investigated similar cyclic storage in a brackish, clayey-sand aquifer, experiencing serious well plugging which they attributed to geochemical reactions. However, they were able to arrest further loss of well production and injection capacity through treatment of the well with calcium chloride, prior to termination of the investigation. Neither of these investigations was carried forward into full operation by the participating water utilities at the time, although ASR development is now underway in the vicinity of each of these sites.

Little further operational development of the ASR concept occurred until 1978, the year in which the Southwest Florida Water Management District and Manatee County, FL, initiated an ASR project to seasonally store treated drinking water in an artesian, limestone aquifer. Also in that year, the Goleta Water District in southern California began a program to recharge a local aquifer with treated drinking water using 18 unused irrigation wells and 9 wells equipped with pumps. The latter were to be available for use to redevelop the wells periodically and also recover the stored water when required to meet system demands, augmenting yield of the Cachuma reservoir during droughts. Soon after initiation of the Goleta ASR system, a prolonged drought precluded water availability to recharge the system for several years.

In Florida, meanwhile, the Manatee County project had proved to be successful and cost-effective as a water supply alternative. Other Florida water utility systems began what by then were termed "ASR" programs. These included the Peace River project and then others for the City of Cocoa and Port Malabar, all of which successfully stored treated drinking water in brackish limestone aquifers and were carried forward into full operation. The utility of ASR as a viable, cost-effective, and operationally acceptable water supply alternative was therefore established in New Jersey by 1972 and in Florida by 1988, when all four of these Florida ASR systems were in operation, three of which had already been expanded or had expansion plans.

It remained, then, to determine whether ASR operational success was limited to the New Jersey coastal plain deposits and certain Florida lime- stone aquifers. These could be characterized as karst, with moderate to high transmissivity, and with total dissolved solids concentrations less than 1300 mg/L in the native groundwater. Alternatively, ASR might have broader applicability.

Based upon the Florida and New Jersey successes with ASR, new projects were initiated in several states that would eventually expand the acceptance of ASR as a widely applicable water supply alternative. ASR projects were initiated in Seattle, WA; Kerrville, TX; Chesapeake, VA; Marathon, FL; Calleguas, CA; Swimming River, NJ; Highlands Ranch, CO; and Tucson, AZ; and were soon followed by many others.

As of October 1994, 20 ASR projects were operational in the U.S., compared to three in 1983. These are listed in Table 1.1. All are storing treated drinking water underground in suitable aquifers through wells, with recovery to meet a growing variety of needs. About 40 additional projects are currently in various stages of investigation, design, permitting, construction, testing, or operational startup. The future for ASR technol- ogy and implementation appears bright, with new technology develop- ments, implementation in other countries, and broader applications for storage of water from different sources such as reclaimed water and agricultural supplies. These are discussed in Chapter 8, Future Directions.

1.5 ASR APPLICATIONS TO MEET WATER MANAGEMENT NEEDS

Although most ASR applications are for seasonal, long-term, or emergency storage of drinking water, many other applications have been considered or implemented at ASR sites across the U.S. and overseas. An ASR system can usually be designed and operated to meet a primary objective and one or more secondary objectives. However, such a system must be carefully planned to ensure that facilities are situated, designed, permitted, and operated in such a way as to achieve these multiple objectives. Twenty-two ASR applications that can be used to guide future planning at potential ASR sites are now briefly described.

1. **Seasonal Storage.** Water is stored during wet months, or months when it is available, and is recovered during dry months, or months when it is needed. Where water is plentiful, it may be stored during times when quality is best and recovered during times of poor quality. The duration of seasonal recovery periods may be several days to several months. Storage zones can be confined, semi-confined, or unconfined aquifers containing fresh, brackish, or saltwater.
2. **Long-term Storage.** Water is stored during wet years, or during years when new supply, treatment, and distribution facilities have spare capacity, and is recovered during drought years, or years when the capacity of existing facilities is inadequate to meet system demand. This type of storage is sometimes referred to as "water banking."
3. **Emergency Storage.** Water is stored, when available, to provide an emergency supply or strategic reserve to meet demands when the primary source of supply is unavailable, whether due to accidental loss, contamination, warfare, or natural disaster. This type of storage is particularly appropriate for water systems that rely heavily on a single source and a long transmission pipeline.
4. **Disinfection Byproducts (DBP) Reduction.** ASR storage can reduce the concentrations of DBPs, such as trihalomethanes (THMs) and haloacetic acids (HAAs), and also their formation potential. For some water utilities, an ASR system providing underground treatment may therefore be an inexpensive method of meeting pending standards for THM and HAA concentrations in drinking water.
5. **Restore Groundwater Levels.** Continuing trends of water level decline can be reversed by incorporating an ASR system into a regional water management plan. A small percentage of the stored water can be left in the aquifer each year, or increased storage during wet years can be accumulated, to eventually bring water levels to within a target range of elevations.

6. **Reduce Subsidence.** By restoring groundwater levels, ASR systems can help reduce subsidence in areas where it is occurring because of continued water level decline.
7. **Maintain Distribution System Pressure.** Aquifer storage at those locations within a utility distribution system that experience seasonal low pressures can help to maintain these pressures by recovery during peak demand months. Such locations are frequently at the ends of long transmission or distribution pipelines that are undersized to handle existing or projected flows. ASR location in conjunction with small elevated or ground storage facilities can be a cost-effective means of meeting seasonal peak demands at adequate pressures.
8. **Maintain Distribution System Flow.** Aquifer storage at those locations within a utility distribution system that experience seasonal low flows can help to maintain adequate disinfectant residuals and other indicators of water quality. This is probably an alternative to flushing pipelines to waste.
9. **Improve Water Quality.** ASR systems can provide many water quality improvements at different sites, including pH stabilization or adjustment, THM and HAA reduction, iron and manganese reduction, hydrogen sulfide reduction, possible arsenic reduction, blending with native groundwater, and softening. Nutrient and coliform reduction may also occur where these constituents are present in the recharge water.
10. **Prevent Saltwater Intrusion.** Placing ASR wells in a line parallel to saltwater-intruded portions of an aquifer can prevent further movement of the saltwater intrusion front, while also meeting seasonal peak demands.
11. **Reduce Environmental Effects of Streamflow Diversions.** The reliable safe yield of surface water supplies has historically been established according to allowable diversions during dry weather and low flow months in order to protect aquatic and terrestrial ecosystems while maintaining downstream water quality. ASR systems primarily divert water during high flow months when the percentage of streamflow that is diverted is frequently small. This reduces the environmental effects of streamflow diversions and thereby facilitates environmentally sound use of surface water sources.
12. **Agricultural Water Supply.** Seasonal storage of water for agricultural irrigation purposes is possible in many areas where aquifer hydraulic characteristics are such that high yield wells can be developed for recharge and recovery purposes. Fresh or brackish aquifers are potentially useful for such purposes. Regulatory and technical issues must be addressed where recharge water quality may not meet all potable standards.
13. **Nutrient Reduction in Agricultural Runoff.** Nutrients are frequently present in agricultural runoff, causing eutrophication of lakes and reservoirs. Storage of agricultural runoff can reduce nitrogen concentrations through bacterial denitrification. In addition, some aquifers can reduce phosphorus concentrations through physical-chemical and bacteriological mechanisms.
14. **Enhance Wellfield Production.** Wellfields are usually designed and operated to produce water at rates within their long-term safe, sustained yield. When these same wellfields are converted to an ASR mode, it is frequently

possible to produce water at higher rates during peak demand months, counting on artificial recharge during off-peak months to restore water levels before the following peak season.

15. **Defer Expansion of Water Facilities.** Water system components with ASR are sized differently than those without ASR. In particular, it is frequently possible to make more efficient use of existing investment in treatment and conveyance capacity by operating these facilities at full capacity throughout the year and throughout the life of the facility, except for scheduled maintenance periods. Excess treated water is stored for later recovery. With this approach, expansion of water facilities can be deferred and downsized, with substantial cost savings.

16. **Compensate for Surface Salinity Barrier Leakage Losses.** In south Florida and possibly other areas, salinity barriers are located on major drainage channels discharging to saltwater. Leakage around these structures through the adjacent shallow aquifer during drought periods can be substantial. With ASR, wells would be located adjacent to these barriers, recharging water into deep brackish aquifers during wet months. Stored water would be recovered during drought months to compensate for leakage losses.

17. **Reclaimed Water Storage for Reuse.** High quality reclaimed water may be stored seasonally in brackish aquifers for recovery to meet irrigation demands. This eliminates the need for expensive aboveground storage that is often required for those periods when the seasonal demand for irrigation water is reduced. In concept, the same principle could be used for reclaimed water storage in potable aquifers. The stored reclaimed water would be hydraulically controlled within a small radius around each ASR well, rather than dispersing freely in the storage zone. It would be recovered from the ASR well, having undergone some improvement in quality during seasonal aquifer storage.

18. **Soil Aquifer Treatment.** Aerobic and anaerobic bacterial treatment processes occur in both the saturated and unsaturated zones of an aquifer system. In addition, physical-chemical processes are effective in treating water stored in ASR wells. While much remains to be learned regarding soil aquifer treatment in ASR wells, this appears to be a beneficial application of ASR technology.

19. **Stabilize Aggressive Water.** Aggressive water is frequently treated with calcium carbonate to achieve stability of product water from water treatment plants. In limestone storage zones, stabilization can also be achieved at lower cost during aquifer storage.

20. **Hydraulic Control of Contaminant Plumes.** In portions of aquifers that are threatened by movement of contamination plumes, it is sometimes possible to control movement of these plumes through the appropriate use of injection or recovery wells. However, use of ASR wells can also achieve seasonal production from these aquifers while controlling movement of the plumes. With such applications, legal liability issues must be addressed carefully.

21. **Diurnal Storage.** ASR wells have been used in some areas to store water at night for recovery during the day, in situations where daytime demands exceed supply capacity.
22. **Fish Hatchery Temperature Control.** Seasonal variations in source water temperature can be used to advantage by recovering and blending ASR water to meet temperature control requirements.

Other applications of ASR technology may become evident in future years, supplementing the above list. Each of these applications entails associated hydraulic, hydrogeologic, and water quality requirements for ASR system development and operation.

1.6 OBJECTIVES OF THIS BOOK

The purpose of this book is to present the author's vision of ASR as a powerful and cost-effective water management tool that can alleviate growing water supply problems around the world. If the vision can be conveyed clearly, then those who choose to follow along this path will be able to make reasonable judgments and decisions as they encounter different circumstances, needs, and opportunities at new ASR sites.

The book endeavors also to provide useful guidelines, procedures, design concepts, and other pertinent information to guide those who may be interested in establishing new ASR systems. With a clear understanding of the ASR vision and a grasp of the procedures that have proven sufficient to ensure ASR success at many sites, the reader will hopefully be well prepared to proceed with a new ASR program.

Chapter 1 presents the ASR concept as a relatively new global water management tool with many different potential applications. The historical development of ASR is also discussed. A proven, successful approach for ASR system development is presented in Chapter 2, including a series of development phases leading to system operation and expansion. Chapter 3 discusses the design of ASR wells, wellhead facilities, and wellfields. Several design features are unique to ASR and are highlighted in this chapter. Design considerations that apply to all wells and are not unique to ASR are generally not covered. An understanding of many technical issues unique to ASR constitutes the body of knowledge comprising ASR technology. These issues are discussed in detail in Chapter 4. Chapter 5 presents a practical guide for understanding ASR geochemistry issues. Geochemistry is a very important technical element of the ASR development program at most sites. Other non-technical issues are equally as important to ASR success, such as economics, water rate impacts, legal and regulatory issues, environmental impact, and public involvement. These are discussed in Chapter 6. Chapter 7 presents alternative ASR applications for storage of water from sources such as reclaimed water and untreated or partially treated surface water. Chapter 8 discusses future directions for ASR, including technical and regulatory developments that can be reasonably foreseen, and also expected global applications. Chapter 9 includes several case studies selected from among existing operational ASR systems, or those that are in development. These are presented to illustrate the range of potential applications, the successes, and the problems that have arisen at these sites, and how these problems were resolved.

As the first book to be published on the subject of aquifer storage recovery, it is hoped that this will be perceived as a useful and complete reference to guide professionals, water managers, water users and others with a common interest in achieving more efficient, sustainable, and cost-effective utilization of our global water resources.

CHAPTER 2

ASR Program Development

The winds blow forth; to earth the quivering lightnings fall,
The plants shoot up; with moisture streams the realm of light.
For all the world abundant nourishment is born,
When by Parjanya Earth is fertilized with seed.
The rain of heaven bestow, O Maruts, on us,
Of your strong steed pour forth the streams abundant.
With this thy thundering roar do thou come hither,
And shed the waters as our heavenly father.
With roar and thunder now the germ deposit,
Fly round us with thy water-bearing chariot.
Turn well thy water-skin unloosed downward,
Make, with the waters, heights and hollows level.
Draw the great bucket up and pour it downward,
And let the liberated streams flow forward.
On all sides drench both heaven and earth with fatness;
Let there be for the cows fair pools for drinking.

from the *Rig Veda*, Book V, 83, *Parjanya* (c. 1300 BC)

2.1 INTRODUCTION

By following a logical process for the planning and implementation of ASR projects, the probability of ultimate success can be maximized. Such a process has been developed as a result of experience at several existing operational recharge sites, and familiarity with others, and is presented in this chapter. Although each ASR project tends to have important and site-specific issues that determine the nature and direction of activities, common themes emerging from these different projects form the basis of a recommended process for consideration at potential new ASR sites.

An essential element of the process is a phased approach in which the level of effort and associated financial investment is related to the degree of risk, both technical and non-technical. A minimum of three phases is normally appropriate:

Phase 1: Preliminary Feasibility Assessment and Conceptual Design
Phase 2: Field Investigations and Test Program
Phase 3: Recharge Facilities Expansion

In some cases, the level of risk is higher than usual, justifying a greater number of phases along the path to implementation. An increased number of phases may also be necessary as a result of funding constraints, particularly during field investigations. Certain regulatory, legal, or water rights issues may be sufficiently important that a separate phase may be devoted to full investigation of these issues before remaining portions of the program can continue. In other cases, the level of risk is known to be quite low as a result of nearby successful ASR experience, justifying moving immediately into field investigations following an initial conceptual design effort.

The natural tendency to forego initial investigations and move immediately into field testing at a selected site tends to be risky. Some ASR projects have developed in this fashion and have been successful. Others have encountered significant problems. The problems usually lead to loss of confidence in the concept of well recharge, as a result of which the project is halted. The resulting loss of momentum can be hard to overcome. More careful attention to initial planning details can identify and resolve many of these issues in advance, thereby minimizing the risk of failure. Where the penalty associated with failure is low, the higher risk may be worth taking. However, where the penalty of failure is high, possibly leading to the need for development of a far more expensive water supply alternative, it is usually wiser to invest more in the proper initial development of the ASR program to maximize the chance of success.

The probability of successfully implementing an ASR program can be enhanced by assembling a multi-disciplinary technical team that includes a balance of engineers and hydrogeologists with capabilities in the areas of geochemistry, water treatment processes, utility operations, hydraulics, aquifer simulation modeling, economics, water chemistry and design of pipelines, pumping stations, and related elements of a water utility system. Support may also be necessary to address environmental, legal, regulatory, political, and possibly other issues. Failure to consider these issues at the planning and conceptual design stages of the program can lead to costly mid-course corrections at a later date, or possibly to failure of the program.

Phase 1 usually culminates in a preliminary feasibility report that presents the ASR program in some detail to support issuance of permits for Phase 2, and to obtain financial, political, and environmental support for the subsequent phase. Phase 2 completion usually occurs when the first ASR well is fully operational, tested, and permitted, providing a firm basis in some detail to obtain regulatory, financial, political and environmental support for Phase 2. Upon completion of Phase 3, the ASR program is ready to move into a long-term operation and maintenance mode. Furthermore, sufficient confidence in the role of ASR as a valuable water management tool has usually been developed that long-range water supply plans can be formulated that incorporate ASR as a key component. These elements of an overall ASR program are presented in the remainder of Chapter 2.

2.2 PHASE 1: FEASIBILITY ASSESSMENT AND CONCEPTUAL DESIGN

Phase 1 frequently provides the key to ASR success. By planning the project in detail, and evaluating all of the many varied factors that comprise successful implementation, this phase provides a sound basis for the remainder of the program. Success requires satisfactory performance for both technical and non-technical issues. In recent years, many of the technical challenges that have adversely affected performance of recharge systems have been resolved. All too frequently regulatory, legal, political, environmental, and other issues have become the most significant factors determining recharge program success or failure. The Phase 1 assessment can provide not only the conceptual design of ASR facilities but also a broader evaluation of how the program fits into the larger water management picture, thereby constituting an effective tool for gaining a consensus of regulatory and public support for the overall ASR program.

Several elements comprise an ASR preliminary feasibility assessment. Not all of these require significant effort in every case; however, consideration of all of the following issues is most likely to ensure ultimate success.

Recharge Objectives

It is important to carefully consider the range of recharge objectives for any proposed project and to select and prioritize those that are applicable. While this may sound intuitively obvious, it is a step that is usually ignored, frequently leading to projects that are situated in the wrong location or that fail to provide the degree of benefit that could otherwise have been achieved. There is usually a primary objective that everyone can agree on; otherwise, the project does not reach the Phase 1 stage. There are also frequently one or more secondary objectives, early consideration of which can broaden the base of support for the recharge program while also affecting the conceptual design. Chapter 1 presented a list of potential ASR applications derived from projects currently in operation around the world or in various stages of development:

- seasonal storage and recovery of water
- long-term storage, or "water banking"
- emergency storage, or "strategic water reserve"
- disinfection byproduct reduction
- diurnal storage
- restore groundwater levels
- reduce subsidence
- maintain distribution system pressure
- maintain distribution system flow
- improve water quality
- prevent saltwater intrusion
- reduce environmental effects of streamflow diversions
- agricultural water supply
- nutrient reduction in agricultural runoff
- enhance wellfield production
- defer expansion of water facilities
- compensate for surface salinity barrier leakage losses
- reclaimed water storage for reuse
- soil aquifer treatment
- stabilize aggressive water
- hydraulic control of contaminant plumes
- fish hatchery water temperature control

Other ASR objectives are undoubtedly possible. This list may be used to assess potential primary and secondary objectives. For example, a community may have a primary objective of storing potable water to meet seasonal peak demands and thereby defer the need for expansion of water treatment facilities. Location of the associated aquifer storage recovery wells could be at the water treatment plant or at any suitable location in the water transmission and distribution system. Many communities have portions of their distribution system that require flushing to waste during low demand months, in order to maintain a disinfectant residual. In peak demand months the same areas may experience low pressures. Location of the ASR wells at the problem areas in the distribution system can meet both the primary and secondary objectives since excess water would be stored during low demand months, maintaining adequate line velocity. During peak demand months, the water would be recovered, helping to maintain adequate local pressures while also augmenting system peak supplies.

Water Supply

Careful consideration of alternative sources of water for recharge is essential. Each source should be evaluated as to the average flow available, monthly or other variability in flow rate, and any trends in flow. Water may be available for recharge at a higher rate initially, declining with time as other, higher priority demands for that source arise. Monthly variability is also common, based upon raw water source availability, seasonal variations in quality competing demands, legal or regulatory constraints, or other characteristics of each system. It is usually insufficient to know the average and peak rate of water availability for recharge, although this is helpful. A knowledge of monthly variability is essential.

Figure 2.1 shows a typical situation for water supply variability for the Peace River ASR site in Florida, which stores treated drinking water from a river source that is highly variable in terms of both flow and quality. For each month, the maximum, minimum, and average streamflows are shown. The minimum regulatory flow is also shown, above which up to 10% of the flow can be diverted. No diversions are allowed below the regulatory low flow for each month.

Figure 2.2 shows the drought duration frequency analysis for this same site, based upon over 50 years of records and assuming the regulatory diversion rule was in effect for the full duration of the records. It is apparent that, despite the high average flow from this source, drought durations of up to 7 months would have occurred, while those lasting two months are common. While not shown in this figure, a long-term trend analysis of the streamflow records at this site indicated a fundamental shift in streamflow characteristics occurring approximately at the middle of the record. As a result, monthly hydrologic parameters that were later used for ASR simulation modeling were obtained from the more recent portion of the hydrologic record.

This type of hydrologic analysis is essential for the Phase 1 ASR feasibility assessment because it defines water supply and variability.

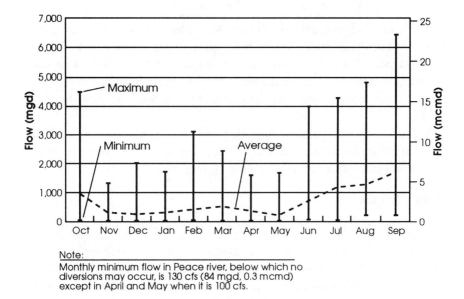

Note:
Monthly minimum flow in Peace river, below which no
diversions may occur, is 130 cfs (84 mgd, 0.3 mcmd)
except in April and May when it is 100 cfs.

Figure 2.1 Peace River, Florida, monthly flows at Arcadia, 1932–1991.

Where multiple sources are available, the results can then be compared
between sources to assess their relative suitability for ASR purposes.

Recharge Water Quality

Recharge water quality also has to be addressed carefully. Frequently,
average values mask an underlying seasonal cycle or long-term trend,

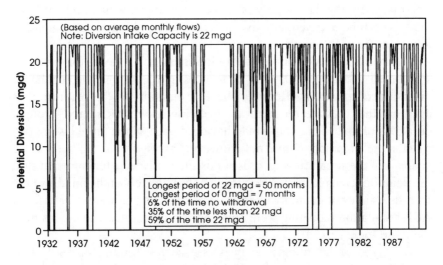

Figure 2.2 Peace River, Florida, drought-duration-frequency, 1932–1991.

which can affect recharge activities. Months when high flows are available for recharge can also be months when significant water quality issues are prevalent that would create water treatment difficulties or cause ASR wells to plug. A thorough scan of recharge water quality records for each source is required in order to properly assess the potential for problems. An initial scan on at least one sample obtained during typical recharge months and flows can be quite helpful in guiding the future direction of the project. Chapter 9 includes several examples of water quality data for specific ASR sites, showing the parameters deemed to be of significance at these sites.

Comparison of recharge water quality constituent concentrations with applicable water quality standards is an important element of the preliminary feasibility assessment. Where treated drinking water is to be stored, it is usually sufficient to show that primary and secondary standards are being met during recharge months. Where other water sources are under consideration for ASR storage, such as untreated or partially treated surface water, untreated groundwater, or reclaimed water, then comparison with applicable water quality standards can provide a basis for regulatory evaluation of the proposed ASR project, as discussed further in Chapter 7, Alternative ASR Applications. Appendix A includes 1993 water quality standards for the U.S. Environmental Protection Agency, the European Community, and the World Health Organization.

For the Peace River water supply example discussed above, potential diversions from the river were further constrained by the algal content of the river water. Analysis revealed that there are several months during the year when water may be available for diversion, considering only quantity criteria; however, the quality is so poor that it cannot be readily treated. The probability of not diverting due to algal content varied from zero in September through December to one-third in March. Furthermore, experience at that site suggests that this is not easily predicted, occurring somewhat randomly during certain months of the year and averaging about 17% of the potential diversions for the entire year. This analysis had a pronounced effect upon the initial conceptual design of the ASR system and the other facilities at the Peace River site. Subsequently, facilities were constructed and additional facilities are planned so that this water can be diverted and treated rather than lost.

An important water quality consideration is the suspended solids content of the recharge water source. Invariably this data is not available for Phase 1 investigations. As discussed in Chapter 4, well plugging and redevelopment is an important ASR technical issue, the resolution of which includes field data collection to gather information regarding total suspended solids content in the recharge water, how these solids vary with time and with flow, and what materials contribute to these solids. In some cases particle counting and a detailed analysis of particle size may be appropriate in order to provide a basis for understanding and resolving plugging issues. Where this data is not available for Phase 1, it may be obtained early during Phase 2 investigations.

Once recharge quality and quantity issues have been addressed, it is possible to combine the two and thereby evaluate those times of the year when recharge water is available in a useful quantity and with suitable quality. This provides the basis for determination of annual recharge volume potentially available in the initial and subsequent years.

Water Demand

In most situations it is important to evaluate water demands, including average demands, monthly variability, and trends. Records of monthly maximum day, average day and minimum day demand for a period of three or more years are frequently helpful in showing the degree of seasonal, monthly and daily variability. Ratios of maximum day to average annual demand, maximum week to average annual demand, and/ or monthly demand as a percent of average annual demand are frequently helpful in assessing the duration of peak demand periods when recovery of stored water would provide maximum benefit. Such ratios also help in assessment of the amount of idle supply, treatment, and transmission capacity to be expected during each month, as a result of normal monthly variability in system demands. Another useful output of this analysis is the annual volume of water required during recovery to meet system demands.

Sometimes the duration of the peak demand period is quite short. An analysis of monthly water demand records for the Alexander Orr Water Treatment Plant at Miami, Florida, indicates little seasonal variability in demand, with a ratio of 1.2 for maximum to average day demand. The relatively low ratio partly reflects the large size of this urban area. However an analysis of daily records shows that peak demand durations of up to 18 days per year are associated with the highest 20 MGD of peak demands. The highest 10 MGD occurs for only about 3 days per year. Similarly, the peak demand for Wildwood, New Jersey, occurs during July 4 weekend each year when large numbers of people visit this coastal resort community. In such situations, the volume to be recovered from storage is therefore not that great. The rate of recovery in such situations may determine the number of ASR wells required to meet system demands. In other situations, the number of wells may be determined by the recharge rate necessary to store the volume of water required to meet demands during the recovery period.

Municipal water systems are usually designed to meet peak day demands during some future year. The typical ratio of peak day to average annual demand is about 1.3 to 2.0, although ratios as high as 5.0 are known. Consequently, it is not uncommon for water systems to have a substantial amount of idle capacity during much of the year. This capacity can be utilized for treatment and storage of water during off-peak months, using ASR and other recharge facilities.

Figure 2.3 shows the record and projection of average and maximum day water demand for Evesham, NJ, showing the effect of adding one ASR well on extending the useful capacity of existing water supply and treatment facilities. Typically, an ASR system enables a water utility to meet maximum day demands with water supply and treatment facilities sized to

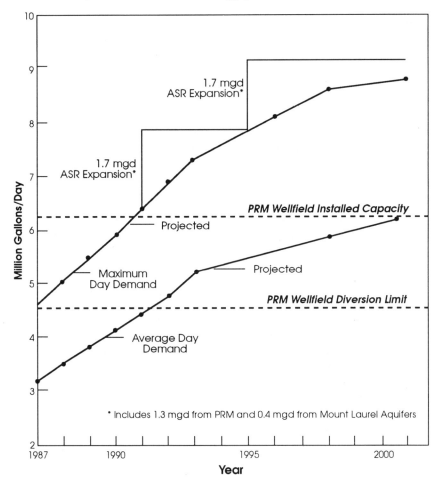

Figure 2.3 Supply–demand relationship with new ASR facilities, Evesham Municipal Utilities Authority, New Jersey.

meet close-to-average demands, and ASR facilities sized to provide the difference. Usually treatment and piping facilities are much more costly than are ASR wells.

Quality requirements for the recovered water also need to be evaluated to aid in assessment of the potential treatment requirements. Usually it is only necessary to disinfect the recovered water prior to distribution. In some cases pH adjustment may also be required, either to maintain stability within a desired range or to maintain disinfection effectiveness.

Hydrogeology

This is frequently the most time-consuming element of the Phase 1 feasibility assessment. Careful evaluation of area hydrogeology can lead to the selection of suitable storage zones, recharge water sources, and treatment requirements, and usually affects the location and design of ASR facilities. Depending upon the amount of information available for review, it is frequently possible to complete this task without substantial field investigations. On the other hand, such an evaluation may indicate important technical unknowns that can only be addressed through drilling and testing. Whether such field investigations are performed during the first phase or deferred to the second phase is a site-specific decision; however, identification during Phase 1 of what is known and what is unknown can lead to more enlightened decisions regarding project planning and funding.

The hydrogeologic evaluation during Phase 1 should consider the following technical issues to the extent possible with available data and resources. Where the data is not available and is deemed to be significant to the program, plans should be made to obtain this information through supplemental investigations, whether during Phase 1 or at the beginning of Phase 2:

- stratigraphy, including geologic cross-sections
- aquifers (areal extent, thickness, and depth)
- confining layers or aquitards (aerial extent, thickness, and depth)
- lithology of aquifers and confining layers
- potential availability of cores
- hydraulic characteristics (transmissivity, storativity, leakance, hydraulic conductivity, porosity, etc.)
- typical well construction and production rates
- mineralogy of clays, sands, and other soil components
- geophysical logs
- water quality of each aquifer
- geochemical compatibility of recharge and native water with formation minerals

- structure (unconsolidated, consolidated, fractures, bedding planes, solution features, fissures, etc.)
- recharge and discharge boundaries
- water table levels or potentiometric surface
- local gradient of the potentiometric surface
- natural groundwater velocity and direction
- well inventory within a reasonable radius
- groundwater withdrawals within the surrounding area
- proximity of potential sources of contamination
- proximity of potential contamination plumes that may be affected by recharge operations

In many cases the selection of the aquifer to be recharged appears obvious. However, it is important to consider all possibilities, since ASR operations can also occur in deep, brackish, or otherwise undesirable aquifers containing water quality too poor for normal consumption. It is quite common for attention to focus upon shallow aquifers for which considerable data is already available, when deeper aquifers at the same site may have equal or better potential for ASR storage but may be less well documented. With appropriate design and operation, these aquifers can frequently be used for storage purposes, and all of the water stored can be recovered without the need for retreatment. "Stacking" the storage zones is frequently highly cost-effective, using multiple zones for storage at a single site and thereby saving construction and operation costs for an extensive piping network.

The level of effort appropriate for assessment of geochemical issues should be carefully evaluated. The attention given to this part of the project is frequently too small, sometimes resulting in unexpected or unexplained plugging.

It is probably wise to obtain and analyze cores in the potential storage zones for the following situations, so that the risk of geochemical plugging can be properly assessed:

- where insufficient data is already available to perform such an assessment
- clays, silts and other fine materials are possibly present in the potential storage zones; or
- there is no local recharge experience to consider

The cost of conducting this work can be weighed against the perceived risk and cost of potentially plugging and losing use of a recharge facility. Where the risk is low, or the value of the facilities at risk is not too great, then this portion of the assessment can probably be deferred or minimized. Otherwise, coring and geochemical analysis should be conducted during

Phase 1 or at an early stage in Phase 2. Further consideration of coring and core analysis is included in Chapter 5, Geochemistry.

Selection of Recharge Processes

An informed decision can usually be made at this point regarding the recharge process or processes that appear appropriate. Consideration of water supply and demand factors will usually indicate the annual volume of water available for recharge or required for recovery. If the hydrogeologic evaluation indicates that surface recharge is probably feasible, it is then possible to conduct an evaluation of possible sites for instream or offstream surface recharge facilities. Once potential sites are identified, a preliminary screening can usually indicate whether sufficient recharge capacity is likely to be present to fully utilize available recharge flows. If land availability and hydrogeology are favorable, surface recharge is usually the most cost-effective recharge approach if the objective is limited to getting the recharge water into the ground. Where either of these factors become limiting, then well recharge should be considered.

As discussed in Chapter 1, aquifer storage recovery wells are more likely to meet overall recharge objectives than are conventional single-purpose injection wells. In some cases, a combination of the two recharge approaches (surface and well recharge) offers operating flexibility while also fully utilizing the available recharge flows and storage capacity in the area.

As part of the selection process, it is very important to consider potential recharge objectives. Getting the water into the ground is usually only part of the process. Equally important is the ultimate potential use and value of the stored water at the point of recovery. Storage and recovery of higher value potable water may, in many cases, be more cost-effective and useful than storage and recovery of lower cost raw water from rivers, especially if all of the stored water can be recovered without the need for retreatment. Recharge economics should therefore consider not just the cost of getting the water into the ground, but also the overall cost for achieving the local water management objectives.

Site Selection

For most ASR projects, the site for the Phase 2 test facilities is best located either at or close to the water treatment plant, or at some point in the distribution system where major facilities are already in place, such as a pumping station or a ground storage reservoir. This provides increased likelihood that qualified personnel will be available during the test program to aid in gathering hydraulic and water quality monitoring data. Sampling and data collection can be intensive during portions of the test program, so it is helpful to have operations staff close to the site to assist with the program. Conversely, a site that is distant from operations staff will have greater difficulty and cost in obtaining the requisite data, and correspondingly higher potential for occurrence of problems that are not noticed until considerable time has elapsed.

Site selection should also take into account the need for disposal of water produced during drilling, testing, backflushing, or well redevelopment operations. In some residential areas, water disposal can be a major constraint unless an adequate surface drainage system is in place. In other areas characterized by extensive wetlands, disposal of brackish waters during initial testing of brackish water storage zones can be a significant challenge. It should be possible to pump the well to waste at its design production rate, or somewhat higher if pumping to free discharge, and to maintain this rate for several hours or possibly several days.

Hydrogeologic considerations may affect site selection, such as a significant trend of well yields, water quality, or well depth within the potential storage aquifer across the utility service area. By affecting well depth and yield, the presence of geologic faults may affect the suitability of some areas for ASR purposes, either positively or negatively. Frequently site selection is dominated by the availability of suitable existing wells for testing purposes. Proximity of recharge water pipelines with adequate transmission capacity is also an important factor, especially in more remote areas.

Although very little land area is required for ASR facilities, this can be a constraint in urban areas. Consequently, land availability and site access have to be considered. In some states, regulatory requirements stipulate the need for a radius around ASR wells, within which no existing or potential contamination sources are allowed. Typical radius values are about 31 m (100 ft). This can control location of ASR wells near property boundaries, potentially affecting the site selection process.

Conceptual Design

Assuming that ASR is selected as the recharge process and one or more alternative sites have been identified, advantages and disadvantages of each site can be considered, and criteria can be developed and applied for recharge site selection. Conceptual design of facilities to achieve recharge objectives can then be conducted. Usually it is sufficient to develop a preliminary conceptual design for Phase 2 test facilities, with some consideration for the probable layout of the ultimate ASR site to ensure that test facilities are located and constructed reasonably consistently with an ultimate plan. The conceptual design would include the location and sizing of the ASR test well, pipelines, building, controls, and other facilities, in sufficient detail that an order-of-magnitude cost estimate can be developed. If desired, a similar cost estimate for the ultimate recharge plan can also be prepared to provide a basis for consideration of unit capital or operating costs for expansion of peak water supply capacity.

Existing Well vs. New Well

A key element of the conceptual design is the decision regarding whether to utilize an existing available production well retrofitted for ASR testing purposes or to construct new facilities designed specifically to meet ASR objectives but at increased cost for the test program. Existing operational and abandoned wells are frequently available and should be considered for test program purposes, for economic reasons. However, they are not always situated appropriately, seldom constructed appropriately, and may not be in a condition suitable for ASR purposes.

As discussed subsequently in Chapter 3, ASR wells have a unique design that is different from either production or injection wells. Where storage zone water quality is fresh and potential geochemical problems are minimal, production well design may be very similar to ASR well design. However, when the storage zone water quality differs from recharge water quality, and where geochemical issues are of concern, the ASR well design will tend to be different from the production well design. Hence, the selection between using an existing well or constructing a new well requires some care if optimum well performance is to be achieved.

Where an existing well is selected, it will usually be necessary to verify the condition of the well prior to testing. Whether this is done during Phase 1 or at the beginning of Phase 2 is a site-specific decision. However, this exercise may include video camera logging of the well, followed by such activities as wire brushing the screen and casing while pumping the well; acid treatment; disinfection; and a second video log to confirm results. In old wells that are equipped with oil-lubricated pumps, it is common to find a layer of oil floating in the casing, which should be removed prior to ASR testing operations.

It is important that program objectives not be compromised by a decision to utilize existing facilities that are unsuitable for the purpose. Once an ASR test well becomes severely plugged or has other major problems during testing, the test program tends to lose support. It is then very difficult to regain support by pointing out that further investment in new test facilities would have been more likely to achieve success. Wells are usually abandoned or not utilized for a reason: frequently due to construction, age, sand production, or other water quality problems that can severely hamper the conduct of successful recharge testing. Where existing wells are utilized for testing, great care has to be exercised to identify and, if possible, correct shortcomings in design or construction that could adversely affect success of the recharge program.

Hydrogeologic Simulation Modeling

The application of computer simulation modeling at this point in the recharge program should be considered. In some situations the data available from literature sources may be sufficient to justify hydrogeologic modeling of proposed recharge operations and to provide an improved basis for conceptual design of facilities and planning of the test program. In most cases, however, it is probable that such modeling will be of limited benefit since site-specific data will generally not be available until after construction and testing of ASR facilities during Phase 2. Experience suggests that data uncertainties frequently render such modeling during Phase 1 to be of limited value, although it is certainly of interest. At the end of Phase 2, hydrogeologic modeling of expanded ASR operations is an important and essential tool for planning and budgeting Phase 3 expansion.

Modeling activities are discussed further in Chapter 4, Selected ASR Technical Issues. Generally, ASR hydrogeologic modeling includes three different objectives:

1. hydraulic analysis of wellfield design and operations
2. geochemical simulation to evaluate mixing between stored and native water in the presence of aquifer minerals
3. solute transport modeling to establish the direction and rate of water movement during aquifer storage, to establish constituent concentrations during recovery, and to estimate percentage recovery

Hydraulic analysis is usually best addressed at the end of Phase 2 when site-specific data are available, and results support further consideration of an expanded ASR wellfield. Sometimes this has to be performed during Phase 1, based upon preliminary estimates of aquifer hydraulic characteristics, in order to obtain the permits to conduct Phase 2 testing. Usually, however, such models require estimates of aquifer dispersivity, values for which are determined based upon data collected during Phase 2.

Geochemical simulation is usually conducted early during Phase 2 when formation samples, water quality samples, cores, and core analysis results are available to support a detailed geochemical analysis. A preliminary geochemical assessment is appropriate during Phase 1, based upon available information in the literature, in order to guide Phase 2 planning, site selection, and storage zone selection.

Solute transport modeling is increasingly utilized as a tool in ASR investigations to assess site location relative to other production or ASR wells, and the impact of well interference upon recovery efficiency. As such, it can be a useful Phase 1 tool in accompaniment with hydraulic modeling. In many cases, both of these types of models can be deferred to the end of Phase 2, relying upon site-specific data without incurring significant risks in site selection, design, or operation. As additional experience is gained with the application of solute transport modeling to ASR systems, they will become more useful as prediction tools to guide location and design of ASR systems.

Outline of Test Program

The proposed testing and monitoring program should be developed in some detail. This includes baseline hydraulic testing, water quality sampling, and water level monitoring of the storage zone, followed by several cycles of operation under typical conditions. The location and frequency of data collection and sampling, and the parameters to be analyzed on each sample, should be developed. Typically, this includes water levels, flows, pressures, water quality, and other data at ASR and monitoring well facilities. This is usually a time-consuming and expensive part of the test program and should be planned as carefully as possible beforehand so that it is scoped correctly and budgeted adequately. It is important to collect adequate data to permit interpretation of ASR results at the end of the program.

The next section in this chapter provides greater detail on the design and implementation of test programs. It is important to build considerable flexibility into the Phase 2 test program plan. Flow rates, volumes, storage periods, sampling frequencies, and just about every other monitoring parameter can provide compelling reasons to adjust the original testing plan prepared during Phase 1, in order to meet changing Phase 2 needs.

Regulatory and Water Rights Issues

These issues are different in every part of the U.S. Each state has its own requirements and procedures. The federal government (U.S. Bureau of Reclamation, U.S. Geological Survey) also is involved in recharge activities through funding of local demonstration programs and state water projects, and through regulation of Underground Injection Control (UIC) Class V Well activities by the Environmental Protection Agency.

A common theme among recharge projects is the ownership of the stored water. State legislation and associated case law is increasingly supporting the position that, if the water is already available to a user for consumption or storage purposes, it is also available to that user through recovery from storage. In other words, rights to the water are not lost through underground storage. It is necessary in some states to pass supplemental local ordinances or state legislation to reinforce the water rights associated with a recharge project. In other states, it is necessary to obtain separate permits for recharge and for recovery. During the next few years it is anticipated that a variety of state laws relating to ASR will be passed, hopefully achieving some commonality on critical issues.

Environmental considerations usually are important regulatory concerns. Impacts upon river flows and quality need to be considered, as well as impacts upon groundwater levels, groundwater quality, recovered water quality, and impacted ecosystems. Normally, recharge operations are perceived as benefitting the environment by making more efficient use of water when it is available without causing significant adverse effects. However, each project has to be evaluated on its own merits. Mitigation plans may be appropriate to address potential adverse environmental effects. The potential cost of such plans should be incorporated in the site economic analysis.

Chapter 6, Selected ASR Non-Technical Issues, presents further details on consideration of regulatory, water rights, and environmental issues during the ASR Phase 1 feasibility assessment.

Institutional Constraints

Whether addressed directly in the Phase 1 report, or handled through parallel efforts, this is usually a vitally important element of a recharge program. Water is synonymous with political power, and recharge programs are not exempt from the associated pressures. Sometimes the agency assuming responsibility for the recharge program may have institutional constraints that can hamper effective recharge program implementation. This can take several forms:

- lack of access to suitable recharge sites or water sources
- a range of operations flexibility in the agency charter that is too narrow to encompass effective integration of recharge operations
- higher priority water management programs that can effectively inhibit progress on cost-effective, competing, promising alternatives

- established policies that were developed many years ago without consideration of ASR potential opportunities, and
- numerous other institutional constraints

Failure to carefully evaluate and address these often subtle issues can easily lead to recharge program failure or extended delay. Conversely, use of the Phase 1 feasibility report as a tool to elicit political and institutional support for the recharge program can lead to constructive input and enhanced likelihood of ultimate success.

In some cases institutional or regulatory issues will create the need for public involvement in the planning and implementation of the recharge program. This can occur in several ways but usually involves meetings with advisory committees or at regulatory hearings during development and permitting of the recharge program. The Phase 1 report can be a valuable tool at such meetings to help present the overall program in an unambiguous way and thereby dispel misconceptions that frequently arise.

Economic Considerations

It is important to develop a preliminary estimate of the capital and operating cost of ASR operations as early as possible. It is also frequently useful to develop estimated unit costs for expected increases in peak water supply capacity. Such an analysis would consider the useful life of the recharge project and the anticipated future availability of the recharge water source.

Sometimes it is appropriate to extend the economic analysis to estimate annual costs, including amortization of capital investment plus normal operation and maintenance. For the Phase 1 feasibility assessment, this cost estimate can be provisional, to be confirmed upon completion of Phase 2 testing. It is useful to have some idea of expected overall cost, for comparison with other water management alternatives that may achieve some or all of the same objectives.

ASR operations are invariably cost-effective when compared with conventional water supply alternatives involving development of new water sources, primarily due to the distance to new sources, and the associated environmental constraints and costs. It is not unusual for ASR alternatives to be less than half the capital cost of other water supply alternatives, particularly those involving development of new reservoirs or construction of treatment facilities. Once the feasibility of ASR operations is confirmed through Phase 2 testing, it is usually appropriate to conduct broader economic analyses to assess the optimal plan for overall phased development of water management facilities, including ASR operations.

In some cases, it may be appropriate to extend the economic analysis to include financing and rate-paying considerations, particularly where outside sources of water or funding are to be sought.

As discussed in Chapter 6, Selected ASR Non-technical Issues, off-peak purchase or sale of water supplies can provide the opportunity for substantial reduction in unit costs. Where this approach has been evaluated, unit cost reductions exceeding 50% have been demonstrated for wholesale purchasers.

Project cost-sharing between various contributing sources can be proposed in the Phase 1 report or can be deferred to Phase 2 when feasibility is confirmed through field testing. This can be an important element of an overall ASR program where multiple water users may participate but their individual requirements for supplemental water may vary from year to year.

Final Report

Results of the Phase 1 feasibility assessment and conceptual design should be incorporated in a final document that achieves two key objectives:

1. presents a well-considered technical approach to the ASR program
2. provides a document for use in obtaining necessary permits, environmental support, institutional support, and funding

As the recharge program is implemented, changes will undoubtedly occur in the circumstances surrounding the project, necessitating reevaluation of certain elements of the original plan. Furthermore, as the test program is implemented, evaluation of test data will undoubtedly justify changes in the testing approach or monitoring program. This is normal, and provision for such changes should be incorporated in the original plan and associated contracts and funding arrangements.

2.3 PHASE 2: FIELD TEST PROGRAM

Once the decision is made to move ahead with a Phase 2 test program, the first step usually is to design and construct the ASR test facilities. These should be designed at full scale, capable of recharging and recovering at whatever rates are reasonably available from an efficient well penetrating the selected aquifer. Testing in a small diameter test well necessitates recovery rates that are lower than those that would ultimately be expected from a full size well. Conclusions regarding potential seasonal volumes and associated unit costs therefore tend to be biased on the high side. In some situations, such as storage of freshwater in brackish aquifers, the extrapolation of results from small-sized facilities can lead to incorrect conclusions regarding ASR recovery efficiency and cost-effectiveness compared to results from properly-sized test facilities. Design issues will be addressed in greater detail in Chapter 3, Design of ASR Systems.

Construction issues are similar to other well construction projects, with the possible exception of the greater amount of hydrogeologic data collected during construction. Sometimes this includes collection of cores for detailed laboratory analysis to determine mineralogy, geochemistry, and associated hydraulic properties, as discussed in Chapter 4, Selected ASR Technical Issues, and Chapter 5, Geochemistry. Chapter 6 discusses permitting issues associated with construction and operation of ASR facilities.

Interval pumping tests are sometimes conducted at several different depths during construction of open hole wells, in order to estimate the productive intervals of the aquifer for correlation with geophysical logs and drill cutting descriptions. Geophysical logging is conducted after pilot hole drilling in order to establish casing and screen intervals. For screened wells, caliper logs may be obtained after the pilot hole is reamed and prior to setting the screen, while flowmeter logs are obtained after the screen is installed and developed. For open hole wells, flowmeter and caliper logs are obtained following well completion. The baseline distribution of production with hole depth or screen interval is useful as a reference point against which to compare subsequent logs and thereby estimate changes in the flow distribution due to ASR operations. Upon completion of ASR facilities construction, the test program begins.

Design of the test program reflects a careful assessment of the many issues of concern at each new site. Well plugging is always a primary issue; however, others of importance may include the following:

- geochemical effects such as cation exchange, precipitation, or solution, and their effect upon well plugging
- backflushing frequency required to maintain recharge capacity and control well plugging
- mixing characteristics between stored and native water
- water quality changes for selected non-conservative constituents of interest
- improvement of water quality with successive ASR cycles
- effect of storage time on water quality response
- recovery efficiency
- trickle injection flow rate during periods of no recharge or no recovery, required to maintain a disinfectant residual in the well (this may also be required to maintain a target recovery volume in a highly brackish or seawater aquifer subject to density stratification losses)

- regional and local response of water levels to ASR operations (recharge, storage, recovery)

Other site-specific testing objectives may also occur. Depending upon the relative importance of each of these concerns, the test program design will typically adjust to meeting site-specific needs.

Baseline Testing

The first part of the test program includes baseline hydraulic and water quality testing prior to initiating significant recharge activities. This provides a reference point against which future results may be compared.

Baseline hydraulic testing usually starts with a standard step drawdown pumping test in order to establish well and formation loss coefficients and well efficiency. Following water level recovery, a long duration pumping test should be conducted in order to estimate aquifer hydraulic characteristics in the vicinity of the ASR well. Duration of the test depends upon local experience in obtaining reliable estimates of aquifer hydraulic characteristics. Typical durations are about one day, sometimes longer. If observation wells are present, they should be incorporated in the pumping test to better define aquifer transmissivity and estimate storativity. Observation well data are frequently more useful since it is less affected by variations in the pumping rate that affect drawdowns in the pumping well. The test establishes aquifer hydraulics in the vicinity of the well before recharge commences.

Upon completion of the long-term pumping test and associated recovery of water levels to background, a step-injection test is usually conducted to characterize water level response in the ASR well under reverse conditions from the previous step-drawdown test. Recharge during this step-injection test occurs at three different increasing rates, generally bracketing the expected recharge rate for the well. Each recharge rate step is of the same short duration, such as 2 to 4 hours. Water level response to this test characterizes the baseline water level response of the well in the presumed absence of significant plugging. At any subsequent point in time, this test can be repeated to evaluate whether plugging has occurred between the two tests. This is discussed further in Chapter 4, Selected ASR Technical Issues.

An important element of the baseline step-injection test is that the recharge water should be allowed to flow to waste near the wellhead at the planned recharge rate for a few minutes prior to recharge down the well or until any solids in the recharge water have been flushed from the system. Depending upon the length and the normal flow rate in the tributary piping, flow reversal in the piping during ASR recharge can sweep a substantial volume of solid material down the well, causing immediate onset of plugging if these solids have not first been purged from the system.

Baseline water quality characteristics are also determined during this initial part of the test program. Samples are collected usually at the beginning, middle, and end of the long-term pumping test to fully characterize water quality in the ASR well prior to recharge. The last sample usually receives a complete water quality analysis, including EPA primary and secondary drinking water standards and several other parameters, as suggested in Table 2.2. The initial and middle samples collected during this long-term pumping test are analyzed for a smaller range of parameters in order to estimate whether any trend in water quality occurs during the test. If a trend is apparent from field data such as chloride, conductivity, or pH, it may be appropriate to extend the pumping test until such time as equilibrium water quality is apparent.

Sometimes the situation arises where the storage zone is separated by a thin or poorly defined confining layer from an overlying or underlying highly transmissive aquifer containing water of very poor quality. The adequacy of this confining layer may be of some concern. In particular, if the time required for poor quality water to move through the confining layer under the head differential imposed during seasonal recovery is shorter than the expected ASR operational recovery period, then deteriorating recovery water quality will tend to define the upper limit of ASR recovery duration and volume, regardless of the volume stored.

Figure 2.4 shows the ASR initial test wells and storage zone at Cocoa, FL. Chloride concentration in the storage zone is about 400 mg/L; however, an underlying aquifer, not penetrated by the ASR well, has a chloride concentration of 1320 mg/L. The confining layer separating the aquifers was approximately 36 ft thick and was of uncertain integrity. Following initial pump testing and cycle testing to determine storage zone hydraulic characteristics and water quality response to ASR operations, it was evident that upflow was occurring from the underlying aquifer through the confining layer during recovery. The ASR well and production zone observation well were then plugged back to a shallower depth, with cement, to reestablish the integrity of the lower confining layer. The shorter open hole interval of the ASR well was acidized to restore the production capacity lost when the well was plugged back, following which a 90-day pumping test was performed at the design recovery rate for the ASR well. Chloride concentration during the test remained steady, suggesting that future ASR operations with recovery periods of up to 90 days would not be likely to experience reduction in water quality due to upwelling of brackish water from below.

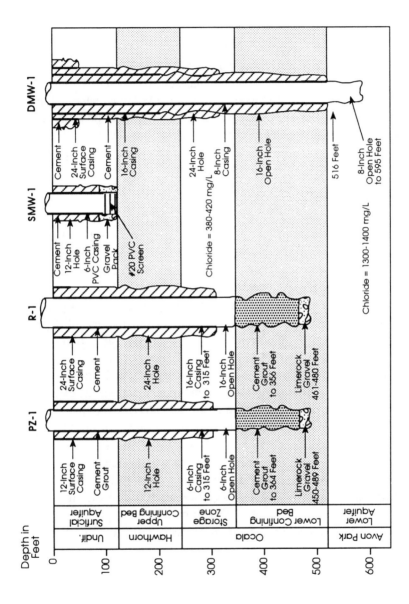

Figure 2.4 ASR initial test facilities and storage zone, Cocoa, Florida.

Another ASR test site at Tampa, FL, showed inadequate lower confinement from a similar test, as evidenced by breakthrough of poor quality water from an underlying highly transmissive aquifer with a total dissolved solids (TDS) concentration of 8000 mg/L. The storage zone had been acidized to improve its low initial yield, resulting in a dramatic increase in specific capacity. Unfortunately, the acid opened up a channel through the lower confining layer, which was about 21 m (80 ft) thick. The breakthrough of poor quality water during recovery occurred after a given volume of recovery, essentially regardless of the volume recharged. In this case, plugging the bottom of the well with cement may possibly help to restore adequacy of the lower confining layer; however, this may also substantially reduce well yield. Relocation of the ASR well would then be necessary.

ASR Cycle Testing

Selection of the appropriate number and duration of ASR cycles during the field investigations entails considerable judgment and the requirement that, whatever initial selection is made, flexibility is retained to make adjustments during the test program in order to respond to changing needs. Certain key points underly the development of the cycle testing plan.

A short initial cycle is advisable, to confirm satisfactory ASR performance at small volumes and to provide a quick appraisal of plugging and geochemical reactions. Usually this has a duration of about 1 to 2 weeks and entails intensive hydraulic and water quality data collection. Except in situations where significant adverse geochemical reactions are expected or identified from water quality data, it is advisable to recover approximately 150 to 200% of the stored water during the first cycle, or until recovered water quality approaches native water quality. This fully defines the mixing characteristic curve for the well at the scale tested in the first cycle and is useful for comparison to comparable data from other sites. As discussed below, recovery should be less than 100% of the stored water volume in situations, where geochemical reactions are of concern.

Recovery efficiency for the initial cycle is typically lower than for subsequent cycles. For storage zones that are brackish or otherwise contain non-potable water, it is important that recovery efficiency expectations are carefully managed and that subsequent cycles at similar larger volumes are conducted.

Data from the first cycle can be used to reasonably estimate performance in subsequent cycles. However, there is no real substitute for actual performance during several cycles if you wish to build confidence in the

ASR system. A brief interval between the first and second cycle is sometime useful to obtain and review laboratory water quality results and hydraulic data, permitting adjustment of the testing plan for subsequent cycles or adjustment of the ASR facilities, if appropriate.

If water quality difference between stored and native water is small and there are no significant concerns regarding geochemical reactions, then a small number of long cycles is appropriate in order to focus upon plugging rates and backflushing frequency required to maintain recharge rates. A minimum of three cycles is usually appropriate in such situations, with the third approximating an operational recharge duration. Recovery of 100% of the stored water volume in each cycle after the first is a reasonable target, although greater or lesser recovery volume may be appropriate in some situations.

If there is a significant water quality difference between stored and native water, a larger number of cycles will be required. After the first cycle, the next three cycles have the same recharge volume and storage period in order to demonstrate improvement in recovery efficiency with successive identical cycles. Subsequent groups of three or more cycles tend to have larger volumes and, in some cases, may incorporate storage periods as discussed below. The total number of cycles may be in the range of 4 to 10. Recovery occurs to a target water quality concentration in each cycle. The water not recovered in each cycle forms a buffer zone to improve quality during the subsequent cycle. The target constituent concentration may or may not be a potable water quality standard, depending upon whether any blending will occur between the recovered water and water from other sources prior to consumption.

If there is a real concern regarding potential geochemical reactions, care should be taken to avoid shocking the formation with a sudden change in quality. Furthermore, storage time should be built into the test program since some reactions such as manganese dissolution require several days or weeks to occur. Geochemical plugging reactions that occur several feet away from the well screen have little effect upon well hydraulics, while those occurring close to the well can have substantial adverse effects. Hence, in these cases it is appropriate to design the ASR test cycles in such a way that stored water is never fully recovered, always leaving a small buffer around the well, which tends to build with successive cycles.

Where potential geochemical reactions are an issue, it is advisable to run a larger number of small cycles to demonstrate control over geochemical issues near the well, before moving on to larger operational cycles. It would not be unusual to run 6 to 10 cycles in situations with complex geochemical issues.

Initial ASR cycles are usually performed with the recovered water discharged either to waste or to retreatment. Water from subsequent cycles during the test program can usually be returned to the distribution system once adequate water quality has been demonstrated and the system is permitted. Frequently the last test cycle is designed so that the volume to be recovered is large, and recovery occurs during a time of peak demand, so that the water meets a local need.

The effect of storage time on system performance is always of interest. However, it is quite rare for this to be a significant parameter in ASR operational performance. One situation where the effect of storage time is significant is where the storage zone TDS concentration is sufficiently high that density stratification is significant. A difference between re-charge and native water TDS concentration of about 5000 mg/L or higher is probably a threshhold above which this should be considered. A second situation would be where the regional or local hydraulic gradient around the ASR well is sufficiently steep, and the aquifer transmissivity suffi-ciently high, that there is real concern that the stored water will rapidly move away from the well prior to or during recovery, such that poor quality native water would then be recovered. Finally, storage time can be significant in situations that are geochemically complex. For example, manganese production requires several days or weeks to develop, as discussed above. Other than these situations, it is better to spend the time more productively, recharging and recovering water and gathering data.

Consideration of these guidelines will reveal that priorities need to be established at the outset of the test program design, since some of the guidelines are incompatible with others. For example, it is not advisable to recover 150 to 200% of the stored water during the first cycle if the initial data suggests that adverse geochemical reactions may be occurring. For low permeability storage zones in brackish sand aquifers with no significant geochemical sensitivity, definition of plugging rates and redevelopment frequencies may take precedence over demonstration of recovery efficiency improvement in successive cycles, moving the test program in the direction of a small number of long cycles instead of a larger number of short cycles.

Tables 2.1, 2.2, and **2.3** present the actual test programs implemented at three sites, illustrating a range of issues to be resolved. Each site is discussed in greater detail in Chapter 9, Selected Case Studies.

In the first case, Table 2.2 shows the test program for Marathon, FL. This ASR site includes a geochemically insensitive sand aquifer contain-ing water with a TDS concentration of 39,000 mg/L. The storage zone is confined, and the ASR test well has a low specific capacity of about 3 G/min/ft. The site provides an emergency water supply for the Florida Keys,

TABLE 2.1 CYCLE TEST PROGRAM: MARATHON, FLORIDA

Cycle No.	Volume (MG)			Duration (Days) (a)			Total
	Recharge	Recovery	%	Recharge	Storage	Recovery	
1	4.5	5.1	33	17	0	19	32
2	9.7	3.5	30	42	34	11	87
3	5.3	4.2	68	27	0	16	43
4	3.6	2.8	72	14	0	9	23
5	15.0	6.5	43	51	39	26	116
6	15.1	7.7	51	56	36	43	135
7	15.8	8.9	55	56	35	30	121
8	15.4	10.1	65	76	21	34	131
9	15.0	10.1	65	54	0	44	98
10	15.3	8.6	56	56	35	31	122
11	14.0	10.4	71	63	81(b)	25	169

Note: (a) Testing conducted 1990 to 1993. (b) Trickle flow of about 50 G/min during last 57 days of storage period. Total recharge plus trickle flow volume = 19.2 MG.

TABLE 2.2   CYCLE TEST PROGRAM: KERRVILLE, TEXAS

	Begin	End	Total Time	Gallons Recharged/ Recovered (millions)	Average Recharge/ Recovery Rate (G/min)
Cycle 1					
Recharge	4/2/91	4/5/91	3 days, 1 hr.	2.93	669
Storage	4/5/91	4/7/91	2 days	—	—
Recovery	4/7/91	4/9/91	2 days, 7 hrs.	2.86	867
Cycle 2					
Recharge	4/15/91	5/14/91	29 days, 3 hrs.	24.9	595
Storage	5/14/91	6/13/91	30 days	—	—
Recovery	6/13/91	7/3/91	20 days	25.0	868

a region that is quite vulnerable to damage during hurricanes. Eleven ASR test cycles have been conducted to demonstrate improvement in recovery efficiency with successive cycles at two different storage volumes, and also to demonstrate the ability to maintain a target storage volume by addition of a trickle flow during storage periods. Different storage durations without any trickle flow showed the adverse effect of density stratification upon recovery efficiency. Plugging was shown to be insignificant.

Table 2.3 presents the test program for the ASR facility at Kerrville, TX. There is little significant difference in water quality between the recharge water and the native groundwater, and no geochemical issues are evident. The potentiometric surface in the Kerrville area is such that all water recharged will ultimately be recovered as long as excessively high water levels are avoided in the ASR well. The aquifer is confined and consolidated and has low-to-moderate transmissivity. The test program needed primarily to focus upon plugging rather than water quality issues. It included only two cycles, neither of which demonstrated any significant plugging.

TABLE 2.3   CYCLE TEST PROGRAM: SWIMMING RIVER, NEW JERSEY

Cycle	Dates	Volume Injected (MG)	Volume Recovered (MG)	% Recovered
1	07-29/08-21-91	9.6	7.2	74
2	08-21/09-13-91	9.8	5.4	55
3	09-13/10-07-91	8.5	7.6	89
4	10-07/11-08-91	10.0	7.0	70
5	11-10-91/06-16-92	52.1	46.8	90
6	06-16/06-22-92	3.4	3.0	88
7	12-02/12-11-92	3.9	3.5	89

Table 2.4 shows the test program for the ASR facility at Swimming River, NJ. The storage zone is a confined, clayey sand aquifer that is geochemically complex and sensitive, including native water iron concentrations exceeding 13 mg/L and manganese sources in the storage zone. The site also has moderate transmissivity of 509 to 658 m^2/day (41,000 to 53,000 G/day/ft). Seven ASR test cycles were conducted to initially resolve plugging issues and then to address iron and manganese geochemical issues. As discussed subsequently in Chapter 4, ASR testing at this site was preceded by a pretreatment phase to condition the aquifer around the well. Initial resolution of these complex issues was successfully resolved through small volume cycles in which reactions occurred reasonably close to the well. It was then possible to move on to larger cycles.

A characteristic of ASR test programs is that utility operational requirements often affect the planning or implementation of the testing plan. Water may not be available for recharge at the rate or time desired. Water may be required from recovery to meet peak demands at a time that may be inappropriate from the limited perspective of the test program. Mechanical and

TABLE 2.4   ASR TYPICAL WATER QUALITY SAMPLING SCHEDULE: SWIMMING RIVER, NEW JERSEY

Parameter	Recharge	Recovery	Frequecy Daily	Frequecy Weekly
* Total suspended solids	X	X		X
Turbidity	X	X	X	
* Total dissolved solids	X	X		X
Sp. conductance	X	X	X	
pH	X	X	X	
Temperature	X	X		X
Dissolved oxygen	X		X	
Chloride	X	X	X	
Total alkalinity	X	X	X	
Calcium	X	X		X
Magnesium	X	X		X
Sodium	X	X		X
Potassium	X	X		X
* Iron		X	X	
Aluminum	X			X
Sulfate	X	X		X
TOC	X			X
* THM	X	X		X
Total coliforms		X		X
Gross alpha (cycle 1 only)		X		X

* = Parameters for which increased frequency required at certain times. During Cycle 1 recovery, DO readings were obtained weekly; temperature, Ca, Mg, SO daily.

electrical breakdowns occur and necessitate delays in the program, which can impact the usefulness of results. Maintaining flexibility is important.

Another important element is the need to continually analyze the hydraulic and water quality monitoring data collected during the test program. This provides a basis for adjustment of the testing plan as new data becomes available, thereby avoiding wasted time and effort. It can be quite frustrating to wait until all of the data is available from the first two or three cycles, and then realize that some important but subtle reactions or hydraulic problems occurred at the beginning of the program and were not detected until too late to adequately respond.

Data Collection

Hydraulic data collection includes the following parameters:

- flow rate during recharge and recovery
- cumulative volume stored
- water level or pressure in the well
- wellhead injection pressure
- water level response in observation wells
- elevation of pressure/water level measurement points

Accurate flow measurement is essential to the success of ASR programs. It is desirable (but usually not implemented) to provide two flowmeters in series during the test program, collecting duplicate data sets with different measurement devices. At such time as the readings may start to diverge, it becomes immediately apparent, and appropriate remedial steps can be taken. While this may appear overly conservative, experience at many ASR sites has included flowmeter failure or loss of accuracy. This is a problem at any time with any system; however, during the middle of an ASR test program, it can have a severe adverse affect upon interpretation of the results.

The same concern relates to water level measurement. Backup manual measurements should be collected along with any pressure transducer measurements, in case transducer failure occurs.

Figure 2.5 shows an example of a typical hydraulic data form utilized during Cycle 3 of testing at the Port Malabar, FL, ASR site. The form is set up in such a way that input data is immediately converted to specific capacity or specific injectivity, and cumulative volume is calculated. This facilitates real-time control of ASR test operations.

Water quality data collection is tailored to the specific needs of each site. Field data collection typically includes conductivity, pH, and temperature. Chloride is frequently selected as a natural tracer constituent, the

Date	Time	Elapsed Time (hrs)	Injection Meter	Volume Injected (gal)	Well Head Pressure (psi)	Injection Rate (gpm) (*calc)	Specific Injctvty (gpm/ft)	Percent Injected	Sample Number	Chloride (mg/l)	Cond. (mmh/cm)	pH	pHs	Alkalinity "P" (mg/l)	Alkalinity "T" (mg/l)	Hardness Total (mg/l)	Hardness Ca (mg/l)	Turbidity (ntu)
08/01/88	10 6	0.00	6839960	0	5.27	0	0	0.00	--	--	--	--	--	--	--	--	--	--
BEGIN INJECTION																		
		New meter installed																
08/01/88	12 15	0.00	0	0		405	0	0.00	--	--	--	--	--	--	--	--	--	--
08/01/88	17 20	5.08	57000	57,000	18.00	360 *	12.2	1.64	C3IR1.1	186	664	8.3	8.08	0	32	180	160	0.60
08/02/88	11 22	23.12	437000	437,000	18.50	351 *	11.5	12.57	C3IR2.1	178	653	8.2	8.14	0	32	168	140	0.32
08/03/88	7 55	43.67	866000	866,000	18.50	348 *	11.4	24.91	C3IR3.1	170	614	7.8	8.26	0	24	160	140	0.36
08/04/88	8 11	67.93	1374000	1,374,000	18.80	349 *	11.2	39.53	C3IR4.1	174	634	8.0	8.21	0	28	164	136	0.47
08/05/88	9 59	93.73	1872000	1,872,000	18.60	322 *	10.4	53.85	C3IR5.1	178	642	8.1	8.21	0	28	164	136	0.46
08/06/88	8 5	115.83	2379000	2,379,000	18.70	382 *	12.3	68.44	C3IR6.1	180	611	8.7	8.11	8	36	164	132	0.54
08/07/88	9 17	141.03	2908000	2,908,000	19.00	350 *	11.0	83.65	C3IR7.1	183	634	8.2	8.22	4	28	164	132	0.87
08/08/88	8 6	163.85	3386000	3,386,000	18.90	349 *	11.1	97.40	C3IR8.1	178	635	8.5	8.16	4	32	164	132	0.49
08/08/88	12 0	167.75	3469000	3,469,000	18.50	355 *	11.6	99.79	C3IR8.2	176	592	8.5	8.38	0	20	168	128	0.34
08/08/88	12 21	168.10	3476250	3,476,250	--	345	--	100.00	--	--	--	--	--	--	--	--	--	--

END INJECTION

NOTE: *Calculated Injection Rate

Figure 2.5A   Cycle testing data collection form, Port Malabar, Florida.

Date	Time	Elapsed Time (hrs)	Recovery Meter	Depth to Water (+=abv. MP / -=below MP)	Recovery Rate (gpm)	Specific Capacity (gpm/ft)	Volume Recovered (gal)	Percent Recovery	Sample Number	Chloride (mg/l)	Cond. (mmh/cm)	pH	pHs	Alkalinity "P" (mg/l)	Alkalinity "T" (mg/l)	Hardness Total (mg/l)	Hardness Ca (mg/l)	Turbidity (ntu)	H2S (mg/l)	Sulfate (mg/l)
08/08/88	14 30	--	0	+15.75	0	0.0	0	0.0	--	--	--	--	--	--	--	--	--	--	--	--
BEGIN RECOVERY																				
08/08/88	15 1	0.00	0	--	0	--	0	0.00	--	--	--	--	--	--	--	--	--	--	--	--
08/08/88	15 2	0.02	300	--	--	--	300	0.01	C3RR1.1	--	--	--	--	--	--	--	--	--	--	--
08/08/88	15 6	0.08	2500	--	550 *	--	2,500	0.07	C3RR1.2	--	--	--	--	--	--	--	--	1.8	--	--
08/08/88	15 12	0.18	5500	--	700	--	5,500	0.16	C3RR1.3	--	--	--	--	--	--	--	--	1.5	--	--
08/08/88	15 17	0.27	10000	--	900 *	--	10,000	0.29	C3RR1.4	--	--	--	--	--	--	--	--	1.25	--	--
08/08/88	15 27	0.43	16000	--	600 *	--	16,000	0.46	C3RR1.5	--	--	--	--	--	--	--	--	1.75	--	--
08/08/88	15 30	0.48	18500	--	-- *	--	18,500	0.53	C3RR1.6	--	--	--	--	--	--	--	--	0.66	--	--
08/08/88	15 40	0.65	25500	--	700 *	--	25,500	0.73	C3RR1.7	--	--	--	--	--	--	--	--	0.38	--	--
08/08/88	15 49	0.80	31500	-39.67	667	12.0	31,500	0.91	C3RR1.8	--	--	--	--	--	--	--	--	0.47	--	--
08/08/88	16 5	1.07	43000	-45.81	700	11.4	43,000	1.24	C3RR1.9	190	614	8.5	8.3	0	24	172	140	0.82	--	--
08/08/88	20 0	4.98	208500	--	710	--	208,500	6.00	C3RR1.10	190	614	8.3	8.0	0	40	188	160	0.25	--	--
08/09/88	7 43	16.70	700500	-48.63	700	10.9	700,500	20.15	C3RR2.1	185	676	8.3	8.0	0	40	190	148	0.50	--	51
08/09/88	14 20	23.32	975000	-48.15	695	10.9	975,000	28.05	C3RR2.2	186	712	8.4	8.0	0	40	192	152	0.22	0.00	52
08/09/88	19 50	28.82	1160000	-46.50	695	11.2	1,160,000	33.37	C3RR2.3	180	712	8.5	8.0	0	40	188	156	0.30	0.00	50
08/10/88	8 20	41.32	1689500	-50.96	725	10.9	1,689,500	48.60	C3RR3.1	195	730	8.2	7.9	0	44	208	164	0.34	0.00	54
08/10/88	13 10	46.15	1870680	--	625 *	--	1,870,680	53.81	C3RR3.2	200	--	--	--	--	--	--	--	0.31	--	--
08/10/88	15 20	48.32	1960000	--	680	--	1,960,000	56.38	C3RR3.3	198	734	8.3	7.9	0	44	204	168	0.32	0.00	55
08/1088	19 26	52.42	2123000	-43.50	660	11.1	2,123,000	61.07	C3RR3.4	216	767	8.2	8.0	0	44	220	160	0.38	0.00	60
08/11/88	8 15	65.23	2672500	-51.43	730	10.9	2,672,500	76.88	C3RR4.1	250	846	8.2	7.8	0	60	252	180	1.4	0.00	67
08/11/88	10 48	67.78	2786000	--	730	--	2,786,000	80.14	C3RR4.2	246	--	--	--	--	--	--	--	--	--	--
08/11/88	12 45	69.73	2873245	-49.21	705	10.9	2,873,245	82.65	C3RR4.3	265	--	--	--	--	--	--	--	--	--	--
08/11/88	14 41	71.67	2910000	--	700	--	2,910,000	83.71	C3RR4.4	265	--	--	--	--	--	--	--	--	--	--
08/11/88	15 45	72.73	2952500	--	700	--	2,952,500	84.93	C3RR4.5	275	--	--	--	--	--	--	--	--	--	--
08/11/88	15 54	72.88	2961000	--	--	--	2,961,000	85.18	C3RR4.6	--	947	8.2	7.9	0	44	268	196	0.48	0.00	--
08/11/88	16 45	73.73	2998000	-45.70	725 *	11.8	2,998,000	86.24	C3RR4.7	263	962	8.2	7.8	0	56	264	184	0.39	0.00	71
08/11/88	17 0	73.98	3001985	--	--	--	3,001,985	86.36	--	--	--	--	--	--	--	--	--	--	--	--

END RECOVERY

NOTE: *Calculated Recovery Rate

Figure 2.5B

concentration of which is analyzed daily during recharge and recovery. Dissolved oxygen may be obtained at a few intervals during recharge and early in recovery. Any significant, rapid change in dissolved oxygen concentration can indicate geochemical or bacterial activity underground. Where possible, field measurements of Eh, or oxidation-reduction potential, are useful to support geochemical analyses, particularly in confined aquifers. Where Eh measurement is not possible, dissolved oxygen measurements provide an indirect indicator of Eh, particularly in unconfined aquifers where groundwaters contain oxygen.

Laboratory analyses typically include a broad range of parameters, a few of which are analyzed daily and the remainder less frequently. Table 2.1 presents a typical range of water quality parameters considered for ASR test program applications. Table 2.4 shows a typical water quality sampling schedule, as planned for the ASR test program at Swimming River. The range and frequency of a water quality sample collection usually changes during the test program, in response to early laboratory analysis results and other observations. However, the first cycle generally is subject to rather intensive sampling, data collection, and laboratory analysis, in order to gain a rapid understanding of hydraulic and water quality issues. Subsequent cycles require less intensive sampling and analysis, unless an important issue is identified in the first cycle that requires such effort in subsequent cycles.

Sampling Frequency

Recharge water quality variability determines the frequency of sampling appropriate during recharge periods. Sometimes recharge quality is sufficiently stable that sampling need occur only at the beginning, middle, and end of the recharge period. Usually more frequent recharge sampling is appropriate. Field data collection should occur daily. Recovery sampling should be sufficiently frequent as to show clearly the system response, both hydraulic and water quality. A minimum of 10 samples during a normal recovery period is usually appropriate except for very short cycles. Daily samples should be collected and analyzed for basic parameters and tracer constituents.

ASR Test Program Duration

Depending upon the range of issues addressed during the test program, the duration may extend for as short as about 3 months and as long as 2 years or more. Short programs are appropriate in situations where no geochemical or water quality issues are involved and where plugging does not appear to be significant in early testing. On the other hand, where the storage zone contains brackish or other non-potable water, or where complex geochemical issues are present, the test program duration may easily extend for two years. Where a short duration is selected, care must be taken to continue monitoring for subtle water quality changes that may not become apparent until the aquifer has reached water quality and geochemical equilibrium. This may require several months.

A typical duration is about 6 to 12 months, at the end of which a report is prepared presenting results of the test program and also a plan for ASR wellfield expansion to meet projected needs, if appropriate. The report provides the justification for issuance of an ASR operating permit in those states where such a permit is required between the end of the test program and the beginning of permanent operation.

2.4 PHASE 3: ASR WELLFIELD EXPANSION

There are a few factors to consider in the design of an ASR wellfield that are quite different than for a conventional wellfield. Failure to consider these factors can reduce the value of an ASR system to meet water demands.

Flow Rate Balancing

Where there is a significant difference in quality between the stored water and the native groundwater and it is essential to minimize mixing between the two prior to recovery, it is important that the wellfield be designed so that recharge flow rate distribution among the wells is proportional to recovery flow rate distribution. If this balance is not maintained, then excessive recharge or recovery in one well can effectively move the stored water bubble away from adjacent wells, reducing overall system recovery efficiency. This requires a design that allows flow rate adjustment at each wellhead.

If recharge and recovery flow rates are unbalanced, the effect may become apparent as a change in recovery efficiency for each of the wells, when compared to results immediately following their construction and initial testing. Some wells may have very high efficiency, while others are very low.

Well Spacing and Arrangement

For an ASR system that primarily meets seasonal water supply needs, the annual volume of water stored will create a bubble of stored water around the well, with a diameter typically in the range of 100 to 300 m (328 to 984 ft). If well spacings are less than this diameter, the storage bubbles from adjacent ASR wells should tend to coalesce. Over a period of several cycles, this should tend to improve recovery efficiency, compared to a wellfield with spacings such that no coalescence occurs. As discussed subsequently in Section 3.3, Wellfield Design, well arrangements including a central well should provide improved recovery efficiency. The tradeoff between increased annual power costs for recharge and recovery in a "clustered" ASR wellfield with closely-spaced wells can be compared with the expected improvement in recovery efficiency, based upon Phase 2 results and modeling, as discussed in Chapter 4. A decision can then be made regarding appropriate well spacing. In general, closer well spacing will be appropriate in more brackish aquifer systems, while conventional well spacing criteria will tend to govern wellfield design in situations where groundwater quality is fresh.

Stacking

In many ASR applications, multiple aquifers or producing intervals are available at the test site. Shallow aquifers are generally the freshest, with deeper aquifers containing water of poorer quality. For ASR purposes, aquifer native water quality is less important than hydraulic characteristics. A deep brackish aquifer that has little or no value for water supply purposes can be quite useful for ASR purposes. This opens up the possibility of storing water in multiple aquifers at the same site. This is termed "stacking." To the extent that storage in the shallowest zone can meet program objectives, this may be the least expensive plan. However, piping and pumping costs to distribute ASR wells over a wide area in such a zone can be substantial. In such cases it is frequently more cost-effective to locate ASR wells in clusters, penetrating multiple zones in order to meet storage volume requirements in a relatively small area.

Figure 2.6 shows a hydrogeologic cross-section of the multiple storage zones at the Peace River ASR wellfield. The water in each of these zones is brackish, with TDS concentrations ranging from about 700 to 900 mg/L in the upper two zones, and about 2900 mg/L in the deepest zone. The shallowest zone, the Tampa formation, stores little water due to well interference with adjacent domestic and agricultural wells. The second zone, the Suwannee formation, stores almost all of the water at this time. The deep zone, the Avon Park formation, is being tested to determine its recovery efficiency characteristics. The high yield of wells in this deep zone suggests low unit costs for water stored in this zone, if reasonable

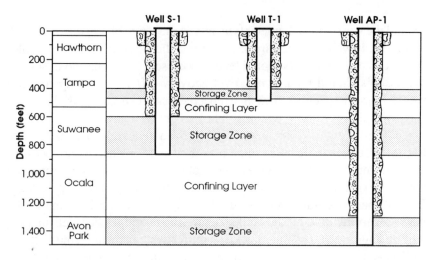

Figure 2.6 Stacking ASR storage in multiple zones, Peace River, Florida.

recovery efficiency can be confirmed. With all three zones, seasonal storage requirements could be met for many years with wells all located on the water treatment plant site. Cost savings would be substantial.

Wellfield Layout

Land ownership and other practical constraints frequently govern the location of ASR wells. As with conventional wellfields, layout can include dispersed individual wells located at key points around the distribution system; linear arrangements, and clusters. As discussed above, more-compact layouts are better suited for brackish or otherwise non-potable water storage zones.

Where the regional hydraulic gradient is such that stored water may move a significant distance away from each ASR well between the time of recharge and the time of recovery, a linear arrangement generally perpendicular to the direction of the regional gradient may be appropriate. Flows during recharge and recovery would be distributed so that more recharge occurs in up-gradient wells and more recovery occurs in down-gradient wells. **Figure 2.7** illustrates this approach, as modeled for a proposed ASR wellfield in Kuwait that was to form part of a peak season water supply and also a strategic water reserve. Regional water movement on a seasonal basis was insignificant. However, long-term water movement over a period of many years would tend to reduce recovery efficiency without provision for downgradient ASR wells. The proposed wellfield arrangement provided for seasonal operation of all wells; however, the distribution of recharge and recovery would be adjusted slightly to achieve

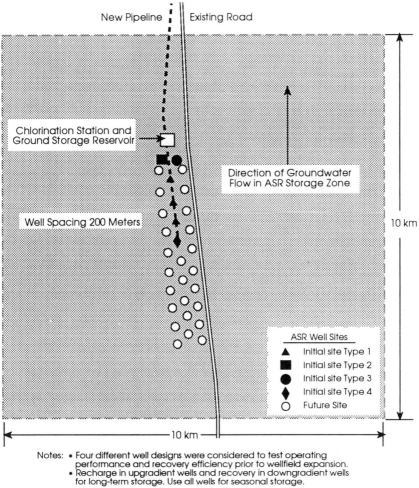

New Pipeline   Existing Road

Chlorination Station and Ground Storage Reservoir

Direction of Groundwater Flow in ASR Storage Zone

Well Spacing 200 Meters

10 km

ASR Well Sites
▲ Initial site Type 1
■ Initial site Type 2
● Initial site Type 3
♦ Initial site Type 4
○ Future Site

10 km

Notes: • Four different well designs were considered to test operating
performance and recovery efficiency prior to wellfield expansion.
• Recharge in upgradient wells and recovery in downgradient wells
for long-term storage. Use all wells for seasonal storage.

Figure 2.7 Strategic water reserve: proposed well layout, Kuwait.

long-term storage objectives. The conceptual design provided for higher recovery rates in the central line of wells so that reasonable flows could be maintained even though saline water may reach the outer lines of wells during emergency recovery.

2.5 OPERATIONS AND MAINTENANCE

ASR wellfield operation and maintenance requirements are marginally greater than for conventional wellfields. Some of the elements that comprise the difference are as follows.

Periodic Change in Operating Mode

This occurs typically two to four times per year as the system changes from recharge to recovery and back again. Where ASR wells are used to meet peak demands in areas where diurnal and weekly variations are more significant than seasonal variations, more frequent changes in operating mode will occur. Automated controls facilitate such operations. The procedure at each well may need to include flushing lines to waste for a few minutes at each wellhead, depending upon the amount of rust, sand, and other solid material in the piping. Proper selection of construction materials can substantially ease this operating requirement.

Backflushing to Waste During Recharge

This procedure is implemented at most ASR sites in order to maintain recharge capacity by purging from the well any solids that may have been carried into the well during recharge. Backflushing frequency ranges from every day at two long-term operational ASR sites in New Jersey, to every few years at the Orange County Water District salinity intrusion barrier injection wellfield in southern California. The two sites in New Jersey are automated and pump to waste for 10 min each day. The Orange County sites require removal of all injection piping from the well, introduction of a packer and jetting assembly, well redevelopment while pumping each of a series of screen intervals, and re-assembly of injection tubing. The effort typically requires about five days per well.

Some ASR wells currently in operation are redeveloped seasonally by extended pumping, as a part of the recovery operation, without any additional backflushing frequency. Some ASR sites backflush to waste every one to two weeks to maintain recharge capacity. Whether automated or not, this increases the operation and maintenance requirements for ASR facilities. Further discussion of well plugging rates and redevelopment frequency is included in Chapter 4.

Disinfectant Residual

It is advisable to ensure a disinfectant residual within the well casing, screen and gravel pack, or borehole. This will control bacterial activity in the formation immediately adjacent to the well, thereby helping to maintain recharge specific capacity and avoid bacterial plugging. This is easily accomplished during recharge if the recharge source is treated drinking water with a small disinfectant residual.

In some situations initial baseline sampling will indicate that no chlorine residual is available at the ASR site, reflecting low flows and long travel times in the water distribution system during recharge months. Operation of ASR facilities for recharge will accelerate local water movement, sometimes achieving a positive chlorine residual.

During storage periods in excess of about two to three days, it is advisable to trickle feed a chlorine, chloramine or other disinfectant solution into the well casing, at a low rate of typically about 10 to 20 L/min (2 to 5 G/min). Alternatively, a periodic flush of recharge water from the distribution system can sometimes achieve the same objective if the volume is sufficient to displace water in the casing and screen or borehole with chlorinated water. Samples obtained from the well at daily intervals following the beginning of a storage period can show the time required for the disinfectant residual to dissipate in the well, thereby providing a basis for calculation of the trickle flow recharge rate or the frequency of casing flushes required to maintain the well free of bacteria.

This is a precautionary step that may not be required in all ASR wells. In particular, it may be unnecessary for storage zones with high transmissivity, karst solution features, or otherwise with low plugging potential due to the low organic and carbon content of the recharge water. However, at most water utility sites it is relatively easy and inexpensive to prevent bacterial plugging, which can be difficult to rectify once it has occurred.

A higher trickle flow rate during storage periods may be appropriate at those sites storing drinking water in aquifers with very high TDS concentrations in the native water. For example, at Marathon, the trickle flow rate required for disinfection during storage periods is under 0.3 L/sec (5 G/min); however, about 2.8 L/sec (45 G/min) is needed to maintain a target recovery volume of 38 megaliters (10 MG) for a typical hurricane season duration of 120 days. The native water TDS at this site is 39,000 mg/L. The trickle flow rate offsets mixing losses due to density stratification at this site.

Monitoring Data and Reports

Monitoring requirements for ASR systems depend upon local needs and regulatory requirements at each site. It is vitally important to accurately keep track of flow rates and cumulative volumes at each well and for the entire ASR wellfield. Regular calibration of flowmeters is therefore essential.

Periodic measurement of water levels in ASR and monitor wells is also important, in order to check for well plugging rates, pump efficiencies, effect of ASR operations upon water levels in adjacent wells, and other criteria. Routine sampling in ASR and monitor wells is also appropriate, in order to keep track of the recovered water quality and, at some brackish zone locations, the lateral extent of the stored water bubble.

For large ASR wellfields, it is advisable to develop a computer-based data collection and analysis system that can provide monthly, annual, and other reports as needed to meet operational and regulatory needs. For automated systems with computer control, this may represent a relatively easy extension of existing hardware and software.

An important element of ASR operation and maintenance is the need for periodic review of the operating data so that performance can be evaluated relative to expectations, and adjustments made as appropriate. This is particularly important in the first two to five years of system operation when the greatest changes in water levels and water quality will tend to occur and normal ranges for various parameters tend to be defined.

2.6 WATER SUPPLY PLANNING WITH ASR

At some logical point in the overall ASR development program, it becomes appropriate to consider how to integrate ASR as a water management tool into the local and regional water supply network. This may occur at the end of the Phase 2 field investigations, as a part of the report preparation. It is more likely to occur after a few ASR wells have been constructed (Phase 3) and are in operation for a year or two. At that point the utility operating the ASR system has gained confidence in long-term performance, justifying reevaluation of future plans for water system expansion.

As discussed in Chapter 4, an ASR water supply system model can be helpful in evaluating the most cost-effective combination of capacities for key water system components (i.e., alternative sources, intake structure, treatment plant, raw water storage, ASR treated water storage, etc.) to meet future demand at different time steps along the expansion path. For many water systems, such a model may be essential in view of the complexity of the economic analysis. The relatively low cost of ASR systems, as discussed in Chapter 6, is usually a strong incentive to provide a careful and accurate analysis of the optimum role that these systems can play in future water supply planning.

CHAPTER 3

Design
of
ASR
Systems

Once there was a great drought. The rain stopped falling and the Earth became dry. Finally the streams themselves stopped flowing. There was a village of people who lived by the side of a stream, and life now became very hard for them. They sent someone upstream to see why the stream had stopped. Before long, the man came back.

"There is a dam across the stream," he said. "It is holding back all of the water. There are guards on the dam. They say their chief is keeping all the water for himself."

"Go and beg him for water," said the elders of the village. "Tell him we are dying without water to drink." So the messenger went back again. When he returned, he held a bark cup filled with mud.

"This is all the water their chief will allow us to have," he said.

Now the people were angry. They decided to fight. They sent a party of warriors to destroy the dam. But as soon as the warriors came to the dam, a great monster rose out of the water. His mouth was big enough to swallow a moose. His belly was huge and yellow. He grabbed the warriors and crushed them in his long fingers, which were like the roots of cedar trees. Only one warrior escaped to come back to the people and tell them what happened.

"We cannot fight a monster," the people said. They were not sure what to do. Then one of the old chiefs spoke. "We must pray to Gitchee Manitou," he said. "Perhaps he will pity us and send help." Then they burned tobacco and sent their prayers up to the Creator.

Their prayers were heard. Gitchee Manitou looked down and saw the people were in great trouble. He decided to take pity and help them and he called Koluscap. "Go and help the people," Gitchee Manitou said.

Koluscap then went down to the Earth. He took the shape of a tall warrior, head and shoulders taller than any of the people. Half of his face was painted black and half was painted white. A great eagle perched on his right shoulder and by his side two wolves walked as his dogs, a black wolf and a while wolf. As soon as the people saw him they welcomed him. They thought surely he was someone sent by the Creator to help them.

"We cannot afford you anything to drink," they said. "All the water in the world is kept by the monster and his dam."

"Where is this monster?" Koluscap said, swinging his war club, which was made of the root of a birch tree.

"Up the dry stream bed," they said.

So Koluscap walked up the dry stream bed. As he walked he saw dried up and dead fish and turtles and other water animals. Soon he came to the dam, which stretched between two hills.

"I have come for water," he said to the guards on top of the dam.

"Give him none, give him none!" said a big voice from the other side of the dam. So the guards did not give him water.

Again Koluscap asked and again the big voice answered. Four times he made his request, and on the fourth request Koluscap was thrown a bark cup half-full of filthy water.

Then Koluscap grew angry. He stomped his foot and the dam began to crack. He stomped his foot again and he began to grow taller and taller. Now Koluscap was taller than the dam, taller even than the monster who sat in the deep water. Koluscap's club was now bigger than a great pine tree. He struck the dam with his club and the dam burst open and the water flowed out. Then he reached down and grabbed the water monster. It tried to fight back, but Koluscap was too powerful. With one giant hand Koluscap squeezed the water monster and its eyes bulged out and its back grew bent. He rubbed it with his other hand and it grew smaller and smaller.

"Now," Koluscap said, "no longer will you keep others from having water. Now you'll just be a bullfrog. But I will take pity on you and you can live in this water from now on." Then Koluscap threw the water monster back into the stream. To this day, even though he hides from everyone because Koluscap frightened him so much, you may still hear the bullfrog saying, "give him none, give him none."

The water flowed past the village. Some of the people were so happy to see the water that they jumped into the stream. They dove so deep and stayed in so long that they became fish and water creatures themselves. They still live in that river today, sharing the water which no one person can ever own.

Mic Mac and Maliseet Indian story, Nova Scotia [1]

3.1 WELLS

ASR wells have certain unique features in their design that differentiate them from production wells or injection wells. When completed, the wells may often be similar; however, the design process is different, and the end results may also be different, depending upon conditions at each ASR site. In this section, the ASR well design approach is discussed, addressing those features that differentiate ASR wells from other wells.

Casing Materials of Construction

ASR wells generate rust from steel casings to a greater extent than either production or injection wells, due to the increased surface area subject to wetting and drying during recharge and recovery. This is particularly true for brackish water storage zones. This rust flows down the well during recharge, contributing to plugging of the well. Solids present in the recharge water are usually more significant causes of ASR well plugging than is rust. However, for low permeability aquifers, the increase in plugging potential due to rust can be unacceptable in some cases, particularly where frequent backflushing to waste is perceived as an operating problem to be avoided if possible. During recovery or backflushing redevelopment, the rust combines with other solids carried into the well during recharge and is pumped from the well, either to waste or into the wellhead piping system.

For ASR wellfields located at water treatment plants, it is not uncommon for the water pumped from the well at the beginning of recovery to be conveyed back to the treatment process for retreatment. Duration of this period may typically range from about 10 min to 2 hours or more. Once the rust and other particulates have been flushed from the well, the water can then be diverted directly into the treated water distribution system following disinfection.

For ASR wells located other than at water treatment plants, the only option is to waste this water to a nearby drainage system or sewer line. The pumping rate at such times may be slightly greater than the design recovery rate for the well, since the pump is usually pumping against a lower head than normal and therefore producing more water. This is good in that it helps to purge solid material from the well; however, disposal of water for an extended period at such rates is sometimes a problem. In residential areas lacking storm drainage networks, ASR backflushing operations can cause temporary, localized flooding of streets and homeowner opposition due to inconvenience and apparently wasted water. In other areas with adequate drainage networks, regulatory opposition may be encountered due to the ultimate discharge into a receiving stream for this colored water, initially containing considerable solid material. That this event may occur infrequently during initial testing and then perhaps once or twice per year is of little assistance.

One solution to this problem is to utilize casing material that will not contribute to the production of rust. In particular, polyvinyl chloride (PVC) casing offers many advantages in situations where casing length and diameter are suitable. Where a steel casing is required, epoxy coating can substantially reduce or eliminate the surface area of steel that is subject to rusting. Both of these approaches have been used successfully in operational ASR systems. Other potential approaches that may be appropriate in certain cases include fiberglass casings and stainless steel casings. Finally, a frequent solution is to utilize a conventional black steel casing and accept the solids production as a long-term operating issue to be dealt with later. Further discussion of each of these options follows.

PVC Casing

PVC well casing is utilized frequently in the water well industry, particularly for smaller, shallower wells. For larger, deeper wells, greater care is required in order to ensure satisfactory construction. PVC casing diameters are readily available up to 400 mm (16 inches) and are available by special order up to 900 mm (36 inches). Recent introduction of mechanical joint casing connections by CertainTeed Corporation (Phone: 1-800-359-7296) has dramatically speeded up the installation process, thereby reducing both risk and cost.

The deepest PVC casing for an ASR well is at Marathon, FL, which is 400 mm (16 inch) diameter to 118 m (387 ft). For other purposes, the author is aware of 200 mm (8 inch) PVC casing set to 335 m (1100 ft); 300 mm (12 inch) casing set to 293 m (960 ft), and 460 mm (18 inch) casing set to 335 m (1100 ft). All of these are Florida projects. In general, setting a PVC casing to a depth of about 152 m (500 ft) is not too difficult. Below that depth, the technology exists to successfully construct the well; however, considerable time, skill, and great care is required to avoid casing or hole collapse.

For most ASR systems, PVC casing diameters in the range of 200 to 400 mm (8 to 16 inches) will be appropriate. These are outside diameter measurements. Wall thickness for PVC casings is thicker than for steel casings. Schedule 80 casing is frequently selected; however, this is a standard wall thickness and therefore provides decreasing resistance to collapse with larger diameters. Selection of casing according to an SDR (standard dimension ratio) number, such as SDR 17, ensures consistent collapse strength regardless of casing diameter. For SDR 17, the wall thickness is 1/17 times the outside diameter of the casing. For 400 mm (16 inch) casing, Schedule 80 would have a 19 mm (0.75 inch) wall thickness, whereas SDR 17 would have a 24 mm (0.94 inch) wall thickness. Inside diameters of PVC casings are therefore smaller for corresponding pipe sizes.

Couplings add another 40 mm (1.5 inches) outer diameter to the casing string. Since it is important to be able to introduce a 50 mm (2 inch) tremie line into the annulus around the PVC casing for cementing operations, the hole into which the casing is run should be at least 150 mm (6 inches) greater diameter than the couplings. Installation of long runs of PVC casing in a well is comparable to threading a wet spaghetti noodle down the barrel of a rifle. It is therefore advisable to provide a sufficient annular space in order to facilitate installation.

The compressive strength of the PVC casing is weakest during cementing of the casing, when the heat of hydration can be sufficient to raise the casing temperature and thereby reduce its strength. For this reason it is sometimes advisable to cement the casing in a series of stages. Circulating water within the casing while the cement is curing for each stage is another option. Control of temperatures and pressures is important. All of this extra care increases well construction time and therefore cost. The extra cost, combined with the increased installation difficulty, are the principal reasons why deeper, larger ASR well casings have traditionally utilized materials other than PVC. However, successful experience with deep, large diameter PVC casings is becoming more widespread.

It is important that the hole to be cemented stays open during the successive cementing stages. Settling of drilling mud and subsequent hole collapse may preclude staged cementing of the casing to ground surface in some situations where the casing installation and cementing operation requires excessive time. At extra cost, the risk of hole collapse during cementing can be offset through setting additional outer steel casings so that each cementing stage is likely to terminate within the next steel casing string.

For screened wells constructed with cable tool techniques, PVC casing may be inappropriate since the force required to pull back the casing to expose the screen and gravel pack may exceed the tensile strength of the casing couplings, causing them to pull apart. This should be discussed with the well driller where applicable. It is not a problem with rotary drilled wells.

The use of PVC casings at the Cocoa and Marathon, FL, ASR sites has been effective, in conjunction with appropriate selection of wellhead piping materials, in keeping the production of solid materials during recharge and recovery to an acceptable minimum. At each site, water is clear and meets all applicable drinking standards within about 20 min after the beginning of recovery.

A factor to consider during installation of PVC casings is that the density of the PVC is approximately the same as the density of drilling mud. Consequently, it may be necessary to push the casing down the hole prior to cementing, particularly if the drilling fluid weighs more than about 12 lbs/G, or has a specific gravity greater than about 1.44.

Epoxy-Coated Steel Casing

A fusion-bonded epoxy coating can be applied at the factory and is selected to meet applicable American Water Works Association (AWWA) standards for use with public drinking water systems. For welded steel casing, the coating adjacent to each weld will be lost during construction. For threaded and coupled casing, the coating would remain; however, the marginal reduction in surface area exposed to rusting is probably insufficient justification by itself to provide threaded and coupled casing at significantly greater cost.

Care is required during well construction to avoid damage to the coating during well construction activities that occur after the casing is set and cemented in place. Rubber bumpers have been used successfully around the drill pipe to prevent such damage.

Epoxy-coated steel casing has been used successfully at the Peace River, FL; Kerrville, TX; and Chesapeake, VA ASR facilities. Production of solids from the casing during recharge or recovery has not been an operating issue at these sites.

Fiberglass and Stainless Steel Casing

These have not yet been utilized for ASR wells, primarily due to the availability of less-expensive alternatives. It is reasonable to expect that use of these casing materials will occur sooner or later for an ASR system, in order to meet site-specific needs.

ASR system operating costs are discussed in Chapter 6. At such time as we have greater understanding of the annual costs associated with periodic backflushing for well redevelopment, and for initial recovery to waste to remove solids, the tradeoff between investment in higher cost well casing and reduced operating costs will become more clear. Until that time, it seems wise to seek reasonable opportunities to minimize solids production. One of these alternatives is to select an appropriate casing material, even at some increase in well construction cost.

Cost is not the only issue here. Operational complexity is an important factor. An ASR well in a low permeability, unconsolidated aquifer that requires backflushing every few days or weeks represents less of an operating problem if the backflushing duration and pumped volume is minimized. Selection of an appropriate casing material can help to minimize the frequency, duration, and volume of backflushing.

Steel Casing

Uncoated steel casing is utilized in many ASR situations, similar to conventional production wells. When existing production wells are to be utilized for ASR testing or operations, there is usually no choice in the well design or casing material. The only option available is cleaning the existing casing prior to ASR retrofitting.

Operating measures can be implemented to handle the solids produced from the steel casing during recovery and to minimize introduction of solids to the well during recharge. During initial recovery or backflushing operations, water can be returned to the water treatment plant for retreatment. It can also be discharged directly to the storm drainage system, whether piped or by surface conveyance. It may also be discharged to a drywell, pit, or pond, constructed adjacent to the ASR well with sufficient volume to contain a considerable portion of the water to be discharged to waste. This may reduce the peak rate of discharge to the local drainage system, if necessary, and will provide some settling and dilution of the initial solids pumped from the well. This approach is utilized at El Paso, TX, for the Fred Hervey Water Reclamation Facility injection well system. It was also utilized during initial testing at the Riverton Heights ASR test well, Seattle, WA.

One or more injection tubes are sometimes provided inside the casing to control cascading. These tubes also serve to reduce entry of rust into the well during recharge since the surface area of exposed steel is reduced. Water is conveyed into the well through the tubes and therefore is less likely to scour rust from the casing. The tubes may be of steel or PVC. This requires sufficient casing diameter to accommodate the pump and also the injection tube(s). An alternative approach is to recharge down the pump column, achieving the same objective.

Casing Diameter

For ASR wells, casing diameter should be no different than for normal production wells, except when one or more injection tubes are used for recharge. It is advisable to have a small amount of extra space inside the casing to allow easy entry and withdrawal of the pump, injection tubing, air line, and electrical cable for submersible pumps. It is also helpful to be able to lower geophysical logging tools past the pump in order to evaluate changes in flow distribution due to differential plugging of the well during recharge.

However, larger casing diameters increase well construction expense. For the first well at any new ASR wellfield, it may be appropriate to pay the additional expense in order to gain the ability to better understand aquifer response to ASR operations. Subsequent wells could be designed with smaller diameter casings to reduce costs. Frequently well construction costs represent a relatively small component of the overall construction cost for an ASR wellfield.

Due to the increased cost associated with larger casing diameters, the downhole control valve discussed subsequently in Section 4.4, Flow Control and Measurement, is expected to rapidly gain in popularity since it provides needed control of injection flows without requiring larger casing diameters. This valve expedites ASR operations in situations where existing wells with small diameter casings are retrofitted to ASR operations.

Cementing

Casings in ASR wells should be cemented from the bottom of the casing to ground surface to ensure an adequate seal against flow movement outside the casing through possible channels opened during construction. At three known ASR sites in existing uncemented water supply wells, the pressures occurring during recharge caused upward flow around the outside of the casing. At one site this created flow at the surface. At another site the result was formation of a sinkhole adjacent to the well. The third site experienced downward movement of surficial sands into the underlying limestone production interval, causing a severe solids problem in the well. Cementing is normally a desirable practice to prevent production well contamination from adjacent land use activities; however, for ASR wells there are additional hydraulic reasons that apply due to the cyclic operation.

Selection of ASR Storage Intervals

Water Quality Issues

The simplest case is one in which the ASR storage zone under consideration contains water of similar quality to that which will be recharged, and has no potential geochemical issues. In such a case, the ASR well design will tend to be similar to a conventional production well design. If screened, the screen length will tend to be longer to maximize recharge efficiency and to minimize the rate of plugging. If open hole, the hole length will tend to fully penetrate the production interval for the same reason.

In situations where the storage zone is brackish, or contains water of such quality that mixing is to be minimized, the selection of the storage interval requires greater care. Thin intervals that have excellent vertical confinement are best suited for minimizing mixing. **Figure 3.1** shows a geologic cross-section for the Marathon ASR test well, FL, which successfully stores treated drinking water for emergency water supply purposes. Storage is in a confined sand production interval 11 m (40 ft) thick and containing seawater with a total dissolved solids (TDS) concentration of 39,000 mg/L. This site is discussed further in Chapter 9. In less extreme cases of water quality difference, thicker storage intervals with less confinement may be sufficient to provide the desired recovery efficiency.

Where the choice of storage intervals is limited, well and wellfield design can, to some extent, adapt to the limitations imposed by nature. Multiple wells can provide ASR development of a zone that has sufficient storage volume capacity but low yield to individual wells. The cost of additional wells is frequently small when compared to the cost of alternative storage approaches.

Where the storage zone has great thickness or poor confinement and contains poor quality native groundwater, acceptable recovery performance may sometimes be achieved by operating at high rates and long durations during recharge. The volume stored may then be sufficient to displace the poor-quality water away from the well, both vertically and laterally, so that a useful recovery volume can be achieved during each recovery season. This may take several annual cycles of operation, each showing an increase in recovery efficiency. Alternatively, a large initial storage volume may be provided following construction. This may be considered as the formation of a buffer zone, analagous to initial filling of a surface reservoir. Once the buffer zone is formed, or the surface reservoir is filled, ASR operations at the ultimate recovery efficiency can proceed.

The recovery efficiency attainable will depend upon the hydraulic and water quality characteristics at each site. While 100% recovery efficiency is a reasonable target and is obtained in most cases of storage in brackish aquifers, lower recovery efficiency may occur in some situations due to technical constraints or regulatory restrictions designed to promote aquifer recharge. An economic analysis will then indicate whether the lost value of the water not recovered is more than offset by the value of the water recovered when needed. Usually this is the case.

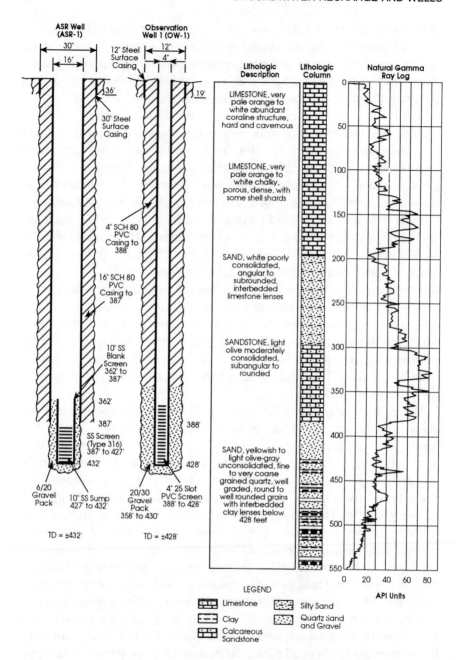

Figure 3.1 ASR construction details and lithology, Marathon, Florida.

Geochemical Issues

The most complex issues pertaining to storage zone selection are with aquifers, or portions of aquifers, that offer geochemical challenges. This is discussed in greater detail in Chapter 5, Geochemistry. However, one solution is to design the ASR well to case out production intervals that contribute severe geochemical problems, if this can be achieved without losing much of the potential production capacity of the well.

Typically, the detailed information needed to make a reasonable judgment regarding well design to avoid geochemical problems can only follow coring, core analysis, and geophysical logging. In the absence of this data, it is difficult to know which intervals are contributing the water quality constituents of concern. Consequently, the design of the second and subsequent ASR wells may benefit from experience gained with the first such well at any new site.

Screen Design

Discussion has occurred during recent years regarding whether the screen and gravel pack design for an ASR well in an unconsolidated aquifer should be any different than for a normal production well. A case can be made that the screen slot size should be slightly larger than normal such that during pumping and redevelopment, the gravel pack will clear of solid particles more readily. In such situations it is necessary to add one or two gravel tubes from the ground surface to the top of the gravel pack, adding gravel as necessary to make up for gravel pack material washed through the casing during recovery. Otherwise the formation material may collapse around the screen during changes from recharge to recovery.

However, some uncertainty will always exist as to whether provision of one or more gravel tubes will adequately protect the well against collapse. The best solution is to design the screen identical to that for a production well so that there is no increased likelihood of gravel pack movement around the screen. Whether used for production or for ASR, a sand-producing well is an operating problem to be avoided.

Pump Setting

One alternative that has not yet been applied in an ASR well is to set the pump below the screen interval at the bottom of the well. For seasonal operation, this practice may enable recovery rates well in excess of normal continuous production rates. The specific yield of unconfined portions of the aquifer will release much larger volumes of water from storage than will the storativity associated with confined aquifers. Consequently, a pump located below the screen may support higher seasonal production rates and longer durations than will a pump in the same well, located at the base of the casing, above the screen. During the subsequent recharge period, water levels would be restored to normal high levels or possibly above those levels. **Figure 3.2** illustrates this design option.

It is probable that this approach would work well at some sites and not at others. In particular, where the uppermost screen interval is close to the top of the aquifer and produces most of the water, the benefits of this approach may not be significant since dewatering the top screen interval would rapidly increase hydraulic losses and reduce well yield. However, where production is distributed more uniformly from the top to the bottom of the aquifer, dewatering the top portion of the aquifer for a brief time may be beneficial in terms of increasing short-term peak well yields.

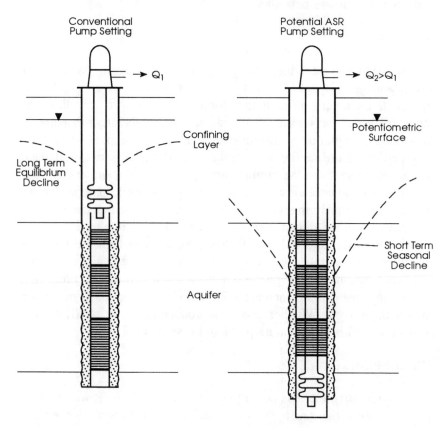

Figure 3.2 Alternative pump setting for seasonal ASR operations.

3.2 WELLHEAD FACILITIES

Several features of ASR wellhead design should be considered for efficient, long-term operation of an ASR system. That is not to say that disregard of these features will cause system failure. Rather, it is probable that consideration of these features will greatly enhance system performance. The various features are discussed in the following paragraphs.

Materials of Construction

As discussed previously in this chapter, materials of construction can play an important part in ensuring efficient and cost-effective ASR system performance. For both wells and wellhead facilities, the same principles apply. Non-ferrous piping systems, such as PVC and cement-lined ductile iron pipe, are preferred for ASR wellhead piping, particularly in systems where aquifer permeability is low to moderate and some concern exists regarding plugging and redevelopment frequency. Where this is implemented, the volume of rust carried into the well during recharge and from the well during initial recovery will be greatly reduced. Furthermore, smooth piping will reduce the opportunity for solids entrapping near the wellhead. These factors, in turn, will tend to reduce the plugging rate, redevelopment frequency, and regulatory issues associated with disposal of backflush water to waste. Any increase in initial capital cost will offset higher long-term operating costs associated with handling of initially rust-colored, turbid recharge and recovery flows.

For downhole piping, such as injection tubes and pressure transducer tubes, PVC or other non-ferrous materials are also recommended. Galvanized tubing should be avoided, reflecting adverse experience with failure due to electrochemical corrosion. In particular, galvanized surfaces create a corrosion cell with non-galvanized surfaces, causing reduced service life and screen plugging. Where such corrosion is anticipated, it can also be partially inhibited with use of protective coatings and cathodic protection. However, this requires maintenance for continued operation and also requires additional space within the casing to accommodate the cathodic protection system.

In general, use of non-metals, epoxy coatings, cement-lined piping and special alloy materials of construction in ASR wells is usually a wise investment. The economic savings usually attributable to implementation of ASR justifies reasonable investment in the design of the wells and wellhead facilities.

Pipeline Flushing and Waste Flow Discharge

Regardless of the materials of construction of the well and wellhead piping, the transmission and distribution system conveying water to the wellhead may have deposits of solid material trapped in places such as connections, valves and fittings. Pressure surges and reversal of flow in this piping, due to startup of recharge operations, can resuspend these particulates and carry them down the well, causing rapid plugging.

For this reason, it is advisable to flush both the well and the wellhead piping to waste immediately prior to recharge, at a flow rate at least as high as the flow rate expected into or from the well. Wellhead piping must be designed with the ability to discharge waste flows at high rates, frequently to a ditch or sewer line near the wellhead or to the water treatment plant for retreatment. The duration of this flushing period may range from 10 min to as long as 2 hours or more, depending upon the site and the amount of solid material in the recharge water or in the well.

In many situations where the ASR well is supplied by a long, dead-end transmission pipeline, disposal of the resulting water from line-flushing operations can sometimes be difficult due to the large volumes and high rates involved. Furthermore, disinfecting new pipelines by filling them with chlorinated water and then draining them to waste at a low flow rate does not remove solids from the pipeline. The solid material will be carried into the well during the first recharge operation. Mechanical cleaning of the pipeline, such as by "pigging," can be helpful to remove solids in such situations.

Where the ASR well is supplied by a new long, cement-lined pipeline, high pH values may occur during initial recharge testing, due to grout and cement curing in the pipeline. At one site, this caused pH values to exceed 9.0 in the recharge water. Such an effect should be considered transitional until cement curing is complete.

Provision should be included in the wellhead design to isolate each well from the system and flush it to waste at the wellhead, so that remedial work can proceed while the remainder of the system is in normal operation. Otherwise the entire wellfield may need to be shut down during periodic backflushing operations. This is particularly important in situations where backflushing is expected to occur more frequently than a few times per year. For large ASR wellfields, it may be sufficient to be able to isolate groups of adjacent wells for backflushing operations.

Trickle Flows

Whether in the surface piping, wellhead piping, or well casing and screen, stagnant water is to be avoided. During periods of neither recharge nor recovery, it is advisable to maintain a trickle flow of chlorinated water into the well. This can be provided through small-diameter tubing conveying typically 2 to 5 G/min (8 to 19 L/min) down the well. In addition to the tubing, a small flowmeter and a valve are suggested, bypassing water around any isolation valve at the wellhead that prevents recharge flows. The required rate of this trickle flow can be easily calculated by monitoring the rate at which a chlorine residual in the recharge water dissipates in the well or in the wellhead piping at the end of a recharge period. Typically, a residual is maintained for up to one or two days, seldom longer. Maintaining a small chlorine residual in the well during storage periods prevents bacterial growth in and adjacent to the well, thereby reducing the potential for bacterial plugging.

A second reason for providing a trickle flow of recharge water during idle periods is applicable particularly in very brackish and saline aquifers. Maintenance of freshwater in the casing permits use of pump materials that are less expensive than those that would be necessary for a well in which water quality can change from fresh to brackish or seawater.

For long lengths of transmission piping, the trickle flow at 2 to 5 G/min (8 to 19 L/min) may be insufficient to maintain a chlorine residual in the transmission piping to the well, in which case the residual is also lost in the well. In such situations, one alternative is to provide a small chlorine feed using low flows from chlorination facilities provided at or near the wellhead for treatment of recovery flows. Another approach is to periodically slug the well with a large volume of recharge water during storage periods, sufficient to provide a residual of chlorinated water in the well and surface piping. The frequency of such an operation would be site specific, but probably every few days. For short storage periods, this practice may be considered acceptable. For longer potential storage periods, provision for wellhead chlorination of recharge flows may be wiser.

Sampling Taps

Sampling taps should be provided in the piping at the wellhead to permit sampling of both recharge and recovery flows. They should be suitable for collection of bacteriological samples and therefore should utilize non-ferrous material such as bronze. For the same reason, they should also be unthreaded at the discharge end. They should be installed on the side of the pipeline. Addition of a sampling tap at the bottom of recharge piping may also be advisable in situations where there is concern for suspended solids in the recharge water.

At some locations special provisions may be necessary to convey sample flows away from the wellhead and piping in order to avoid ponding, iron staining, rusting, or algae formation. A short drain is usually sufficient.

Care must be taken to ensure that the sampling tap is located at a point of positive pressure. During recharge, negative pressures can develop at the wellhead if flow is insufficient to maintain positive pressure in the well or injection piping. If the sample tap is located downstream of the last control valve at the wellhead and negative pressure develops, it will not be possible to obtain samples. This can be resolved by installing the recharge sampling tap upstream of the last control valve on the recharge line.

Pressure gauges should not be connected to sampling taps. Where both are connected to the same tap in the pipeline, pressure gauge readings will tend to be erroneous during sample collection. Although proper valving can remedy this problem, it is wiser to avoid potential human error by installing separate taps for each purpose.

If the recovered flows will be disinfected at or near the wellhead, or if other chemical addition is planned such as pH adjustment, an additional sampling tap may be required sufficiently downstream of the chemical feed point that a representative sample is obtained.

Disinfection of Recovered Flows

Water recharged to and recovered from most ASR facilities meets drinking water standards and can be used following disinfection. However, there are certain disinfection considerations that should be considered during design.

Recovered flows are typically disinfected with chlorine. Chlorine gas is used in many applications, because it is readily obtainable and transportable to most sites. Liquid chlorine can be used; however, larger volumes are required and the disinfectant properties degrade with time. For either approach, adequate facilities for storing and handling the chlorine must be in place. These are usually defined in local and state regulatory requirements.

When chlorine gas is added to water, the pH will decrease to some extent, dependent upon the chlorine dosage, the alkalinity of the water, and the blend ratio between ASR recovered flows and those from the water treatment plant or other source. In some cases the pH decrease can be sufficient to produce an aggressive or corrosive water. While the probable extent of the pH decrease can be estimated during design, it is necessary to confirm the actual decrease following construction and operational testing. Provision should be included in the design to incorporate locations for chemical addition, if needed at a later date to raise the pH. Chemicals may include ammonia for chloramine formation, or base chemicals such as sodium hydroxide to raise the pH.

Cascading Control

This is one of the more important elements of ASR wellhead design and requires some care. Cascading occurs when the water level in the recharge piping does not rise to ground surface during recharge. Allowing water to cascade down the well can lead to significant plugging problems due to air binding in the storage zone, and induced geochemical or bacterial activity. Air present in recovered water can cause consumer complaints. Cascading can also cause structural problems due to cavitation damage to pipes, valves, and fittings. Cascading needs to be controlled in order to avoid these problems, each of which causes plugging of the ASR well. Plugging usually can be reversed; however, it requires considerable time and effort.

Water can be introduced into a well through the pump column, the annulus between the pump column and the casing, one or more injection tubes inside the casing, or some combination of these approaches. It can be introduced under pressure or under vacuum, and it can be controlled from the wellhead or from the bottom of the injection piping. Selection among these alternatives is based upon consideration of several factors, principal among which are the following:

- casing diameter
- static water level in the well
- type, size and capacity of the pump
- specific capacity and specific injectivity of the well
- expected production rate and range of injection rates

Some of this information may not be available at the time the design is completed, creating the need for a flexible design approach, capable of accommodating a reasonable range of expected conditions. It is usually wiser to construct and test the ASR well to determine hydraulic performance characteristics before finalizing design of the wellhead facilities. This requires more time; however, it leads to better results. Provision of flexibility is still advisable, since recharge rates can sometimes drop below planned rates, causing unplanned cascading.

Annulus Recharge

High flow rates can sometimes be recharged down the annulus of a well. To maintain positive pressure at the wellhead and thereby prevent cascading, it is necessary to ensure that sufficient flows are always available for recharge. When recharge flows fall below this critical rate, cascading will occur and a vacuum will develop in the annulus and wellhead piping. Air will be drawn into any open air relief or vacuum breaker valves, any leaks in the upper portion of the casing or pump column, or elsewhere in the wellhead assembly and will be carried down the well into the formation, where it will tend to plug the well. This can happen due to reduction in recharge flow rate or due to local or regional lowering of the static water level in the storage zone.

A flexible solution is to seal the annulus at the wellhead and to ensure that any wellhead valves that are connected to the annulus are closed during recharge. In this way recharge can occur regardless of wellhead pressure or vacuum, and under a full range of recharge rates, thereby maximizing recharge volumes and reducing operating requirements.

A disadvantage of this approach is that water flows over a substantial surface area of casing that is alternately wetted and dried. Therefore, annulus recharge in steel casings has a high potential for production of rust that can contribute to plugging the well during recharge and create regulatory problems during backflushing and initial stages of recovery. For new wells, this can be avoided by selection of a non-ferrous casing or coating where possible. For existing wells, particularly those with long steel casings into low or moderate permeability aquifers, annulus recharge can contribute to particulate plugging problems.

A second disadvantage of this approach applies, in particular, to retrofitting of existing wells for ASR purposes. Where the quality of well construction is unknown or suspected to be poor, it is possible that the casing or pump base may not be sealed adequately. Recharge would therefore entrain air even if the wellhead piping and control valves were sealed and closed, respectively. This can be checked by installing a temporary packer in the well and pressure testing the casing to determine if it will hold a given pressure for 30 min. This is sometimes referred to as "mechanical integrity testing." Alternatively, a brief recharge test can be conducted at a low rate in the supposedly sealed annulus, sufficient to create a wellhead vacuum. Recharge is then shut off, and the vacuum is monitored to see if it will hold for 30 min.

A related issue is that pressure surges have been known to occur in the recharge piping of some ASR wells. In other ASR wells, recharge occurs at higher pressures anyway, to overcome high static water levels or to overcome density differences in saline aquifers. In such situations, the pump bases should be designed to withstand expected operating and transient pressures without leaking at the connection to the casing. A flanged connection between the top of casing and the pump base, machined to ensure flat, parallel surfaces, and sometimes provided with a circular groove and o-ring, can provide the required degree of sealing.

Recharge down the annulus of wells equipped with submersible pumps requires care to ensure that the electrical cable port in the wellhead flange is adequately sealed to prevent air entry during vacuum recharge or to prevent leakage during pressure recharge.

Several variations on this annular recharge approach are possible. Recharge could occur down the annulus at sufficiently low velocity below the water level in the well that any entrained air has the opportunity to bubble out before reaching the formation. No known existing or planned ASR sites utilize this approach, but in theory it should work. A downhole water velocity in the casing below the pump would have to be less than the air-bubble rise rate, or about 0.3 to 0.4 m/sec (1 to 1.3 ft/sec) for air bubbles with diameters of 0.1 to 10 mm.

Another variation is to cease recharge at such times as cascading begins, whether due to static water level decline or due to reduction in recharge flows. This assumes that previous testing has shown that cascading causes air entrainment in the well. Such an approach requires a degree of operating attention that is frequently not available. Larger ASR wellfields with computer-controlled operations and telemetered monitoring parameters can build this into their control systems; however, smaller systems are more likely to continue recharge regardless of whether cascading is occurring or not, with resultant reduction in recharge rates, due to plugging.

Methods to increase friction loss in the annulus have occasionally been considered or tried. These have included sizing the pump to minimize the annular space between the pump and the casing; addition of flanges at the couplings in the pump column; inflation of a packer above the pump in the annulus through an air or water line from the surface; and other novel approaches such as adding floating objects in the annulus. Except for the inflatable packer, each of these approaches has the same drawback in that it is sometimes difficult to place in the well, or retrieve from the well, a tightly fitting object. One that is not tightly fitting will probably not provide much resistance to flow. Well casings are not always straight, plumb, or round.

Several ASR sites utilize annulus recharge. Among these are Cocoa, Peace River, Marathon, and Chesapeake. Cocoa and Marathon utilize PVC casings; Peace River has epoxy-coated steel casings on all but the first two ASR wells, and Chesapeake also utilizes an epoxy-coated steel casing. The first two of these sites store drinking water in brackish aquifers within which the depth to static water level is about 3 and −7 m (10 and −22 ft) below land surface, respectively. Marathon utilizes a seawater aquifer with a static water level that varies above or below ground surface, depending upon the water density in the well. At Chesapeake, the aquifer is slightly brackish with a depth to water level of about 20 m (66 ft). The first three sites store water under slight pressure, while Chesapeake recharges either under pressure or with a vacuum in the annulus, depending upon the flow rate.

Injection Tube Recharge

One or more injection tubes are frequently used to control cascading during recharge. The small diameter tubes provide sufficient head loss at high flow rates that the water column is under positive pressure inside these tubes. For instance, a 2-inch (inner diameter) clean, new steel injection tube flowing at 18 L/sec (280 G/min) provides a friction loss of about 1 m for every meter of length. **Table 3.1** shows friction losses for small diameter pipes at several flow rates, assuming a Hazen-Williams friction factor of 160, representative of new, smooth pipe. When recharge flows available exceed the capacity of one tube, a second tube may be opened. Two different sized tubes can cover a broad range of potential flow rates by operating separately or together.

An advantage of this approach is that positive pressures can be maintained at the wellhead over a wide range of flows. Furthermore, the surface area in contact with the water is small, substantially reducing generation of rust during wet-dry cycles associated with recharge and recovery.

Disadvantages are several. Existing wells rarely have sufficient room within the annulus to add one or more injection tubes, in addition to the already existing pump column, power cable (for submersible pumps), air line or other water level measurement device. For new wells, the cost of oversizing the casing in order to provide sufficient room for all of the tubes and pipes can be substantial, particularly for deep casings.

The principal disadvantage of this approach, however, is the operational complexity. While it may be adequate during a test program of limited duration to adjust flow through different injection pipes to meet recharge flow variations, under operational conditions this approach is time-consuming and unforgiving. If available recharge flow exceeds the injection tube capacity for some period of time, this additional flow cannot be recharged without manually adjusting the wellhead to utilize a second injection tube. Conversely, if available flow falls below the capacity of the injection tube in use, cascading will occur, and therefore the well should be isolated from the system. Within the flow range of the injection tube(s), the system will work but requires more operational effort than is usually available. Above or below the flow range, the system will perform improperly by either rejecting available flow or cascading.

Injection tubes are most applicable in situations where adequate operational attention is available to monitor and control flows at the wellhead. They are also quite applicable in multiple ASR well systems where operations are controlled by telemetry and recharge flow variations can be met by adding or deleting ASR wells from operation. In this way the relatively narrow flow range of individual wells is not a substantial constraint.

Such an approach may require careful development of an operating plan for those situations where the storage zone is brackish or otherwise con-

TABLE 3.1   PIPE FRICTION LOSSES

Pipe Diameter (inches)	Flow Rate (G/min)	Velocity (ft/sec)	Head Loss (ft/100 ft)	
			C = 120	C = 160
1.5	50	9	29	16
	100	18	104	61
	150	27	219	129
2	150	15	54	32
	200	20	92	54
	250	26	139	82
2.5	200	13	31	18
	300	20	66	39
	400	26	112	66
3	300	14	27	16
	400	18	46	27
	500	23	70	41
	600	27	98	57
	700	32	130	76
4	600	15	24	14
	800	20	41	24
	1000	26	62	36
	1200	31	87	51

Note: 1. Based upon Hazen Williams formula:

$$Q = 1.318 \, C \, R^{0.63} \, S^{0.54} \, A$$

where Q = flow in ft^3/sec, C = roughness coefficient, A = pipe area (ft^2), S = slope of total head line = head loss/length, R = area/perimeter for round pipe.

2. Head loss is based upon water at 60°F. Viscosity at 32°F increases head loss by 20%.

3. Range of C factors:
   Welded or seamless steel   80 corroded
                              150 new
   PVC pipe   140 design value
                              160 new

tains water of unacceptable quality. Design of ASR wellfields should usually attempt to balance water storage among the ASR wells so that hydraulic interference does not move the storage bubbles away from each well. Adding or deleting ASR wells in such situations to respond to variations in recharge rates may contribute to reduced recovery efficiency unless the wells added and deleted are selected with care.

When injection tubes are utilized, they should extend below the lowest expected static water level in the well. They may utilize an orifice plate at the bottom to increase friction loss. Changing the orifice plate therefore changes the head loss in the injection tube if this becomes desirable. To further dissipate head and to protect the well and screen from the effects of prolonged high speed jetting, it may be desirable to install a short screen or bucket assembly at the base of the injection tube in order to deflect flows laterally or vertically. A recent alternative is installation of a downhole control valve to adjust flows as needed in order to maintain water levels in the recharge piping. This option is discussed further in Section 4.4, Flow Control and Measurement.

An example of injection tube ASR applications is at Kerrville, TX, where the injection tube is stainless steel to a depth of 61 m (200 ft) in order to remain below the static water level in the well. This is the primary method of recharge; however, the well is also equipped for annulus recharge at higher flow rates.

Pump Column Recharge

Vertical Turbine Pumps. Recharge through vertical turbine pumps has been implemented at several ASR sites. The head loss generated by reverse water flow through the pump is usually sufficient to control cascading.

A non-reverse ratchet can be installed on the electric motor to prevent backspin of the pump and motor during recharge.

If head loss through the pump column and bowls is insufficient and cascading still occurs, a vacuum will develop in the upper part of the column. This may draw air into the well through the column coupling threads and also at the lineshaft stuffing box, particularly in existing wells where the condition and installation workmanship of the pump column may not be known. The resulting cavitation may potentially damage the column and lineshaft.

The potential development of a vacuum in the column is of particular importance for oil-lubricated pumps, since the vacuum can draw oil into the recharge water, thereby contaminating the water. The best solution is to avoid this potential occurrence by not utilizing oil-lubricated pumps in

ASR wells. Alternate approaches include the use of food grade oil for lubrication, or making special precautions to avoid vacuum development in the pump column. Use of new column pipe is also advisable to minimize the likelihood of leaks at the threaded connections.

For vertical turbine pumps, cast iron discharge heads are available in various sizes, and fabricated steel heads can be made to accommodate most configurations. Standard cast iron discharge heads can be machine surfaced to fit a steel sole plate grooved for an O-ring. This approach may be useful for retrofitting an existing well for ASR purposes. It also provides a reasonable pressure seal for situations where recharge water levels may rise above land surface during continued ASR operations or pressure surges.

Examples of recharge through the columns of vertical turbine pumps include Goleta Water District, CA; Calleguas Municipal Water District, CA, Las Vegas, NV and one of the ASR wells at Kerrville. All four sites utilize existing wells retrofitted for ASR purposes.

Submersible Pumps. Water may also be recharged through the columns of submersible pumps. These pumps typically include a check valve at the base of the column to prevent water from running backwards through the pump. This valve can be removed to provide for recharge; however, it is then necessary to provide a motor restart delay to the pump controls to avoid severe pump damage during power failures and emergency restarts, or during normal ASR operations. Another consideration is that reverse spin of the motor will generate electricity. Resistive loads wired to the motor leads at the motor starter could be used to dissipate this generated electrical energy. Recharge flow rates through the submersible pump should not exceed design production rates for the pump, as excessive rotational speeds may develop.

Regarding development of a vacuum in the submersible pump column, the same concerns apply as for a vertical turbine pump. Cavitation can damage the upper part of the column, potentially leading to structural failure, as well as drawing air into the column through threaded connections or the wellhead piping.

A better approach for cascading control in wells where the static and recharge water levels would otherwise be below ground surface is the use of a downhole control valve. This was first developed and tested during 1992 at Highlands Ranch, CO, in a well with a depth to static water level of about 900 ft (274 m). It has been used successfully at the base of the column of submersible pumps and should also be applicable for vertical turbine pumps. This valve is discussed in greater detail in Section 4.4, Flow Control and Measurement.

The type of well head seal will depend upon the type of pump in the well. For submersible pumps, a flanged surface plate should be used. Alternatively, a blind flange bored and welded to the column pipe can be fitted to a ring flange on the well casing.

Combinations

In general, pump column recharge is likely to provide the greatest degree of head loss for recharge flows, while annulus recharge is likely to provide the least head loss. Where it is desired to maximize recharge rates, and water levels during recharge should be at or above land surface, annulus recharge may be most applicable. Where water levels during recharge will probably be below ground level, pump column or injection tube recharge may be most applicable.

Flexibility to utilize more than one method of recharge is sometimes useful, particularly in situations where a wide range of recharge flows or static water levels may be encountered, or where considerable uncertainty exists as to the ultimate operating conditions. For example, the ASR system at Chesapeake, includes the provision to recharge down the pump column and also down the annulus, or both if high flows are available for recharge. At Kerrville, the ASR system includes one well equipped to flow down the vertical turbine pump column in a retrofitted production well, while a new ASR well is equipped to recharge down the annulus or two injection tubes.

Air and Vacuum Relief

All ASR wells experience a greater degree of water level change than typical production or injection wells. This change in water level results in air being drawn into, or released from the well during different phases of operation. Adequate venting on the casing and on the wellhead discharge piping should be provided in the form of air/vacuum release valves or other form of vented opening. However, it is essential that these valves be closed during recharge to prevent entry of air during potential vacuum recharge. This is an important operating requirement, the omission of which can entrain substantial quantities of air and plug the well.

Air relief valves are usually designed to vent air under relatively high operating pressures. ASR wells usually recharge under much lower operating pressures at the wellhead. Sometimes under these lower pressures the air relief valves will leak slightly. Provision for drainage of this leakage water will avoid a problem that may be aesthetically unappealing (rust), inconvenient (ponding), or sometimes slippery and dangerous. An easy solution is to provide a low pressure seat for the air release valve on the recharge piping.

Pressure and Water Level Measurement

Accurate pressure and water level measurement is important to ASR success. While recharge and recovery may occur without collection of this data, there would be no way to determine whether plugging is occurring until such time as the water level rise begins to inhibit recharge rates. By that time, the severity of plugging may preclude easy redevelopment by pumping. Instead it may be necessary to pull the permanent pump and any additional tubing from the well; clean the casing and screen with scraping, jetting, brushing, or other redevelopment methods while pumping the well with a temporary pump; acidize the well and surrounding formation; and disinfect it prior to reinstalling the permanent pump. This is time-consuming, expensive, and risky since the recharge specific capacity may not be fully restored. More cost-effective would be periodic redevelopment by backflushing to maintain recharge capacity. The need for backflushing is usually based upon pressure and water level measurements.

Pressure Gauges

Pressure gauges should be both durable and accurate. Sealed cases filled with glycerin or silicone stand up well to harsh, outdoor conditions. The fluid-filled gauges also provide needle damping if vibrations are present.

Pressure readings are useful in many places on ASR wellhead piping. Consideration should be given to installing taps for pressure gauges at the distribution system piping supplying the ASR well, upstream and downstream of any pressure control valves, upstream and downstream of any wellhead filters, and on the wellhead recharge and recovery piping. If vacuum or negative pressures may occur, particularly at the wellhead, a combination vacuum/pressure gauge should be provided.

Gauges should provide the level of accuracy necessary for each location. Generally, a gauge with 0.5% accuracy is desirable for the wellhead but is not necessary at other locations.

To protect the gauges against damage during pressure surges, spikes, or fluctuations, dampening devices can be installed for each gauge. These range from a fitting provided by the gauge manufacturer to a simple, small petcock.

Water Level Measurement

A variety of systems are available for obtaining accurate measurement of water levels in a well, among which are the following:

- casing access tube for direct measurement
- air lines and bubbler systems
- portable electronic sounders
- electronic pressure transducers

It is important to provide a direct means of measuring water levels through a casing access tube, even if other indirect means are also provided for convenience. The selection of the measurement system should reflect the probable frequency of water level measurement and other operational needs and opportunities.

Water levels fluctuate over a larger range in ASR wells, from recharge pressures attained at the end of the recharge period to drawdowns at the end of recovery or during backflushing to waste at high rate. The range can sometimes exceed the design range for pressure transducers, causing their failure. Pressure transducers are also vulnerable to failure due to lightning strikes.

Air lines and bubbler systems work well. A small diameter tube is installed in the well with the end of the tube submerged below the static water level. Air or nitrogen gas is pumped down the tube at a low rate until the gas bubbles out of the end of the tube. The pressure required to pump the gas out of the end of the tube is equal to the depth the tube is submerged below the well water level. However, at some ASR installations the range in water levels exceeds 30 m (100 ft), which is the approximate limit for use of a bicycle pump to purge an air line. In this case a small air compressor can purge up to about 75 m (230 ft) of air line, or a nitrogen bottle can meet higher pressure needs. It is possible to use two separate air lines with the appropriate valving for applications involving large water level changes.

Providing a small diameter PVC casing access tube with a cap on the bottom and a perforated section near the bottom is advisable, regardless of the measurement method. For recharge down the well annulus, this is probably the only way to measure water levels since cascading in the annulus will otherwise preclude accurate water level measurement. Cascading, whether under a vacuum or not, creates a column of water expanded with air bubbles so that it is difficult to determine the true water level in the well unless it is measured at greater depth.

Flow Measurement

A very important aspect of ASR operations is measurement of flows and volumes of water recharged and recovered. This is important for both technical and regulatory reasons.

Flowmeters used on ASR projects have included propeller, turbine, magnetic, venturi, and ultrasonic meters, as well as orifice plates and other approaches. Selection of the appropriate flowmeter should reflect project needs as well as meters currently in use at other locations operated by the same water agency or utility. Accuracy of these meters ranges from ±2% of actual flow, down to ±0.5% of actual flow. For example, an ASR site recharging at a rate of 4 megaliters/day (1.1 MG/day) for 90 days would store about 360 megaliters. A flowmeter with 2% accuracy may be off by 7.2 megaliters, or almost 2 days' pumping. For a site storing treated drinking water in a storage zone containing freshwater and with no real risk of geochemical reactions, this would be quite adequate. However, for a site storing treated drinking water in a storage zone with very poor water quality and/or high potential for geochemical reactions near the well that may cause plugging, a more accurate flowmeter may be important in order to ensure that an adequate buffer zone of stored water is maintained around the well at all times. This is discussed in greater detail in Chapter 4.

Selection of the appropriate flowmeter range is important since it is quite common for recharge flows to vary over a broad range during initial testing and subsequent operations. An ASR system may be designed to recharge at a high rate. However, water may not always be available for recharge at this rate due to operational constraints such as increasing distribution system demands or maintenance of minimum distribution system pressures in the vicinity of the ASR well. The alternatives include continued recharge at whatever lower flow rate may be available, or stopping recharge until flows are available at a rate within the range of the flowmeter. A flowmeter with an accuracy range of 10 to 120% of the design flow would probably be sufficient to permit continued recharge for most of the time until the system is switched over to recovery. Added operational flexibility at the low end of the operational recharge flow range can extend the usefulness of ASR in situations where there is a need to store as much as possible of a limited supply of seasonally available water.

Flowmeter accuracy depends on appropriate location in the wellhead piping, requiring an adequate distance of straight pipe upstream and downstream. For new ASR wellhead facilities, the flowmeter selected and the associated piping distances can be easily incorporated in the design. However, for retrofitting existing wells for ASR purposes, it is frequently necessary to select a flowmeter type that will provide the desired accuracy within the piping distance available. Straightening vanes are sometimes used to straighten flow lines upstream and downstream of the meter.

For larger ASR systems, or those involving automated control systems, it may be appropriate to obtain a certificate of proper flowmeter installation from the manufacturer.

For all ASR systems, consideration should be given to providing dual flow measurement capability, at least for the duration of the test program. Meter failure or loss of calibration during the test program has occurred at several sites for a variety of reasons. Loss of calibration is difficult to detect at the time, and usually only becomes apparent late in the program when it is too late to repeat the tests. The resulting data can be difficult to interpret. It is desirable to have two different types of flowmeters, one of which is the primary meter. Any trend of increasing difference in measurements between the two meters would signal the need for calibration or meter replacement before proceeding further with the test program. These problems appear to be more common with propeller meters that are used widely in the water industry. Having a standby propeller meter or replacement parts on hand can be helpful, available for rapid substitution if necessary. A venturi tube or similar device incorporated in the wellhead piping can provide the backup flow measurement during testing, and can easily be removed when the system changes into long-term operation, if desired.

Bi-directional flowmeters have been used at some ASR sites where it was desired to convey both recharge and recovery flows through the same pipe. However, bi-directional propeller meters have proven much less reliable than corresponding venturi or magnetic meters.

Flowmeters utilized on ASR systems should include totalizing measurement in order to monitor cumulative volumes during both recharge and recovery. This is typically provided with propeller type flowmeters, which are readily available, relatively inexpensive, and have been used widely on ASR projects. Propeller meters are usually accurate to within ±2% of the actual flow rate.

Turbine meters are similar to propeller meters; however, they use a turbine instead of a propeller. The turbine spins at a higher velocity and subsequently requires a more precise bearing and mechanism. For this reason, turbine meters are more sensitive to sand and particles in the water flow. Upstream screens should be installed with turbine meters. They typically provide higher accuracy and a wider operating range than does a propeller meter. Typical accuracy is about ±1.5% of actual flow. Cost is usually about 30% greater than the corresponding propeller meter.

Venturi meters offer the advantage of having no moving parts. They place a smooth constriction in the flow stream and then measure the reduction in pressure at the throat of the constriction. The difference in pressure between the meter throat and the adjacent pipe is related to the flow rate. The actual meter tube can usually be installed between two pipe flanges and therefore requires little space. However, adequate upstream and downstream pipe distances must still be provided. These meters result in relatively low head loss through the meter. They require a mechanism to read the differential pressure and a separate totalizer to integrate the flow signal. Reading the differential pressure requires a fairly sensitive gauge. Typically, a differential pressure transmitter is mounted at the venturi tube and sends a signal to a remote flow rate indicator and totalizer. These meters are accurate within ±1% of full scale.

Magnetic flowmeters also have no moving parts and have the advantage of compact size. The meter works by first creating a magnetic field in the pipe. When the water moves through the magnetic field, a voltage is induced that is proportional to the flow rate. Flow rate indicators and totalizers are available with either local or remote mounts. Magnetic meters are bi-directional, with no loss in accuracy. The required upstream pipe distance is usually low, as a result of which these meters are particularly useful in retrofitting existing wells for ASR purposes. However, these meters are typically more expensive than other meter types. They can be obtained with accuracies of ±0.5% of the actual flow rate.

Ultrasonic flowmeters are portable, and can be moved easily from one length of pipe to another. The meter mounts to the outside of an existing pipe and requires no moving or other parts in the water flow stream. They operate by electronically measuring the time required for an ultrasonic signal to travel between two or three transducers mounted to the outside of the pipe. The difference in time between signals traveling upstream and downstream is proportional to the liquid velocity. The meters usually consist of several transducers that can be mounted in several configurations and record to a data-logging microprocessor. Pipe material, diameter, wall thickness, and lining type and thickness must be known and entered into the microprocessor. Ultrasonic flowmeters are well suited for checking the performance and accuracy of inline meters and can be obtained with an accuracy of ±1% of the actual flow rate.

Disinfection and pH Adjustment

At drinking water ASR sites, recovered flows usually require only disinfection prior to distribution. Facilities therefore need to include provision for storing and handling the chlorine or other disinfectant that will be used.

Chlorine gas added to water will typically result in a decrease in pH. The magnitude of the decrease will depend upon the chlorine dosage and the alkalinity of the water. Where the recovered water will be blended with a much larger flow of water, the effect may be negligible. However, where little or no blending will occur prior to consumption, the pH drop following chlorination can be sufficient to produce an aggressive water, capable of causing corrosion of pipes and fittings, and associated "red water" complaints from consumers. The need for pH adjustment following recovery is usually determined following construction and initial testing of ASR facilities. Consequently, it is desirable to equip ASR wellhead facilities with locations for chemical addition, if later required.

Adjustment of pH may also be advisable for recharge flows. Where manganese is present in the storage zone, recharge at pH of less than about 8.0 may tend to cause the manganese to go into solution during an extended storage period. Recovery of the stored water may then create a problem with excessive concentrations of manganese and associated black discoloration of wetted surfaces. Adjustment of the recharge water pH to levels of about 8.5 or above will help protect against recovery of water with high manganese concentrations.

Depending upon the potential for formation of disinfection byproducts such as trihalomethanes and haloacetic acids when the recovered water is disinfected, it may be necessary to add ammonia to the recovered water to form a chloramine residual. Where ammonia is present in the recharge water, its presence in the recovered water should be tested before making a determination as to whether re-ammoniation is necessary. Typically, ammonia is substantially reduced during aquifer storage. Reduction of disinfection byproducts during ASR storage is discussed in greater detail in Section 4.5, Disinfection Byproduct Reduction.

Pump Considerations

Selection of a pump for an ASR well includes a few features not normally considered in pump selection for a normal production well. Pumping water levels may vary depending upon the degree of well plugging. At the beginning of recovery, pumping water levels may be lower than those occurring following redevelopment. Hence, it is frequently advisable to set the pump deeper in an ASR well, which requires additional column pipe. Pump hydraulic characteristics should be selected so that operation occurs over a reasonable range around the design point for flow and head. The additional column pipe provides operating flexibility, since the range of pumping water levels is usually not known until after a few cycles of operation. Net positive suction head (NPSH) and motor electrical horsepower should also be sufficient to match the full range of expected pumping water levels.

Pump setting has been within the casing, or within a blank section between screen intervals, in all ASR wells to date. However, it is anticipated that some future ASR installations may set the pump *below* the producing interval in a bottom section of casing that serves as a sump. In this way, seasonal production may be conducted at rates higher than those associated with normal well operation, causing rapid seasonal lowering of water levels and potential partial dewatering of confined or semi-confined aquifers.

The volume of water available from dewatering a confined aquifer is defined by the specific yield, which typically ranges from 5 to 35% of the volume of aquifer material dewatered. In contrast, the volume of water released from lowering of water levels within a range above the top of the same confined aquifer would be defined by the storativity, which typically ranges from 0.1 to 0.001%. Hence, in situations where hydrogeological, geochemical, and bacteriologic considerations permit, it may be very desirable to better utilize the large volume of water stored in an aquifer by producing at a high rate for a short period of a few weeks or months. The aquifer would then be recharged during the following low demand season. This is illustrated in Figure 3.2. The pump design would then entail additional column pipe and, for submersible pumps, a shroud around the pump and motor to ensure that water flows around the motor during pump operation, to provide adequate cooling.

To date, ASR wells have been equipped with vertical turbine, submersible, and horizontal centrifugal pumps. All have proven adequate for their specific applications.

In a few situations where storage zone permeability is very low, plugging potential is deemed to be high, or discharge of initially turbid water is a significant concern, consideration should be given to coating the column pipe, both inside and outside, in order to reduce the surface area subject to rusting during alternate wetting and drying periods associated with recharge and recovery.

Normally it is wise to utilize the same pump manufacturer utilized for other wells and pumping installations operated by the owner of the ASR well. However, certain submersible pump manufacturers have indicated that they will not honor the pump warranty if the pump is used for injection. In this case, alternate manufacturers or types of pumps should be considered. Injection through a submersible pump entails removal of the check valve normally provided at the base of the pump column. With the check valve removed, there is greater risk of premature re-start of the pump after recharge or after a power failure, at a time when water is still draining down the column pipe. The resulting additional torque can damage the pump. Hence, a restart delay may need to be provided to protect the pump.

A related consideration pertinent to the use of large submersible or vertical turbine pumps is that large motors should not be cycled on and off repeatedly without an intermediate period for heat dissipation, as well as for cessation of flow in the column pipe. Turning these large motors on and off causes considerable wear and tear, which should be minimized. Each manufacturer will have its own criteria for acceptable pump operation. ASR well redevelopment and backflushing sometimes includes pumping the well to waste at a high rate for a few minutes, resting the well, then pumping the well again for a few minutes. This cycle is sometimes repeated one or two times to surge the well and thereby remove solids from the screen and gravel pack, or surrounding formation. The redevelopment operation may occur as frequently as every day or two, or as infrequently as once every year at the beginning of seasonal recovery. For vertical turbine pumps, such a redevelopment sequence is less of a problem. However, short cycle operation of large submersible pumps for redevelopment and backflushing may be inadvisable, depending upon manufacturer requirements. Where the need for frequent cycling of large submersible motors becomes apparent, it may be advisable to reconsider well design and operation, to reduce the generation of solids and the associated frequency of backflushing.

For vertical turbine pumps, a non-reverse ratchet should be included to prevent impeller rotation during recharge and also following pump shutoff. With a non-reverse ratchet, the torque on the impeller is in the same direction during recharge as during normal pump operation, so there is no tendency for the pump to unwind during recharge. Without the non-reverse ratchet, situations can develop that can unwind the shaft.

Oil-lubricated vertical turbine pumps should generally be avoided in ASR wells, where possible. Under certain operating conditions, the potential may exist for a vacuum to form in the column pipe where this is used for recharge. The oil would then be pulled into the recharge water, plugging the storage zone and contaminating the water. If the annulus is used for recharge, any floating oil in the annulus will be carried into the formation. The problem can be minimized through use of a separate injection tube. In vertical turbine pumps, shutting off the oil reservoir supply to the pump shaft at the beginning of each recharge period and opening it at the beginning of each recovery period can also work; however, this is somewhat risky as a long-term operational requirement since it would be easy to overlook the adjustment.

To date, no ASR wells have been provided with variable frequency drives, providing for adjustment of recovery rates over a wide range. However, some wells in California have been provided with two-speed motors to facilitate energy recovery during recharge. One site in the planning stages in southeast Florida is considering a two-speed motor to enable recovery at normal rates to meet distribution system diurnal variations in demand, and higher rates if needed to meet fire flow requirements. At this site, the ASR system is under consideration as a cost-effective alternative to a new above-ground storage tank in an existing residential area. Residents of this area oppose construction of the above-ground tank.

Other ASR Well Site Considerations

Pressure control valves may be required on either the recharge line, the recovery line, or both. This provides operational flexibility in situations where recharge pressures may fluctuate, where available flow may be limited during certain hours of the day or months during the year, or where recovered flows may interfere with system head curves at certain times.

A permanent survey benchmark should be provided at each ASR site, showing the elevation. This will provide a reference point for measurement of water levels.

ASR projects typically require substantial onsite testing during both day and night. Consequently, it is important to provide adequate lighting, not only for the ASR well but also for any observation wells that will be measured or sampled at night. Electrical outlets at the site facilitate use of test equipment, power tools, and other activities, the need for which may not be apparent at the time the wellhead is designed.

If observation wells are to be sampled, consider how the samples will be taken and if dedicated pumps should be installed in these wells.

Adequate site access is important. Delivery of chlorine cylinders and suitable access for pump trucks is important. Adequate road access should be provided so that cars can get to the wellhead, rather than just four-wheel drive or track vehicles.

Provision of telemetry control is frequently desirable, particularly with larger ASR systems, in order to reduce operational labor requirements. Not only does this facilitate routine operations, it can also simplify data collection, monitoring, and reporting requirements. ASR systems may be changed from recharge to recovery mode, typically once or twice a year. However, adjustment of flow rates may occur more frequently, during both recharge and recovery. Telemetry control may include some or all of the following functions:

- pump on-off
- pump failure alarm
- recharge pressure control valve setting
- recovery pressure control valve setting
- water level in ASR and observation wells
- chlorine residual
- recovery flow rate
- butterfly valve operation
- conductivity probe
- turbidity probe

The telemetry control system should provide adequate capability for data storage and processing, and preparation of monitor program reports to track cumulative storage volume, water quality, and operational performance. It should also include a physical or software lock to prevent inadvertent discharge of turbid water into the treated water distribution or collection system upon initiation of recovery.

Energy Recovery

Where depth to static water level is substantial, the opportunity for energy recovery may be considered. Modifications would probably be required to the pump and bowl assembly in order to accommodate reverse rotation and power generation in an ASR well. In particular, conversion to a two speed motor would probably be required, the higher speed for pumping and a lower speed for power generation. A less desirable alternative is to design the system so that the pump motor is disengaged during recharge while a second motor is connected through a right-angle gear drive. Electrical modifications would be required for both approaches.

The kilowatt output capability of a typical turbine is approximated by the following formula:

$$kw = (1.88 \times 10^{-4}) \times Q \times H \times \text{Turbine Efficiency}$$

where: Q = flow in G/min and H = net head in feet.

Typically, the expected energy produced by well pump/turbines of this type is approximately 30% of the well production brake horsepower.

Energy recovery has been implemented at the Calleguas Municipal Water System in southern California.

3.3 WELLFIELDS

Design of an ASR wellfield differs from design of a production wellfield whenever mixing between stored and native water is to be minimized. Mixing can occur due to two situations:

- mixing due to dispersion around each ASR well
- mixing due to advective movement of stored water away from the well

Where no significant difference in water quality occurs, or where the intended use of the recovered water is such that any mixing is acceptable, then conventional wellfield design procedures relating to spacing and arrangement of wells are applicable.

Dispersive Mixing

Clustering of ASR wells provides the opportunity to create a bubble of stored water from the center of the bubble outward, thereby displacing poor quality native water away from the wellfield and avoiding trapped areas of this poor quality water. When designed and operated in this manner, ASR system performance can exceed that which would occur as a result of conventional wellfield design.

The difference lies primarily in the ASR well spacing, which tends to be closer than for conventional wellfield design. The spacing tends to be related more to the lateral extent of the stored water around each well at projected cyclic operational volumes, rather than short-term well interference effects during recharge and recovery. For example, the ASR wellfield for the City of Cocoa, includes six ASR wells around the periphery of the water treatment plant site on 60 acres (25 hectares) of land. The spacing between ASR wells averages about 183 m (600 ft), or approximately the theoretical radius of the stored water bubble around each well at its planned seasonal operating volume of about 61 Mm^3 (160 MG). This spacing is somewhat closer than would be appropriate for a conventional wellfield in the same aquifer. Native water at the ASR wellfield site in the storage zone beneath the water treatment plant has a chloride concentration of about 400 to 1200 mg/L and a total dissolved solids concentration of about 1000 to 3000 mg/L.

In addition to spacing, well arrangement also affects ASR recovery efficiency in situations where mixing between stored and native water is to be minimized. To date, no ASR wellfield has been operated in such a way as to attempt to displace native water potentially trapped between ASR wells. However, this situation has been addressed theoretically by Merritt (2) for water storage in a brackish aquifer using different wellfield arrangements. The situation is similar to the centuries-old practice of the bedouins of the Kara Kum Plain, as discussed in Chapter 1.

Whether the storage zone contains brackish water, high nitrates, or some other deleterious compound, it will not be long before a situation arises where an ASR wellfield is designed and operated to minimize mixing through radial recharge and recovery of the stored water. Recharge would commence in the center of the wellfield and proceed outward, adding wells as the stored water front displaces native water past these wells. During recovery, the opposite procedure would be followed. Central wells may be designed to recharge and recover at equal and also higher rates than peripheral wells, in order to stabilize wellfield operational flow rates during recovery.

Where a radial wellfield arrangement is incompatible with available site constraints or with local geology, a linear arrangement incorporating some of the same design considerations may be appropriate. For example, a central row of higher yield ASR wells could be paralleled with two adjacent rows of lower yielding ASR wells. Initial recharge and late recovery would occur in the central wells, while other ASR operations would occur in all wells. This is shown in Figure 2.7.

Figure 3.3 shows different wellfield design arrangements, as discussed above [2]. These results are theoretical, since wellfields are seldom designed without paramount consideration of available land area and shape. However, incorporation of these principles into wellfield design can potentially improve overall ASR performance in situations where this is important. For the wellfield arrangements modeled in Figure 3.3, the total volume of water stored was identical; however, the number and arrangement of wells varied and also the recharge approach. In some situations, recharge commenced at the center and subsequently commenced at surrounding wells when the freshwater front reached these wells. This was termed "sequential" recharge. In other situations, recharge occurred simultaneously in all wells. This was termed "simultaneous" recharge. For all arrangements, recovery occurred simultaneously in all wells. All recovery efficiency results were compared to a baseline, single well recovery efficiency of 83.1%.

Twelve different arrangements were modeled, with two to nine wells in each. Recovery efficiencies ranged from 74.0 to 82.5%. In each case, an

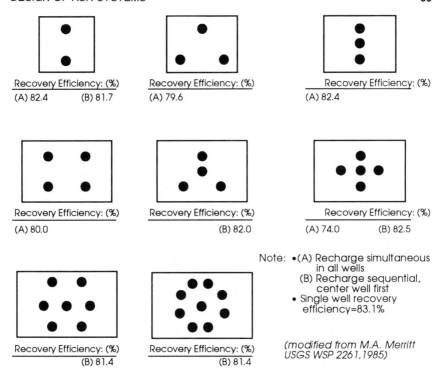

Recovery Efficiency: (%)
(A) 82.4   (B) 81.7

Recovery Efficiency: (%)
(A) 79.6

Recovery Efficiency: (%)
(A) 82.4

Recovery Efficiency: (%)
(A) 80.0

Recovery Efficiency: (%)
                        (B) 82.0

Recovery Efficiency: (%)
(A) 74.0   (B) 82.5

Recovery Efficiency: (%)
                (B) 81.4

Recovery Efficiency: (%)
                (B) 81.4

Note: •(A) Recharge simultaneous
           in all wells
       (B) Recharge sequential,
           center well first
     • Single well recovery
       efficiency=83.1%

(modified from M.A. Merritt
USGS WSP 2261,1985)

Figure 3.3 Alternative ASR wellfield designs and recovery efficiencies.

arrangement with a central well that was recharged sequentially achieved higher recovery efficiency than an arrangement with the same number of wells but no central well, which was recharged simultaneously in all wells. No analyses were performed at the same volume to compare results when recovery also occurs sequentially, first in all wells and then in the central well. By inference, this should lead to improved recovery efficiencies.

Advective Mixing

ASR wellfields are subject to advective movement of the stored water away from the well at a rate that is usually very slow, depending upon the regional gradient and the aquifer hydraulic characteristics. The lateral distance that the stored water moves between recharge and recovery is usually insignificant when compared with the radius of the stored water bubble during a typical recharge and recovery cycle. It is not unusual for the cyclic volume stored to occupy a theoretical radius of a few hundred meters around the ASR well, whereas the advective movement of the storage bubble may typically displace this volume at the rate of only a few meters per year. Consequently, the loss in recovery efficiency is slight, and usually difficult to detect.

Some ASR wellfields may potentially store water in aquifers for which the background advective rate of movement is significant relative to the radius of the stored water during a typical ASR cycle. For example, the cycle may entail water storage for several years to bridge drought/flood periods or to meet emergencies. Alternatively, the storage zone may be an unconfined aquifer, which typically has a greater rate of groundwater movement than does a confined aquifer. In these situations, improved recovery efficiency should be possible by elongating the wellfield design in the direction of expected regional groundwater flow and providing for a greater portion of recharge in upgradient wells and a greater portion of recovery in downgradient wells.

Figure 2.8 shows an example of this kind of situation. It is a conceptual layout of an ASR wellfield to store drinking water in a brackish, confined limestone aquifer in Kuwait; it is designed to help meet seasonal peak demands during summer months, and also to provide a strategic water reserve for emergency purposes. The regional gradient would not be a significant factor affecting recovery efficiency for annual ASR cycles; however, that portion of the potable supply in long-term storage to meet emergency needs would be subject to advective losses. Hence, the wellfield is arranged in a linear fashion in the direction of regional groundwater flow.

CHAPTER 4

Selected ASR Technical Issues

And God said, Let the waters under the heaven be gathered together unto one place and let the dry land appear; and it was so. And God called the dry land earth and the gathering together of the waters He called the seas; and God saw that it was good.

Genesis 1:9–10

ASR is not high technology requiring skills beyond the capability of all but a few specialists in the field. But neither is it low technology. It is somewhere in the middle of this range. The body of knowledge that differentiates ASR from other water management and recharge technologies has been developed since about 1970 through investigations and operating experience at several sites. Design considerations were discussed in Chapter 3, for wells, wellheads, and ASR wellfields. In this chapter, several key technical issues are discussed in greater detail to provide a broader understanding of the technology. Geochemistry issues pertaining to ASR systems are discussed in Chapter 5. Taken together, the information presented in these three chapters comprises the current status of ASR technology development. Chapter 8 presents some probable future directions for ASR technology.

4.1 RECOVERY EFFICIENCY

The two most frequently-asked questions following ASR presentations are "How much of the water that I inject will I get back?" and "What is to stop someone else from drilling a well to pump out the water that I store?" The first question is addressed in this section, while the second question is addressed in Chapter 6, Selected ASR Non-Technical Issues.

Recovery efficiency usually has little significance where both stored water and native water are potable. In such situations the main concerns are usually aquifer plugging and redevelopment frequency. However, to the extent that the difference in water quality between stored and native water is significant so that mixing has to be controlled, recovery efficiency can become an increasingly important factor in the assessment of ASR feasibility.

Definition

Recovery efficiency is defined as the percentage of the water volume stored that is subsequently recovered while meeting a target water quality criterion in the recovered water. If 1 Mm³ (264 MG) of drinking water are stored in a brackish aquifer and, subsequently, 0.8 Mm³ (211 MG) are recovered before the total dissolved solids (TDS) concentration of the recovered water exceeds a target criterion of 500 mg/L, then the recovery efficiency for that ASR cycle is 80%.

A key element of this definition is that it is based on volumes stored and recovered. It may be of theoretical interest to some individuals to evaluate recovery efficiency based on percentage recovery of a tracer in the recharge water. Sometimes referred to as "counting the molecules," this approach will always lead to a lower estimate for recovery efficiency, since it eliminates any allowance for mixing between stored and native water. Such mixing can occur without any adverse effect upon use of the recovered water, so long as the degree of mixing is within the limitations of the water quality criteria for the recovered water. However, most people interested in ASR are less concerned about whether the same molecules are recovered that were injected, and are more interested in knowing the volume of water that is recovered that is useful for their intended purpose. An illustration of this difference is as follows:

Assume for the example above that the average recharge TDS concentration is 200 mg/L; background TDS concentration in the aquifer is 1000 mg/L; the drinking water standard for TDS is 500 mg/L; and during recovery, TDS concentration increases as shown on **Figure 4.1,** reaching the target criterion of 500 mg/L TDS at 80% recovery.

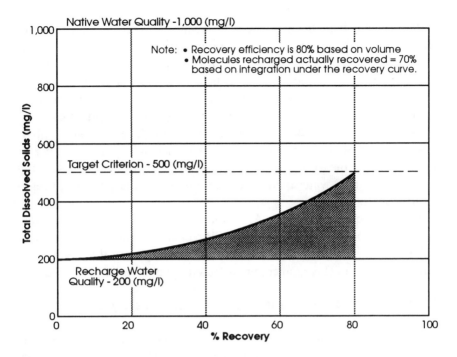

Figure 4.1 ASR recovery efficiency example.

The recovery efficiency is 80%. However, at the beginning of recovery the water is 100% recharge water, while at the end of recovery the water is a blend of 62.5% recharge water and 37.5% native water. Integrating beneath the recovery water quality curve suggests that about 70%, or 185 MG of the actual stored water was recovered during this cycle, while the balance (79 MG) was from native water in the aquifer.

In practice, the difference in analytical approaches is sometimes more significant than this example would suggest. By suggesting a lower percentage recovery, the second approach illustrated in the example can cause confusion among non-technical decision-makers trying to understand and evaluate the results from an ASR test program. The confusion can easily contribute to some loss of confidence in the program. It is much simpler to follow the recommended definition of recovery efficiency consistently, while being aware that individuals with a theoretical rather than an operational interest may occasionally ask valid questions regarding recovery efficiency calculated as performed in the example.

A second key element of the definition of recovery efficiency is that the target water quality criteria can easily vary from site to site, depending upon hydraulic and other factors. Most ASR sites are located at water treatment plants or at locations in the water transmission or distribution system, where blending can occur between recovered water and water flowing through the plant or distribution piping. So long as the water quality of the blend meets applicable drinking water standards, regulatory criteria are met. Consequently, it is usually not necessary to terminate recovery when drinking water standards are reached. Recovery can continue until such higher concentration is reached that the blend going to the consumer approaches but does not exceed applicable standards. Obviously the target water quality criteria will depend upon a number of factors such as the available blend with water from other sources during recovery periods, water quality for these other sources, and local regulatory constraints.

For the situation where the ASR well is located within the distribution system and consumers may receive ASR recovered water directly without any blending, then drinking water standards will govern the target water quality criterion. This is uncommon, based upon experience to date.

Water Quality Improvement with Successive Cycles

Recovery efficiency tends to improve with successive cycles when the same volume of water is stored in each cycle. "The more you use it, the better it works." This is because the residual water not recovered in one cycle becomes a transition or buffer zone of marginal quality surrounding the stored water in the next cycle. This is illustrated in **Figure 4.2**, which is based upon data from several operational ASR sites.

Building the buffer zone around each ASR well is usually completed over a series of cycles, typically about three to six, at the end of which the ultimate recovery efficiency for the site is achieved. However, in theory it can be completed at one time by storing an initial large volume of water in the well immediately after construction, and then proceeding with the expectation of achieving the ultimate recovery efficiency of water stored from that point on. This is closely analogous to filling a reservoir following dam construction, before using the reservoir for water supply, recreation and other purposes. The problem with the latter approach is that the required initial volume is usually not known until considerable ASR investigations, testing, and operations have been performed in an area. Consequently, it is usually only appropriate when an ASR wellfield is being expanded and there is reasonable confidence that new wells will perform similarly to existing ASR wells.

The financial investment in stored water that is required to achieve the ultimate recovery efficiency at a site is usually quite small relative to the cost of the ASR facilities. This investment is made with water generated

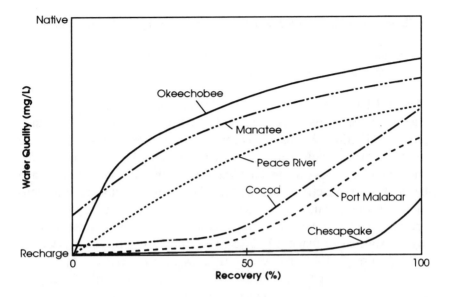

Figure 4.2 Water quality during initial cycle recovery.

during off-peak months and therefore has relatively low marginal costs, reflecting only electrical power, chemicals, and a small amount of operation and maintenance. The investment may be made over a period of several years through successive full scale cycles in which increasing volumes of water are recovered each year. Alternatively, it may be made up front during several months of continuous recharge and no recovery. The value of the buffer zone water invested is invariably quite small relative to the savings achieved by proceeding with an ASR solution to water supply needs.

The ultimate recovery efficiency attainable at any site has to be determined through testing and operations. At most ASR sites, 100% recovery efficiency is attainable; however, the number of cycles of operation to achieve this level may vary, as may the volume of buffer zone water invested. Where 100% recovery efficiency is not attained after several cycles, several factors may contribute to this result:

- inappropriate ASR well or wellfield design or operation
- testing at too small a scale for the storage zone
- insufficient number of cycles to develop the storage zone
- increasing volumes on successive cycles
- density stratification in highly saline aquifers
- high transmissivity of storage zone, particularly with more brackish or poorer water quality aquifers

- advective loss of stored water due to regional hydraulic gradient in the storage zone
- regulatory constraints designed to achieve aquifer recharge by requiring that a certain percentage of the water remain underground

Table 4.1 presents selected results for improvement in recovery efficiency with successive cycles for several ASR sites in brackish artesian aquifers in Florida. All show improvement in recovery efficiency with successive cycles; however, not all have attained 100% recovery efficiency. Those that have not include Marathon in the Florida Keys and Lake Okeechobee. As discussed below, Marathon utilizes a storage zone containing seawater, while Lake Okeechobee ASR utilizes a very transmissive, thick storage zone containing very brackish water. Neither of these sites is expected to reach 100% recovery efficiency. Boynton Beach is expected to reach 100% recovery efficiency over a few more cycles. As discussed below, recovery efficiency below 100% may still represent a wise and cost-effective water management decision.

The Marathon site has a storage zone in a sand aquifer containing seawater with a TDS concentration of 39,000 mg/L, causing substantial tendency for density stratification. As shown in **Figure 4.3,** ultimate recovery efficiency is primarily related to storage time at this site, and secondarily related to storage volume. Expected recovery efficiencies under long-term operating conditions are in the range of 50 to 75%. The annual investment in water not recovered is small compared to the cost of other alternatives to supply water during emergencies that may occur, such as loss of water treatment or transmission facilities during a hurricane. The operating strategy at this site is to store a given volume immediately prior to the hurricane season, maintain a target recovery volume during the

TABLE 4.1
ASR RECOVERY EFFICIENCY IN BRACKISH AQUIFERS

Site	Native Water TDS (mg/L)	Recovery Efficiency (%) (a)
Peace River, Florida	700–920	100
Cocoa, Florida	1000–2000	100
Port Malabar, Florida	1320	100
Boynton Beach, Florida	5000	80+ (b)
Marathon, Florida	37,200	40–75 (c)

Note: (a) Ultimate recovery efficiency after initial formation of underground reservoir. (b) Recovery efficiency approaching 100% expected once underground reservoir formation is complete. (c) Range reflects duration of trickle flow of about 50 G/min to offset losses due to density stratification.

hurricane season by adding a trickle flow of water to offset density stratification losses, and then recover the water during the following peak demand season.

Lake Okeechobee has a storage zone TDS concentration of 7000 mg/L. The aquifer transmissivity is very high, about 60,000 m²/day (4.5 million G/day/ft). Furthermore, the well was designed for disposal, not recovery, and all testing to date has been at a scale too small to properly draw conclusions regarding attainable recovery efficiency. Nevertheless, any recovery efficiency greater than about 40% at this site represents a net gain to the water management system since evapotranspiration and seepage losses associated with surface reservoir storage and canal conveyance are at least 60%. Water not recovered from storage in this aquifer will ulti-mately benefit the region, since the aquifer is increasingly being relied upon for brackish water supply to desalination treatment facilities. The lost ASR water will recharge the aquifer and may eventually tend to reduce the TDS concentrations.

Boynton Beach has a storage zone TDS concentration of 5000 mg/L. ASR testing to date has included seven cycles, five of which were at a volume of 227,000 m³ (60 MG), while the first and fourth cycles were at smaller volumes. Recovery efficiencies have climbed to 80% on the seventh cycle and are expected to approach 100%. On the fourth cycle,

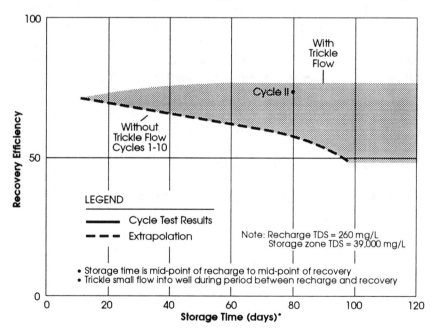

Figure 4.3 ASR recovery efficiency, Marathon, Florida.

recovery efficiency reached 95%. However, the smaller volume recharged and recovered compared to previous cycles helped to achieve this high recovery efficiency.

Two other operational Florida ASR sites shown in Table 4.1 have achieved 100% recovery efficiency in aquifers that have TDS concentrations in the range of 1000 to 1320 mg/L. A fourth site, Peace River, is still building the buffer zone volume. However, operating results to date suggest that full recovery efficiency should be attained. The storage zone TDS concentrations at Peace River range from 700 to 900 mg/L.

Where water supply is quite limited, or prices are already high due to major capital investments in treatment and transmission facilities, public reaction to the apparent loss of water can be a more difficult problem to handle than the actual value of the "wasted" water. Hence, it is always advisable to strive for as high a recovery efficiency as possible, using whatever tools are available to achieve this end. Careful site selection, well design, and operation are major factors in achieving this goal. Other important factors include careful control of expectations of ASR program early results, particularly in higher risk situations.

ASR testing in storage zones containing brackish or poor quality water usually includes at least three cycles with the same volume stored, in order to evaluate the trend in recovery efficiency improvement with successive cycles. Where storage volumes in successive cycles vary, different recovery efficiencies will result in each cycle and may or may not show an improvement with successive cycles. For example, following a series of equal, larger volume test cycles with a smaller volume cycle may substantially increase recovery efficiency in the smaller cycle due to the relatively large buffer zone available from the earlier cycles. Conversely, a series of test cycles each of which is larger than the one before will tend to reduce or eliminate any increase in recovery efficiency between cycles, since the buffer zone formed from the previous cycle is small relative to that required for the larger subsequent cycle.

Water Quality During the Initial ASR Cycle

The first ASR cycle at a new site provides a unique opportunity to gather useful data that can provide an early indication of ultimate recovery efficiency and ASR performance. Once the first cycle is completed and residual water in the aquifer around the well no longer reflects background water quality, then evaluation of performance in subsequent cycles is more complicated because it has to be interpreted with careful consideration of previous operations. Consequently, the first cycle should be planned carefully and implemented under conditions that are as well-controlled as possible.

In theory, it is always possible to repeat the first cycle after recovering to backgroundwater quality. This would provide the opportunity to correct problems that may have arisen, or to vary some of the test conditions, such as altering the volume stored to determine the effect of operating scale upon recovery efficiency on the first cycle. In practice, this is not easy to implement. There are invariably practical, operational constraints that provide a strong incentive to build the buffer zone and achieve ultimate recovery efficiency at the earliest possible date, usually in time for the next anticipated operational recovery period.

The first cycle is usually designed to confirm that wellhead facilities are operating correctly, to gather preliminary data regarding aquifer hydraulic response and geochemical and biological changes, to assess the recovery water quality response due to mixing in the aquifer, and to revise the remaining test program, if appropriate. The volume is usually small relative to subsequent cycles. Typically, recharge will occur for about a week, followed immediately by recovery.

Figure 4.4 shows that the initial cycle recovery water quality results for several ASR sites in brackish aquifers. Of some interest is the difference in the shape of these curves. Sites such as Marathon and Port Malabar, FL, and Chesapeake, VA show very little mixing with surrounding brackish water until late in the recovery portion of the initial cycle. These three sites utilize relatively thin, confined aquifers for ASR storage. Conversely, Lake Okeechobee shows substantial mixing at the beginning of the cycle, reflecting the relatively small volume utilized for testing, the high transmissivity of the aquifer, the substantial aquifer thickness, and the high TDS of this zone, as discussed above.

The shape of the recovery-water quality curve on the first cycle is an indication of the mixing or dispersion characteristics of the aquifer in the vicinity of the ASR well. Curve shapes that are initially flat, showing little or no mixing close to the well, are encouraging signs that successive cycles are likely to form a buffer zone that will support higher ultimate recovery efficiencies. Curves that are initially steep, showing mixing close to the well, are indicative of lower ultimate recovery efficiencies. Where the storage zone is fresh or only slightly brackish, the curve shape may not be very significant. However, where the storage zone is very brackish and little mixing can be tolerated in the recovered water, the curve shape needs to be reasonably flat at the beginning of recovery, in order to sustain expectations for ASR storage zone development to achieve high recovery efficiency.

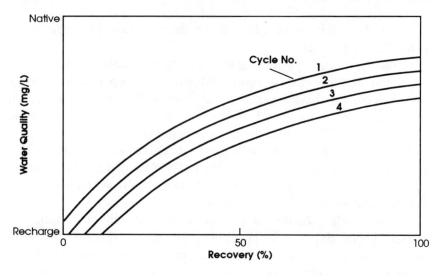

Figure 4.4 Water quality improvement in successive cycles.

The shape of the recovery curve is only determined following expenditure of much time, money, and effort for preliminary investigations, design, permitting, and well construction. As a result, there is only so much that can be done to improve recovery efficiency once the facilities are constructed and initially tested.

Occasionally initial test results may indicate that the well requires partial backplugging to eliminate a zone of poor water quality at the base of the storage zone or to improve lower confinement. This was discussed in greater detail in Section 2.3, Phase 2: Field Test Program. Following such corrective action, the test program proceeds.

Assuming that the best available zone for the intended use has been selected, and that the well is designed and constructed appropriately, the remaining variables that can be used to improve recovery efficiency are primarily operational: storage volume, recharge and recovery rates, and storage time between recharge and recovery. Results from the first cycle can then be used to adjust the planned test program so that recovery efficiency is enhanced. This may entail use of larger storage volumes; higher recharge and recovery rates; shorter storage times than perhaps originally planned; or addition of a trickle flow to the well during the storage period to compensate for losses due to density stratification. Through data collected during the test program, reasonable ranges for these operating variables can be determined to support ASR feasibility assessment and to guide subsequent planning, operations, and ASR expansion.

4.2 WELL PLUGGING AND REDEVELOPMENT

Artificial recharge of groundwater through a well usually results in increasing resistance to flow, or head buildup near the well, which is referred to as "plugging" or "clogging." The primary sites of plugging are the gravel pack (if present), the borehole wall, and the formation immediately surrounding the borehole wall. Increased head buildup in the well due to plugging changes the hydraulic characteristics of the well. Plugging during recharge can result in a decreasing rate of recharge or the need to continually increase the recharge head to maintain a constant recharge rate. Plugging that occurs during recharge and remains during recovery, otherwise known as "residual plugging," will have a negative impact on pumping. Residual plugging increases drawdown during pumping (decreased specific capacity) and thus reduces the pumping rate and/or efficiency during pumping. Residual plugging is probably aggravated by increasing recharge pressure or water mounding to excessive levels in order to maintain recharge rates.

To mitigate the effects of plugging, ASR wells are periodically redeveloped by pumping. Single purpose injection wells are typically redeveloped by installing a vertical turbine pump, by air lift pumping (sometimes with packer systems), or by swabbing and bailing with cable tool drilling equipment. ASR wells with a permanent pump can be redeveloped at more frequent intervals (daily, weekly, monthly, seasonal), whereas single purpose wells are typically redeveloped at long intervals of one year or longer. ASR wells are more suitable for the majority of applications where annual redevelopment is insufficient to maintain recharge capacity.

The preferred method of redevelopment is periodic pumping to minimize plugging and to prevent any lasting effects of residual plugging. Such redevelopment of a well is easily managed where a permanent pump is installed and where redevelopment flows can be conveniently discharged. Difficulties with discharging the redevelopment water require consideration of longer intervals between redevelopment activities. In situations where the number of ASR wells to be used is large and conditions are not ideal, the frequency of redevelopment can be a key issue in determining project feasibility and cost.

Since the rate of plugging during recharge ultimately determines the required frequency of redevelopment, it is appropriate to investigate the factors affecting the rate of plugging. With an increased understanding of plugging mechanisms, predictive tools can be used during the planning stages of ASR programs to estimate redevelopment requirements. An

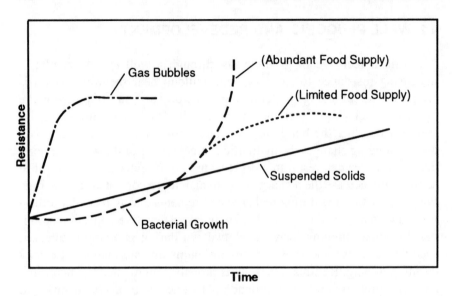

Figure 4.5 Typical clogging processes.

increased understanding of plugging will also be useful during operations in diagnosing the magnitude and origin of plugging, and in developing operations and maintenance guidelines.

Plugging Processes

Previous researchers have documented a list of processes that are primarily responsible for plugging of recharge wells [1, 2, 3, 4]. These processes include entrained air and gas binding, deposition of total suspended solids (TSS) from the recharge source water, biological growth, geochemical reactions, and particle rearrangement in the aquifer materials adjacent to the well. Site-specific conditions, such as aquifer and groundwater characteristics, well construction, recharge facilities design, and source water quality determine the influence of these processes on well plugging.

Each plugging mechanism or process is briefly described below, followed by a discussion of its relative importance. **Figure 4.5** illustrates the typical relationship between time and resistance to flow for plugging caused by suspended solids, entrained air, and biological growth. During recharge, an increased resistance to flow results in an increase in the water level in the well. Comparing a graph of water level rise due to plugging with these typical curves can be a useful tool for diagnosing the cause of the observed plugging.

Entrained Air and Gas Binding

During recharge, air bubbles may be entrained by free fall of water inside the well casing or by allowing air to enter the recharge piping where negative pressures occur. If recharge water with entrained air is allowed inside the well, there is a danger that these air bubbles will be carried downhole; through the well screen, perforations, or open hole; and out into the aquifer formation. For entrained air bubbles to move downward in the well casing, the downward velocity must exceed 0.3 m/sec (1.0 ft/sec), which is the rate at which 0.1- to 10-mm bubbles rise in still water [4]. When the entrained air enters the formation materials, the bubbles tend to lodge in pore spaces. This increases resistance to flow, resulting in higher water levels within the well.

Air entrainment is characterized by a rapid increase in the resistance to flow, which levels off in a matter of hours. Air entrainment effects stabilize because the rate at which bubbles redissolve into the flowing water equalizes with the rate of bubble formation. The plot described as "gas bubbles" on Figure 4.5 illustrates a typical well response to entrained air.

Typically, the possibility of air entrainment is prevented by proper wellhead design and operation. Maintaining positive pressure in the injection tube or pump column prior to discharge below the water level in the well is the most common method of preventing entrained air. Another method is to recharge with a wellhead designed to be airtight. Even though the recharge water may cascade within the well's annular space, injection tube, or pump column, preventing air from entering the well eliminates the possibility of air entrainment.

A plugging mechanism related to air entrainment is caused by a release of dissolved gases within the aquifer formation after injection, which also causes gas binding. This results in reduced permeability. Dissolved oxygen (DO) is an indicator of the concentration of gases in solution. Generally, gas dissolution is not a concern unless DO concentrations exceed 10 mg/L. If dissolved gases are present, they may be released due to an increase in temperature or a decrease in pressure, causing a dissolution of gases contained in the recharge water. An increase in temperature is more likely in northern climates where cold, oxygenated water may be available during winter months for storage in seasonally warmer aquifers. However, a decrease in pressure is unlikely in ASR operations, particularly during recharge. On the contrary, an increase in pressure tends to occur as the water moves down the well and into the storage zone. This pressure increase tends to keep dissolved gases in solution.

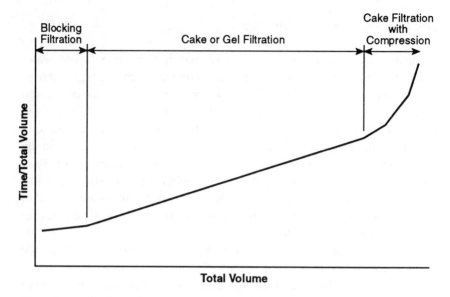

Figure 4.6 Stages of clogging on a membrane filter.

Microbial activity may also release gases as a metabolic byproduct that can result in reduced permeability. Although microbial activity is increasingly recognized as a being prevalent in ASR operations, no evidence of plugging due to release of gases from microbial activity has been noted in ASR wells to date.

Suspended Solids

In unconsolidated formations (typically sands and gravels with minor silts and clays), suspended solids are removed from the recharge water as it flows through the gravel pack into the formation. Resistance to flow near the well increases as the filter cake accumulates due to filtration.

A theoretical analog of this process is plugging of a membrane filter, which has been described as a three-phase progression: blocking filtration, cake or gel filtration, and cake filtration with compression. A typical filtration curve is presented in **Figure 4.6.**

Blocking filtration is characterized by particles physically blocking pore spaces in the filter medium. The duration of this process is typically short and the magnitude of plugging is minor compared to the later stages of plugging of membrane filters. Blocking filtration may be more consequential in ASR and other recharge wells because the pores in the formation and the filter pack are larger than the pores in a membrane filter. The filter pack surrounding the well screen may trap larger particles before they reach the borehole wall, thus reducing the long-term plugging rate by acting as a coarse pre-filter. It is possible that blocking filtration in the filter pack continues while caking filtration progresses at the borehole wall.

The next stage of plugging is cake or gel filtration. Cake filtration begins when the layer of filtrate on the filter begins to thicken. The resistance is directly proportional to the thickness of the filtrate. Cake filtration in an ASR well is evidenced by a linear increase of injection head over time while maintaining a constant injection rate. This linear response conforms with the response of a membrane filter during the caking stage of filtration.

Cake filtration continues until the filtrate thickness increases enough to allow compression of the filtrate, thus initiating the final stage of plugging: cake filtration with compression. Cake filtration with compression is characterized by a sharp increase of resistance to flow, which is dependent on the compressibility of the suspended solids. If this stage of plugging occurs at an ASR well, continuing injection after this point may not be practical due to the associated high plugging rate and/or resulting increased difficulty of redeveloping the well. Identifying the beginning of this stage of plugging during recharge may provide the signal for redevelopment of the well.

Suspended solids are present in the recharge water for virtually all ASR wells constructed to date. While data on turbidity is readily available for potable water sources, data on total suspended solids is not commonly available. Experience at many different ASR sites has shown the presence of an interesting range of solids in the recharge water, including sand, rust, diatoms (single cell algae), alum floc, twigs, dead mice, live shrimp and slugs. Accordingly it is wise to assume that solids are probably present and take steps to quantify their occurrence and typical concentrations. This provides a basis for remedial design and operational measures. Solids typically occur in short intervals, probably associated with pressure transients and flow reversals in adjacent portions of the water distribution system. Discrete, small volume samples are less likely to define the solids loading in the recharge water than long-term composite sampling. Similarly, samples from the bottom of a pipe are more likely to be representative than samples from the side.

Biological Growth

Plugging that occurs due to biological growth during recharge is not well understood. The plugging mechanisms include an accumulation of impermeable slimes, development of a mat of dead cells and byproducts, and the dispersion or alteration of colloidal particles in the soil-aquifer matrix. The degree of biological growth is directly related to the amount of carbon and nutrients present. Although the concentration of nutrients in the source water may be low, the process of concentrating suspended particles near the well, due to filtration, often provides the substrate needed to foster biological growth.

A common method of controlling biological growth during recharge is to maintain a chlorine residual of 1 to 5 mg/L in the source water. However, even with chlorination during recharge, a pause in operations for more than about two days can allow biological growth to form [1]. Continuous addition of chlorinated water at a trickle flow rate between periods of recharge and recovery is frequently practiced to maintain a chlorine residual and thereby control bacterial activity in the ASR well. The trickle flow rate can be estimated by monitoring chlorine residual in the well, following the end of a recharge period, to determine the number of days before the residual has dissipated. Typically this is 1 to 3 days. The trickle flow rate is then determined so that the volume of water in the well is displaced in about half of that period. Typical trickle flow rates for disinfection purposes range from about 0.1 to 0.3 L/sec (2 to 5 G/min).

A drawback of chlorination during recharge and storage is potential formation of disinfection byproducts (DBPs) such as trihalomethanes and haloacetic acids. However, as shown subsequently in this chapter, disinfection byproducts generally decline in concentration during aquifer storage.

Reports from operators of the salinity barrier injection wells for Los Angeles County, CA, indicate that in early years chlorine was added prior to injection. Later, the practice of injecting Colorado River water with only the residual chlorine remaining from the water treatment plant was adopted, resulting in satisfactory operations for over 20 years. However, since the water treatment plants have switched to chloramines (a combination of chlorine and ammonia) for disinfection to reduce DBPs, operators have observed an increase in plugging and an increased difficulty in redevelopment of the injection wells [5].

In a few European countries, well recharge is practiced using water with little or no residual chlorine since water treatment includes chlorination followed by dechlorination. Pretreatment to reduce total organic carbon (TOC) is sometimes practiced to remove undesirable organic constituents and also to control bacterial activity in the well. While this is effective, it is also expensive. Where undesirable organic constituents are absent in the recharge water, it may be more cost-effective to control bacterial activity in the well with chlorine rather than with TOC removal treatment processes. As discussed in Section 4.5, Disinfection Byproducts, natural processes occurring in the aquifer may reduce or eliminate disinfection byproducts in the ASR well during storage.

Geochemical Reactions

During recharge, geochemical reactions can occur that adversely affect aquifer permeability or cause changes in the quality of the recovered water. These chemical and physical changes are a function of:

- recharge water quality
- native groundwater quality
- aquifer mineralogy
- changes in temperature and pressure that occur during recharge and recovery

The most notable of the possible adverse geochemical reactions are precipitation of calcium carbonate (calcite); the precipitation of iron and manganese oxide hydrates; and the formation, swelling, or dispersion of clay particles. Geochemistry is discussed subsequently in Chapter 5.

Particle Rearrangement

Repeated cycles of recharge and recovery may result in rearranging and settling of the aquifer materials in the annular vicinity of the well (mechanical jamming), which may lead to a decrease in pore spaces and produce a reduction in permeability [6]. This effect may extend to a maximum distance of several feet from the well bore. After the initial settling of particles, no further reduction in the permeability is likely to occur. Rapid plugging, which occurs during initial startup of injection, may be caused by particle rearrangement.

Reductions in permeability caused by particle rearrangement are small and are not likely to be an important mechanism in plugging. After the initial settling of formation particles occurs, plugging due to particle rearrangement is not likely to have an appreciable effect during recharge.

Most ASR sites have experienced a difference in recharge and recovery specific capacities, with the recharge specific capacity invariably being lower than the recovery specific capacity. Exceptions include some sites in highly transmissive limestone aquifers where little or no difference occurs. The ratio of recharge specific capacity to recovery specific capacity, for comparable flows and durations, typically ranges from 25 to 100%, with 50 to 80% being a reasonable range for unconsolidated aquifers. The reason for this difference has prompted considerable discussion over the past few years. Particle rearrangement, or a "skin effect," is usually postulated as the reason.

An alternate hypothesis that has been proposed by the author is a "balloon effect." It is easier to let the air out of a balloon than to inflate it. This would imply a hysteresis effect, or differential response of the formation and overlying formations to the stress imposed during recharge and the release of that stress during recovery. It would also imply that lower specific capacity ratios would be expected for more compressible formations and for deeper storage zones. Further work on this issue may be helpful to provide an improved basis for planning and design of ASR wellhead facilities, which are frequently designed and constructed before data is available to determine the specific-capacity ratio.

Measurement Methods for ASR Well Plugging

The drawdown in a pumping well is a function of (1) the aquifer parameters, (2) the design and construction of the well and pumping facilities, and (3) the pumping (discharge) rate. When water is recharged through a well, the same three items (with discharge rate modified to become recharge rate) plus the changes that occur from plugging determine the water-level rise in the well at any given time.

Water level data collected at any ASR well during recharge operations can provide a basis for evaluation of plugging characteristics and comparison to similar data from other sites. Such data may be adjusted to reflect barometric variations and regional groundwater level changes. Three methods have been developed for evaluation of plugging from the adjusted data: (1) specific time of injection, (2) water-level difference, or (3) observed vs. theoretical water level rise.

Specific Time of Injection Method

The specific time of injection method is particularly useful when an observation well is not available, since only the water level in the ASR well is used for the analyses. The theory behind this method is that for a selected recharge rate, held constant during the test, the water level rise in the well since the start of recharge is repeatable, assuming no plugging has occurred. If the water level rise is only attributable to well losses (laminar and nonlaminar) and aquifer response, then repetition of the same recharge rate over the same period of time should produce the same resulting water level rise. Therefore, a comparison of two specific time of injection measurements taken at the same recharge rate and time interval indicates whether plugging has occurred. Any time interval could be chosen for this analysis, but, typically, an elapsed time of 2 to 4 hours since start of recharge is used so that the water level measurement is taken when the rate of water level change is reduced.

A drawback to the specific time of injection method is lack of control over recharge rates. Between tests, these often vary due to factors beyond the control of the operator. Rising water levels in the well can affect the hydraulics of the recharge system and thereby change recharge rates substantially. One way around this disadvantage is to conduct step injection tests over a range of flows at the beginning of testing. Presumably these tests are conducted during "pre-plugging" conditions. Later, step injection tests conducted over the same time period can be compared with an interpolated value from the "pre-plugging" step injection tests to determine the magnitude of plugging. Interpolation of the "pre-plugging" data is done by fitting the data to a power curve ($y = ax^b$). The type curve is then used to obtain a calculated water level rise at a specific injection rate. In this instance, the equation for a power curve was considered more appropriate than the Theis equation because non-linear well losses are the dominant factor in water level rise for the short test period.

Difference in Water Level Rise Method

The difference in water level rise method to determine *in situ* plugging rates uses data from both the ASR well and one or more observation wells. The accuracy of this method is predicated on the assumption that the recharge rate is kept constant and that the wells are perforated or screened in the same interval. When the flow regime in the aquifer system has reached a quasi-steady state, the difference in the water levels in the ASR well and the observation wells theoretically will remain constant. An increasing difference in water levels indicates plugging.

Observed vs. Theoretical Water Level Rise Method

The water level rise observed in the ASR well is a combination of aquifer response and well losses. It is assumed that for a constant recharge rate, the well losses should remain constant, and therefore any water level rise in the well, without plugging, would be due solely to aquifer response. Therefore, using estimates of aquifer parameters, transmissivity, and storativity or specific yield, the water level response in the aquifer is estimated and compared to the observed change in water level in the well. The difference between the calculated, or theoretical, water level and the observed water level is presumably due to plugging. The term theoretical water level is used as a reminder that theoretical aquifer conditions (i.e., homogeneous, isotropic, and infinite in areal extent) are some of the assumptions when using a groundwater flow equation to calculate water level response during recharge or pumping. When using this method, it is best to calculate differences in theoretical water levels and compare them with observed changes in water levels. Avoiding the variability that can occur early is accomplished by choosing a beginning time that occurs several hours after the start of recharge. The average rate of recharge for the beginning time and ending time, along with estimates of aquifer parameters, is used to calculate a theoretical change in water level, for comparison with the observed change in water level in the ASR well. For this method to be valid, recharge rates at the beginning and ending times must be the same so that well losses will be similar. Varying recharge rates between measurements is not a concern and is factored into the average recharge rate used to calculate the rise at the ending time.

If the recharge rate is held constant during injection and the plugging rate is low, a graphical procedure can be used. With the graphical method, water level rise vs. time is plotted on a semi-log chart, and a straight line is drawn through the moderate time data points (i.e., greater than 2 hours but less than 24 hours). Theoretically, the water level rise would plot along a straight line, assuming no plugging is occurring and boundary conditions within the aquifer are not reached. Therefore, the variance from the straight line can be an indication of plugging.

Normalization of Plugging Rates

Common factors affecting long-term plugging rates during recharge include (1) velocity or hydraulic loading (herein referred to as "flux") at the borehole wall, which is a function of the surface area through which the water is entering the aquifer and the rate of recharge; and (2) viscosity of the recharge water, which is a function of temperature. The flux of water entering the aquifer could be likened to the hydraulic loading rate of filters. Higher hydraulic loadings cause faster plugging because of the greater amount of total solids load over a given time interval. Previous studies of recharge wells and filters, performed in the Netherlands, have demonstrated the effects of suspended solids and temperature on plugging rates.

Use of a standard flux at the borehole wall and a standard temperature to normalize recharge well plugging data allows for a more meaningful comparison of plugging rates. Normalized rates are not necessarily estimates of actual plugging rates under those conditions, but are meant to adjust the relative plugging rates of various recharge and ASR well tests for comparison purposes. The following formula (modified from Formula 3.22, Ref. 6) was used to calculate normalized plugging rates:

$$\Delta\Phi_{norm} = \Delta\Phi\left(\frac{q_s}{q}\right)^2\left(\frac{\mu_s}{\mu}\right)$$

where $\Delta\Phi_{norm}$ = rate of plugging normalized for a recharge flow velocity (flux) of 3 ft/hour at the borehole wall over a period of one year at a temperature of 20°C; $\Delta\Phi$ = rate of plugging (feet of head per year); q_s = standard flux (loading rate or velocity) at borehole wall of 3 ft/hour; q = calculated average velocity (flux) at the borehole wall in ft/hour; this is the injection rate/infiltration surface area (over the effective saturated thickness or perforated/screened interval); μ_s = viscosity at a standard temperature of 20°C (centipoise); μ = viscosity at temperature of injection water (centipoise).

Source Water Characterization

Samples of recharge water, native groundwater, and recovered water have to be collected and analyzed to investigate possible physical, chemical, or biological factors that contribute to well plugging. Where chemical and biological factors can be eliminated from consideration, and where air binding is controlled through appropriate ASR well design, construction, and operation, particulate plugging frequently remains as a significant issue requiring evaluation. Particulate plugging is common to almost all ASR sites, even when recharging treated drinking water. The important issue is the rate at which it will occur, and the associated backflushing frequency required to maintain acceptable recharge capacity.

The two basic concerns with quantifying the suspended solids concentration of the recharge water are (1) obtaining accurate measurements for low concentrations and (2) obtaining measurements that account for the changes in concentration that may occur during recharge. The possibility of flushing sediments contained in the pipelines is a concern, since recharge rates often create high flow velocities in a direction opposite to the normal flow pattern. Periodic sampling may miss "slugs" of sediment-laden water entering the well.

Several direct and indirect measurements of suspended solids in potable water for the purposes of ASR testing have been unsuccessful. These unsuccessful test methods have included turbidity measurements, standard laboratory suspended solids measurements, and Rossum Sand Tester measurements. The measurement of turbidity, a common measurement of drinking water quality (which indirectly measures suspended solids), has been shown to have limited value for ASR purposes. The correlation between turbidity and suspended solids concentrations is poor, and the range of turbidity measurements is small. Typical measurements of turbidity for potable water range from 0.1 to 0.7 NTU (nephelometric turbidity units).

TSS measurements conducted during routine laboratory testing of potable water samples have typically indicated non-detectable results for potable water. Since suspended solids were often the suspected clogging mechanism during injection, it was determined that the TSS detection limit of 0.4 mg/L for *Standard Methods* [7] was inadequate to measure suspended solids in the recharge water. Attempts were made to increase the detection limit for the laboratory analysis of TSS, by increasing the volume of filtered samples from 1 L to as much as 10 L. Laboratory experience indicated that filtering large sample volumes tended to erode the filter material, resulting in inaccurate measurements.

The Rossum Sand Tester is a standard device commonly used during well development or when a production well is suspected of producing sand, to measure suspended materials in the discharge water. Data collected with Rossum Sand Testers during injection have indicated a lack of sensitivity when used with potable water. Typically, the test results show a low concentration of suspended materials in the water during the initial startup of an ASR well, and non-detectable results during the remainder of the testing.

Membrane Filter Index

A method that can be used to define the plugging potential of potable water is the membrane filter index (MFI). The theory, equipment, and methodology for MFI testing was developed in the Netherlands. Originally, MFI testing was developed for measuring the potential of waters to plug membranes during reverse osmosis water treatment. Later, MFI was adapted for use on injection and ASR wells. MFI testing equipment and methods, derived from work by Schippers and Verdouw (1980) [8], were first used by Huisman and Olsthoorn (1983) [4] during the early 1970s. The testing procedures have subsequently been refined for ASR purposes, based upon experimentation and field experience.

The basic theory behind MFI testing is to assume that the rate at which a filter becomes plugged at a constant pressure can be used to define a "plugging index" for a specific water at a given temperature. The membrane filter tests were conducted by directing recharge water through a 0.45-μm, 47-mm diameter membrane filter at a constant measured pressure of 15 to 30 pounds per square inch (psi). A temperature measurement was made during each test. The filter operated initially at 0.2 G/min or less. A single test typically required 15 min to 1 hour of field time.

The membrane filter tests were used to develop an MFI for each water source. The MFI is represented by the slope of the straight portion of the plot of time/volume (t/V) vs. volume (V), on a linear scale. Because of the small amount of water and the short times used in the test, the reporting units for MFIs are sec/L/L.

MFIs, as determined by plotting, were normalized to standard conditions so that MFIs measured with different pressure and temperature conditions could be compared. The standard conditions used were a pressure drop of 30 psi and a temperature of 20°C. The following equation was used to normalize the measured values to standard conditions:

$$MFI_{norm} = MFI \times \frac{\mu_{20}}{\mu} \times \frac{P}{30}$$

where MFI = slope of the straight portion of the plot of individual values (sec/L/L); μ_{20} = viscosity of water at standard temperature of 20°C (centipoise); μ = viscosity of water at measured temperature in °C (centipoise); P = pressure drop across filter (psi).

Bypass Filter Test

Bypass filter test (BFTs) are conducted on the source of recharge water, to measure the average concentration of suspended solids over periods of time ranging from a few hours to a week or more. The source water is directed through a 5-μm, 10-inch-long, spun polyester cartridge filter at pressures ranging from 5 to 30 psi. Cartridges with smaller pore sizes (0.45 μm or 1 μm) are available, but have higher costs and shorter life expectancy due to rapid plugging. A flowmeter similar to those used by utilities for household water use is installed in the filter piping, to measure the volume of flow through the filter at each site.

Analysis Method

The bypass filters are used to measure the suspended solids concentration in the recharge water, over extended periods of time. The filters are dried and weighed to the nearest 0.1 g in the laboratory. The totalizer on the flowmeter is read prior to putting the filters into service. The filters are operated during injection until the flow rate through the filter decreases to about 25% of the initial flow rate. When a filter is taken out of service, the flowmeter is read, and the spent filters are put in plastic bags and delivered to the laboratory for drying and weighing. The polyester filter material cannot withstand the 105°C temperature of the standard drying oven.

Therefore, the filters can be placed on top of the ovens and dried for several days. The difference in filter weights and the meter readings are used to calculate the concentration of suspended solids in the recharge water. Filters utilized in tests to date were operated between 2 to 20 days, with 10 days as the average service life.

Well Plugging Relationships

If long-term plugging is assumed to be a function of suspended solids in the recharge water, the rate of plugging will primarily be a function of recharge water plugging potential and aquifer conditions (defined in terms of hydraulic conductivity). The plugging rate can be normalized for rate of recharge and water temperature, by a normalizing procedure that accounts for the flux (also referred to as "velocity") of recharge water at the borehole wall (a function of injection rate, well diameter, area of perforation/screen, and effective saturated thickness) and the viscosity of the source water. The flux at the borehole wall is analogous to the hydraulic loading rates applied to filter media. The adjustment for viscosity accounts for the increment of head buildup created by recharge source waters of different temperatures.

It seems reasonable that, with enough data points from operating facilities, a family of type curves could be developed that relate normalized plugging rates to hydraulic conductivity for source waters of different suspended solids concentrations. The objective is to use these type curves during Phase 1 ASR feasibility investigations, to estimate well plugging potential and probable frequency of well redevelopment required.

These type curves could also be used to determine whether an ASR well is operating within a "normal" range of plugging. Determining whether well plugging is excessive could provide a signal to investigate other causes of plugging.

Plugging Rate Site Investigations

Data was collected during testing at nine ASR sites, including information regarding treatment and conveyance of the water prior to recharge, well construction, recharge rates, pumping rates during redevelopment, hydrogeology, and aquifer parameters. **Table 4.2** summarizes well construction and hydrogeologic conditions, while **Table 4.3** summarizes the ASR well testing characteristics and general information about the source water. General information regarding seven of the nine sites is included in project descriptions included in Chapter 9, Selected Case Studies.

The water level data from the ASR test wells and, when available, the data from nearby observation wells were used to estimate plugging rates.

TABLE 4.2 PLUGGING RATE SITE INVESTIGATIONS: WELL CONSTRUCTION AND HYDROGEOLOGY

Name of Agency and Injection Well(s)	Well Construction			Hydrogeologic Conditions			
	Casing Diameter (inches)	Perforated Interval (ft)	Surface Area (ft²)	Type of Aquifer and Description of Materials	Depth To Water (ft)	Transmissivity (G/day/ft)	Hydraulic Conductivity (G/day/ft²)
Centennial Water and Sanitation District, CO Well A-6	10	932–1,354 (total 200)	524	Confined, partially cemented sandstone	950	8,500	40
Las Vegas Valley Water District, NV Well 11A	20	360–980	3,246	Leaky confined, sand and gravel basin-fill deposits	175	250,000	400
Calleguas Municipal Water District, CA Well 97	14	670–930	953	Confined, sand and gravel marine deposits	220	140,000–70,000	250–500
City of Pasadena, CA Garfield Well	26	192–629 (total 160)	1,089	Confined, sand and gravel	300	200,000–300,000 300,000	1,000–1,500
Seattle Water Department, WA Metro Test Well	16	285–335	293	Confined, sandy gravel outwash deposits	165	370,000	2,500
Sonoma County Water Agency, CA Occidental Road Well	12 and 16	400–800	1,466	Unconfined, sand and gravel	56	20,000	50–100
Tucson Water, AZ Well B-44B Well C-14B Well C-26A	16 16 10 and 12	141–480 260–600 128–480	1,324 1,424 775	Basin fill with interbedded Basin fill with interbedded Basin fill with interbedded	200 220 250	65,000 25,000 50,000	200 60 180

TABLE 4.3 PLUGGING RATE SITE INVESTIGATIONS: WELL HYDRAULIC CHARACTERISTICS

Name of Agency and Injection Well(s)	Testing Characteristics			Monitor Well (distance in feet)	Source Water		
	Rate of Injection (G/min)	Rate of Pumping (G/min)	Wellhead Design		Raw Water Source	Treatment Process
Centennial Water and Sanitation District, CO Well A-6	260	410	Pump column pipe with downhole control valve	None	McLellan Reservoir	Chemical addition, flocculation, filtration and chlorination
Las Vegas Valley Water District, NV Well 11A	1,460	2,000	Pump column pipe	870 and 1,200	Lake Mead (Colorado River)	Chemical addition, flocculation, filtration and chlorination
Calleguas Municipal Water District, CA Well 97	605	680	Pump column pipe	80	CA state project water	Chemical addition, flocculation, filtration and chlorination
City of Pasadena, CA Garfield Well	1,550	1,600	Pump column pipe and well annulus	None	CA state project water	Chemical addition, flocculation, filtration and chlorination
Seattle Water Department, WA Metro Test Well	700	800	Conductor pipes in the well annulus	70	Cedar River, diverted via intake screening	Chlorination and addition of lime to reduce corrosivity
Sonoma County Water Agency, CA Occidental Road Well	590	1,500	Pump column pipe	75	Russian River, pumped from collector well system	Chlorination
Tucson Water, AZ Well B-44B Well C-14B Well C-26A	1,210 1,220 1,060 } Bailed to redevelop		3- and 4-inch conductor pipes }	149 60 50	Pumped groundwater }	Chlorination

The three methods of analysis of plugging previously described were used where applicable. Due to ease of use, the most predominant method was the observed vs. theoretical water level rise method. Only a few sites had observation wells nearby, which are necessary for the water level difference method to be used. The specific time of injection method requires testing and data collection procedures designed specifically for this method and was therefore only performed at a few of the sites. Plugging rates varied widely, from undetectable to 220 ft/month. The results of these analyses are presented in **Table 4.4.**

While in most cases extensive water quality monitoring of chemical and bacteriological parameters was conducted during the test program, only the data related to suspended solids concentration or plugging potential (MFI measurements) are presented in **Table 4.5.**

As shown on the table, the amount of data collected varies widely between testing programs. Where numerous tests were performed, the testing results often covered a wide range of values. Therefore, interpretations and conclusions drawn from these data should be tempered by the amount of data collected on a case by case basis. Each of the source waters tested were considered potable water, yet the testing results indicate a wide range of suspended solids concentrations/plugging potentials that can affect ASR well performance. Where water was delivered from an existing water distribution system, the sediment loads in the water were always higher at the beginning of recharge, and sometimes sediment loads would increase for short periods of time during recharge. These data suggest that reversing the flow through existing pipelines often results in sweeping sediments contained in the pipes into the ASR wells.

A summary of the ASR well plugging data is presented in **Table 4.6.** The normalized plugging rates shown in the table are typically less than the observed plugging rate, mostly due to the flux at the borehole wall being greater than the standard of 3 ft/hour. The relationship between normalized plugging rates, hydraulic conductivity, and suspended solids concentration is shown in **Figure 4.7.** The size of type used to label the data points is intended to be roughly representative of the magnitude of suspended solids in the source water. Generally, the data points appear to follow a logical pattern, such as a comparison of Well 97 and Well 11A. These wells have similar hydraulic conductivities but the well with the highest suspended solids concentration has the highest plugging rate. An anomalous data point is the Garfield Well which, in comparison with the other data, should have a plugging rate that is considerably lower.

The results of this testing indicate that the relationships between ASR well plugging, source water quality, and aquifer permeability generally follow intuitive reasoning. The instances where the ASR well performance

TABLE 4.4 PLUGGING RATE SITE INVESTIGATIONS: INJECTION WELL CLOGGING RATES

Name of Agency and Injection Well(s)	Test Interval			Water Level Changes				Clogging Rate	
	Start Injection (date)	End Injection (date)	Injection Interval (days)	Initial Water Level Buildup (ft)	Ending Water Level Buildup (ft)	Water Level Chg. Due to Clogging (ft)		Observed Clogging Rate (ft/day)	Observed Clogging Rate (ft/month)
Centennial Water and Sanitation District, CO Well A-6	11/30/92	1/14/93	45	120	465	325		7.22	220.3
Las Vegas Valley Water District, NV Well 11A	1/0/00	6/1/91	135	15	34	15		0.11	3.4
Calleguas Municipal Water District, CA Well 97	3/19/91	5/1/91	43	60	230	160		3.72	113.5
City of Pasadena, CA Garfield Well	10/29/92	1/13/93	76	55	122	62		0.82	24.9
Seattle Water Department, WA Metro Test Well	3/4/91	4/16/91	43	16	24	6		0.14	4.3
Sonoma County Water Agency, CA Occidental Road Well	7/7/92	7/23/92	16	34	50	Undetected		0.00	0.0
Tucson Water, AZ									
Well B-44B	11/17/89	3/26/90	129	50	70	6		0.05	1.4
Well C-14B	11/28/89	3/13/90	105	100	145	10		0.10	2.9
Well C-26A	10/11/89	6/26/90	258	77	124	27		0.10	3.2

TABLE 4.5 PLUGGING RATE SITE INVESTIGATIONS: SOURCE WATER QUALITY

Name of Agency and Injection Well(s)	Water Temp. (°C)	Membrane Filter Testing (Membrane Filter Index: s/l 2)				Bypass Filter Testing (Suspended Solids Conc.: mg/L)			
		No. of Samples	Maximum	Minimum	Average	No. of Samples	Maximum	Minimum	Average
Centennial Water and Sanitation District, CO Well A-6	10	N/A	N/A	N/A	N/A	3	3.010	0.130	0.167
Las Vegas Valley Water District, NV Well 11A	20	4	24.4	11.2	16.1	1	—	—	0.083
Calleguas Municipal Water District, CA Well 97	14	N/A	N/A	N/A	N/A	11	2.400	0.100	0.386
City of Pasadena, CA Garfield Well	23	N/A	N/A	N/A	N/A	3	0.023	0.009	0.015
Seattle Water Department, WA Metro Test Well	10	13	165.0	35.0	79.2	2	2.300	1.200	1.750
Sonoma County Water Agency, CA Occidental Road Well	18	Membrane filter would run for hours without clogging		Est. <0.1		1	—	—	0.002
Tucson Water, AZ									
Well B-44B	20	6	0.9	0.5	0.7	9	0.130	0.007	0.060
Well C-14B	22	5	2.2	0.7	1.3	11	0.095	0.011	0.052
Well C-26A	23	6	2.4	0.2	0.9	25	0.123	0.017	0.043

TABLE 4.6 PLUGGING RATE SITE INVESTIGATIONS: WELL CLOGGING SUMMARY

Name of Agency and Injection Well(s)	Source of Water	Water Temp (°C)	Membrane Filter Index (s/l 2)	Suspended Solids (mg/L)	Observed Clogging Rate (ft/month)	Injection Rate (G/min)	Interval Surface Area (ft²)	Flux at the Borehole Wall (ft/hr)	Normalized Clogging Rate (ft/month)	Hydraulic Conductivity (G/min/ft²)
Centennial Water, CO Well A-6	Surface water, treated	10	N/A	0.167	220.3	260	524	4.0	96.6	40
Las Vegas Valley Water District, NV Well 11A	Surface water, treated	20	16.1	0.083	3.4	1,460	3,246	3.6	2.3	400
North Los Posas Basin, CA Well 97	Surface water, treated	14	N/A	0.386	113.5	605	953	5.1	33.9	375
City of Pasadena, CA Garfield Well	Surface water, treated	23	N/A	0.015	24.9	1,550	1,089	11.4	1.8	1,250
Seattle Water, WA Metro Test Well	Surface water, disinfected	10	79.2	1.750	4.3	700	293	19.1	0.1	2,500
Sonoma County Water Agency, CA Occidental Road Well	Surface water, disinfected	18	Est. <0.1	0.002	0.0	590	1,466	3.2	0.0	75
Tucson Water, AZ Well B-44B Well C-14B Well C-26A	Groundwater disinfected	20 22 23	0.7 1.3 0.9	0.060 0.052 0.043	1.4 2.9 3.2	1,210 1,220 1,060	1,324 1,424 775	7.3 6.9 11.0	0.2 0.6 0.3	200 60 180

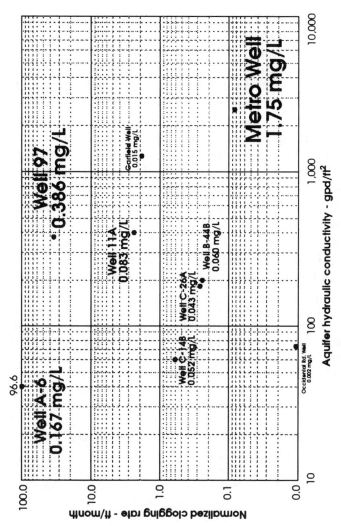

Figure 4.7 Relationship between clogging rate, hydraulic conductivity and total suspended solids.

does not follow the pattern of other wells is possibly due to inability to accurately measure the controlling factors or due to other factors that have not been identified or adequately accounted for. The data from the Occidental Well is possibly the most significant data presented in this study, because it demonstrates conclusively that for low aquifer permeability and low suspended solids content in the source water, plugging does not occur. Well A-6 is another important data point, since it demonstrates that for low aquifer permeability and moderate suspended solids content, the plugging rates are high. The Metro Well demonstrates that low plugging rates can occur with high suspended solids if the aquifer is highly permeable.

Data from additional ASR well sites would further define the relationships presented here. However, this analysis provides a reasonable approach for estimating plugging rates at new well sites prior to well construction and testing, based upon literature values for aquifer parameters, assumed well design, and field measurements of recharge water characteristics. Estimated plugging rates, in turn, can provide a basis for well selection, design, and pretreatment to achieve acceptable backflushing and redevelopment frequency and satisfactory operational performance of recharge facilities.

Following is a theoretical example of how this approach may be used to guide ASR feasibility investigations at a potential new site.

Example: Assume an ASR site is under consideration at a location where the source water would be treated drinking water from a nearby distribution system, the temperature of which varies between 10 and 16°C, averaging 13°C during expected recharge months. The recharge water is tested at a nearby tap in the distribution system and found to have a low suspended solids content of 0.05 mg/L, averaged over three bypass filter tests, each lasting about two days and occurring during representative recharge months. Distribution system pressures at this point during recharge months are typically at least 30 psi. Recharge would typically occur for up to six months followed by a three month recovery period to meet peak demands.

The storage zone under consideration at this site is an unconsolidated, confined aquifer with expected transmissivity of 40,000 G/day/ft and thickness of 100 ft, between 250 and 350 ft below land surface. The ASR well would fully penetrate the aquifer, probably utilizing a 12-inch, .040 slot screen within an 18-inch hole, based upon local experience. Static water level is 50 ft below land surface. Typical well yields in this aquifer are about 1 MG/day.

From this data, which is reasonably available from the literature or from local experience at most sites, the hydraulic conductivity can be calculated by dividing the transmissivity by the aquifer thickness, or 400 G/day/ft^2. The screen open area is about 90 ft^2, considerably less than the borehole surface area of about 470 ft^2. For consistency with the way in which the data in Figure 4.7 were generated, the screen area is utilized for normalizing the data.

Estimated recharge rate is about 0.5 MG/day, or half of the typical well production rates in the area. Recovery specific capacity is estimated from a rule of thumb (transmissivity/2000), or from local experience, at 10 G/min/ft. Recharge specific capacity is expected to be about half of recovery specific capacity.

From Figure 4.7, the normalized plugging rate is estimated at about 0.25 ft/month. This should be multiplied by a factor of 1.2 to account for temperature and viscosity differences. It should also be multiplied by a factor of 31 to account for the flux rate difference at the borehole wall. Adjusting for temperature and flux normalizing factors, the expected plugging rate is about 9.3 ft/month.

At the beginning of recharge, water level in the ASR well would be about 20 ft above land surface. During a typical recharge season, water levels may rise due to plugging to 70 ft above land surface within about five months. This would effectively reduce recharge rates, since the pressure available from the distribution system may be insufficient to overcome further plugging head losses. The need for periodic redevelopment is indicated. A monthly frequency would maintain water levels within an operating range of about 20 to 30 ft above land surface and would be more likely to eliminate residual plugging that can occur when heads build up so high that pumping the well is insufficient to restore its recharge or recovery capacity.

This example illustrates the analysis that might be performed at a proposed ASR site to gain insight regarding probable plugging rates and redevelopment frequency. Such an analysis may suggest the advantages of an alternate site, an alternate water source, treatment of recharge flows to reduce suspended solids, or a different storage zone. As more data becomes available to refine these relationships, the analysis will become more useful.

Redevelopment

Despite the calculations and estimates discussed above, the frequency and method of redevelopment pumping in an ASR well ultimately has to be determined based upon initial testing and operating experience at each site. One of the three methods described above is applied to determine the plugging rate during initial ASR test cycles, following which a judgment is made as to how frequently to redevelop the well in order to maintain recharge rates and also to avoid residual plugging.

A useful starting point is to avoid recharging at a rate, or for a duration, that would cause the water level rise during recharge to exceed the available water level decline in the ASR well during pumping. For the example above, where the recharge specific capacity is estimated at half of the recovery specific capacity, the recharge rate could initially be set at half the recovery rate, or a slightly lower rate with anticipated less-frequent backflushing. The actual plugging rate would then be monitored to compare with the expected value. The well would then be pumped to waste for a few minutes or hours to purge solids from the well. Assuming that recharge and recovery specific capacity are restored, the recharge rate or duration could be extended in small increments in later cycles, each of which would show greater plugging. So long as redevelopment pumping was able to restore specific capacity, the incremental increases in rate or duration would continue until either the desired recharge rate had been achieved, the duration extended to a full recharge season, or signs of residual plugging became evident, such as inability to easily restore recharge specific capacity. The ideal situation is one in which the plugging rate is sufficiently slow that redevelopment only needs to occur at the beginning of scheduled recovery.

Redevelopment pumping or backflushing usually involves pumping the ASR well to waste for anywhere from 10 min to 2 hours. Surging the well by alternately turning the pump on and then shutting it off for two or three cycles in a period of about 3 hours or less is practiced at some sites. This is usually sufficient to restore specific capacity during recharge. However, as discussed in Section 3.2, Design of Wellhead Facilities, care has to exercised to avoid damaging the motor on submersible pumps or the shaft on all pumps by restarting it too soon after shutdown.

The frequency of redevelopment pumping varies substantially between ASR sites. **Table 4.7** lists a number of operational ASR sites and the typical redevelopment frequency. Information is also included regarding

TABLE 4.7   BACKFLUSHING FREQUENCIES AT SELECTED OPERATIONAL ASR SITES

Site	Backflushing Frequency	Lithology
Wildwood, New Jersey	Daily	Clayey sand
Gordons Corner, New Jersey	Daily	Clayey sand
Peace River, Florida	Seasonal	Limestone
Cocoa, Florida	Seasonal	Limestone
Port Malabar, Florida	Monthly	Limestone
Las Vegas, Nevada	Seasonal	Alluvium
Chesapeake, Virginia	Bimonthly	Sand
Seattle, Washington	Weekly	Glacial drift
Calleguas, California	Monthly (approx.)	Sand
Highlands Ranch, Colorado	Monthly	Sandstone

the lithology at each of the sites, to aid in comparison of their operating experiences.

It is usually desirable to pump the ASR wells either to waste or to retreatment during backflushing. Pumping to waste usually provides the opportunity to pump at a high rate, since the discharge head on the pump is substantially reduced or eliminated. This is desirable as it helps to purge solids from the well.

At a few sites, regulatory restrictions on the disposal of water during backflushing operations are sufficiently rigorous that special containment and treatment provisions are required. This is the case for the salinity intrusion barrier injection wells in southern California. Where this is the case, or may be reasonably expected in the future, greater care is needed during design of well and wellhead facilities, to minimize the volume of solids entering the well and thereby reduce the frequency of backflushing as well as to improve the quality of the backflush water.

For unconsolidated aquifers, experience suggests that recharge rates tend to approach an equilibrium level that is lower than the initial recharge rate at the beginning of testing, but can be sustained by periodic backflushing. Some early loss of initial recharge capacity occurs at many such sites, while still maintaining long-term rates at a useful level. For consolidated aquifers, such loss in capacity is less apparent.

It is probably wise to assume that ASR wells will need full redevelopment about every five years, including pulling and setting the pump, cleaning, acidization, disinfection, and possibly other methods to restore its condition. This may not be required at some sites, particularly those in consolidated aquifers; however, in the absence of site-specific evidence to the contrary, the need for redevelopment every few years should be assumed for budgetary and planning purposes.

The ASR system at Manatee County, FL, recharges with water that is diverted from the water treatment plant prior to final pH stabilization. The recharge water is slightly aggressive, but is rapidly stabilized when it comes into contact with the limestone in the storage zone. **Figure 4.8** shows the increase in specific capacity that occurred at this site during the first few cycles of testing. Calculations indicated that the volume of calcium carbonate in the storage zone that was dissolved during this process was very small, and not significant to long-term well operations. This eliminated the need for periodic redevelopment of the ASR wells at this site and also saved the cost associated with stabilizing the water in the treatment plant. During pH stabilization in the aquifer, the total dissolved solids concentration of the recovered water showed an increase of about 25 mg/L. This was evident at the beginning of each cycle, even when the

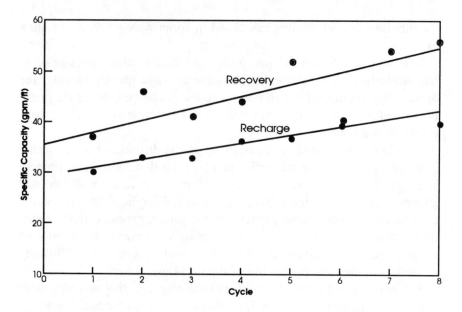

Figure 4.8 ASR well increase in specific capacity, Manatee County, Florida.

storage period between recharge and recovery was on the order of an hour or less.

A similar approach has been considered, but not implemented, for storage of aggressive waters produced from desalination plants in the Arabian Gulf. These plants typically are located over brackish limestone aquifers, some of which would be suited to seasonal, long-term, and emergency storage of drinking water while similarly achieving savings in water treatment costs through pH stabilization in the aquifer.

4.3 WELLHEAD FILTRATION

If the rate of ASR well plugging, or the expected frequency of required backflushing, is perceived as a potential operating problem, a desirable solution is to keep the solids out of the well in the first place. As discussed previously, this involves at least purging the recharge piping to waste prior to initiating recharge.

One solution that has been implemented at an ASR site in New Jersey is to incorporate a short length of large diameter (60 inch, 1500 mm) pipeline into the recharge piping at the wellhead in order to reduce flow velocity and thereby settle out any solids. The primary purpose of this large pipe is to provide detention time for chlorination of recovered flows before they enter the distribution system; however, it is anticipated that it will also serve to settle out any solids in the recharge water.

Gordons Corner, NJ, has operated an ASR system since 1971 that incorporates wellhead sand filtration in order to minimize entry of solids into the ASR wells. It is anticipated that other ASR systems recharging treated drinking water may, in some cases, benefit from providing well-head filtration. Solids in distribution systems often include sand, rust, alum floc, or other constituents capable of plugging an ASR well.

As an ASR operating practice, wellhead filtration is in its infancy with only three sites (Salt Lake County, UT; Salt River Project, AZ; and Gordons Corner) known to be providing these facilities. The Utah test site utilizes a pressure sand filter. The Arizona test site has utilized a drum filter, and the New Jersey operational site utilizes a sand filter. Some new sites recharging treated drinking water into aquifers prone to plugging are expected to include wellhead filtration facilities. In addition, future sites using ASR technology to store surface water containing low levels of suspended solids are expected to incorporate wellhead filtration as a basis of design. This is discussed further in Chapter 7, Alternative ASR Applications.

The technology for wellhead filtration is widely available, as developed for the water utility, mining, and agricultural sectors. Costs of these systems tend to be highest for the water utility applications and lowest for the agricultural applications. However, the particle sizes removed tend to reflect the system costs. The optimum tradeoff between investment in wellhead filtration facilities and system operating costs remains to be determined. For proposed sites where redevelopment pumping is a prob-lem due to cost, potential electric motor damage, water disposal or permit-ting difficulties, the investment in wellhead filtration facilities may be advisable and should be considered during design. A reasonable solution for many sites will be to provide space in the wellhead design to incorpo-rate wellhead filtration at a later date, if required.

Complete sand media filtration systems for agricultural applications are readily available at (1993) manufacturer's costs ranging from $4000 to treat flow rates of 300 G/min, $12,000 to treat 700 G/min, and $30,000 to treat 7000 G/min. Each 300 G/min modular unit has about a 6 ft × 6 ft imprint and stands about 6.5 ft high. Multiple units would be manifolded to achieve the desired filtration flow rate. Such systems are used widely to prevent clogging of spray nozzles in microjet irrigation systems. An alternative approach using a ring filter capable of treating flows of 100 G/min may cost about $7,500. As discussed subsequently in greater detail in Section 7.2, Surface Water Storage, horizontal well technology also achieves significant reduction in the solids content of surface water as a

result of sand filtration. Through patented trenching and backfilling methods, a fabric-covered, slotted pipeline drains water from the water table and
any adjacent surface water sources through a sand formation or backfill,
thereby improving water quality prior to recharge into an ASR well.

For municipal applications, a multi-media pressure sand filter manufactured by the Yardney Company can remove particles down to 10 μm in
size. A system capable of filtering 0.5 MG/day would include two 48-inch
diameter tanks, each 60-inch tall, and would cost about $15,000. Filtering
1 mgd would require four such tanks, or three tanks each 54 inches in
diameter, and would cost about $30,000.

An alternative approach offerred by the 3M Company is to use dispos-
able filter cartridges within an inline pressure filter. Different cartridges
are capable of providing filtration from 15 μm down to 2 μm. Each
cartridge is capable of filtering about 50 G/min, requiring multiple cartridges to achieve flow rates of interest to ASR operations. Vessels capable
of handling 18 cartridges are readily available. The initial pressure drop
across the 2, 5, 10, and 15 micron cartridges is about 0.9, 0.5, 0.4, and 0.3
psi, respectively. Cartridges have to be replaced when the pressure drop
reaches 35 psi.

The cartridges are reported to be able to handle 21 to 24 lbs of suspended material before changeout is required. This data is reported from
tests using silica dust at 40 G/min. Assuming 5 mg/L suspended solids
across a cartridge at 50 G/min, the cartridge would function for 8 to 9 days.
If the loading was 2 mg/L, the cartridges would require changing every 21
to 24 days. This assumes that all of the suspended solids are trapped by the
filter. In reality, the particles trapped will be a function of the size distribution of the particles and the size of the filter pore spaces. The cartridges
may last longer than the numbers presented; however, it is not possible to
estimate the duration without first investigating the water to be filtered.
The 3M Company provides an analysis of particle size and estimated
cartridge life upon request. The 1993 manufacturers' cost of such a system
to handle 0.5 MG/day is about $16,000, while a 1 MG/day system is
estimated at about $31,000. Cartridges cost about $367 each, so frequent
changeout can lead to high operating costs.

Membrane processes may also be used for wellhead pretreatment. In
particular, microfiltration systems can provide a higher level of treatment
but at somewhat higher cost. Since about 1988, Memtec America, manufacturer of the Memcor m³F (continuous microfiltration) process, has
supplied microfiltration units for potable water treatment with capacities
up to 4.4 L/sec (0.1 MG/day), although larger units of 3.8 megaliters/day
(1 MG/day) and 13 megaliters/day (3.5 MG/day) capacity have been

commissioned. The same units have been in use for wastewater renovation since the mid-1980s. Operating pressures for these units range from 25 to 40 psig, while the pressure differential across the membrane varies from 2 psig for a clean membrane to 15 psig when the membrane is fouled. Particle sizes are reduced to below 0.2 µm with this process. The backwash volume is about 2 to 7% of the feedwater volume. The membranes are chemically cleaned when pressure differentials exceed about 15 psig, using caustic-based solutions at pH values above 12. These units retain protozoan cysts such as Giardia and Cryptosporidium, nearly all bacteria of health concern, and turbidity. They also provide between 2- and 4-log removal of viruses.

Installed manufacturer prices for Memcor units range from $1.00/G installed capacity for a unit capable of treating 360,000 G/day; $0.50/G installed capacity for a unit capable of treating 1.1 MG/day, and $0.35/G installed capacity estimated for a unit capable of treating 23 MG/day.

An alternative approach is manufactured by Kalsep, Inc., called the Kalsep Fibrotex System. As applied to alum-flocculated surface water for reverse osmosis pretreatment, this unit is estimated to cost about $300,000 to treat 2.5 MG/day of water with a TSS concentration of 1 mg/L. Result-ant particle sizes in the product water are estimated within the range of 1 to 3 µm.

Considering the range of alternatives presented above, it appears that sand filters, ring filters, drum filters and horizontal wells can filter re-charge water to small particle sizes generally suitable for agricultural applications that would not plug irrigation systems. In some cases, these may also be suitable for ASR wellhead filtration, particularly with storage zones that have high transmissivity. Where aquifers have lower transmis-sivity, other filtration systems are available that can reduce particle sizes down to between 2 and 10 µm using multi-media pressure filters or cartridge filters. Microfiltration systems using membrane filter technology can remove particles down to 0.2 µm. Selection of the appropriate technol-ogy to meet the technical and regulatory requirements for ASR operation has yet to be clearly defined.

It appears that agriculturally oriented systems can be supplied at costs acceptable to the agricultural community, since such systems are already in wide use. At higher manufacturer's costs of roughly $30,000 per 1 MG/day system, wellhead filtration can be provided to reduce particle sizes to levels that would probably be compatible with most ASR systems. Even higher levels of treatment can be provided with membrane and comparable processes, for unit costs in excess of $100,000 per MG/day installed capacity.

Considering the substantial cost savings usually attributable to implementation of ASR technology, it is anticipated that pretreatment costs may, in many cases, have little adverse impact upon the overall cost-effectiveness of ASR systems. As discussed in Section 6.1, Economics, capital costs of ASR systems, including engineering and construction, average about $400,000 per MG/day of recovery capacity ($0.40/gallon). This is frequently less than half the cost of other viable alternatives. Consequently, a capital cost increase of $30,000 to $100,000 per MG/day may not eliminate the cost-effectiveness of ASR in situations where wellhead filtration is desirable.

Wellhead filtration is one of the aspects of ASR technology that is evolving rapidly. Experience is demonstrating the desirability of keeping solids out of the ASR wells. While a higher quality of recharge water will tend to improve overall performance of an ASR system and also expedite regulatory approval, the most cost-effective tradeoff between the investment in wellhead filtration and ASR performance has yet to be established for municipal and agricultural applications.

An important difference between ASR applications and other applications of advanced filtration technology is that ASR wells are provided with a backup capability to remove solids from the well that pass through the filter. This is the periodic backflushing operation. This would not be the case for applications preceding reverse osmosis membrane treatment plants, for instance. Experience at several sites will be required to estimate the most cost-effective combination of wellhead filtration and backflushing frequency for aquifers with different hydraulic characteristics.

4.4 FLOW CONTROL

During recharge, it is not uncommon for wellhead pressure to vary over a very wide range. Initially some low pressure may be required to start flow into the well; however, once flow is established, a full vacuum may immediately develop at the wellhead when the water is allowed to cascade into a sealed well with a few feet or more of depth to water level. As recharge continues, mounding of the recharge water in the aquifer may combine with plugging to cause the recharge water level to rise above land surface. In a few applications, recharge pressure is then increased steadily to compensate for head losses attributable to plugging, up to the maximum available pressure from the source of supply. At that point, redevelopment is required to hopefully restore ASR capacity. Flow control is usually provided to smooth out this extreme range in operating conditions, where they may occur.

During ASR recovery or backflushing, water levels will typically experience a broader range than for a normal production well, reflecting slightly higher initial water levels and slightly to significantly lower pumping water levels at the end of seasonal recovery. This range sometimes presents a challenge in sizing a pump to produce a desired flow rate, since initial flows may be much higher than ultimate flows at the end of a recovery period. One solution to this is to provide a pressure control valve on the recovery piping, so that the pump is always operating within an acceptable range of flow rates.

The broader ranges for operating water levels and pressures have to be considered during well and wellhead design to ensure that the system will operate properly over the expected range. Low pressures at the wellhead tend to cause very high flow rates that can cause operating problems, either in the distribution system during recharge or sometimes in the surface drainage system during backflushing to waste. Flow control is therefore required for many ASR systems.

In Section 3.2, Wellhead Facilities Design, various approaches were discussed for control of cascading during recharge, many of which involved maintaining positive pressure in the well piping. A significant recent development is a downhole water level control valve that is operated by air, water, or oil pressure from the surface, and that throttles recharge flows in ASR wells with substantial depth to water level. The throttling mechanism is a permanent inflatable packer that is connected either in the pump column or at the base of an injection tube. With this approach, flows can be set at a rate that maintains any particular desired wellhead pressure or depth to water level during recharge. Developed by Baski Valve Company, Denver, CO, during 1992, this valve has been applied successfully for the Centennial ASR Well A6 in Highlands Ranch, CO. Depth to static water level in this well is about 900 ft below land surface. Flow control from the end of the recharge piping seems preferable to flow control from the beginning of this piping near the wellhead.

A significant advantage of this downhole water level control valve is its small diameter. As a result, many existing small diameter wells in locations with great depths to water level may be retrofitted to ASR purposes by recharging down the pump column without having to add injection tubes inside the casing to control flow rates and cascading.

Limiting recharge water levels within a reasonable, positive range will tend to stabilize flow rates. Where negative pressures are still expected at the wellhead, it is important that adequate valving is provided to ensure positive pressures a short distance upstream of the wellhead so that flowmeters, sampling taps, and pressure gauges are operating under positive pressures. Otherwise, flow measurements may be erroneous and sampling taps will not function.

The transition from negative to positive pressures at the wellhead during recharge is probably best accomplished with a globe valve at the wellhead with a manual control. Automatic controlled globe valves tend to oscillate under such circumstances. Where an automatic globe valve is used, it should be accompanied with an orifice plate so that the downstream side of the globe valve is always under positive pressure. Butterfly and gate valves do not last long under such operating conditions.

4.5 DISINFECTION BYPRODUCT REDUCTION

Chlorination of water containing color and natural organics, such as humic and fulvic acids produces DBPs, such as trihalomethanes (THMs) and haloacetic acids (HAAs), which are carcinogenic. In recent years, considerable attention has been directed to research into this public health challenge with the result that water treatment practices are changing in response to new regulations. At the same time, data collected from several early ASR sites indicated some reduction in DBP concentrations during ASR storage. The data were generally not conclusive since they were not collected under conditions designed to separate out mixing as a causative factor. However, they were reasonably consistent in showing some DBP reduction between recharge and recovery.

During 1991, a study [9] was initiated to take a closer look at whether or not DBP reduction is occurring during ASR operations and, if so, to evaluate the mechanisms responsible for the reduction. This investigation was performed during a two-year period by CH2M HILL, Inc., consulting engineers, in association with the University of North Carolina. The American Water Works Association Research Foundation (AWWARF) provided funding for the project, while supplemental funding was provided by Thames Water Utilities, Reading, England, and by the Las Vegas Valley Water District, NV. Participating utilities also included the Peace River/Manasota Regional Water Supply Authority, FL; Centennial Water and Sanitation District, Highlands Ranch, CO; and Upper Guadalupe River Authority, Kerrville, TX.

The plan was to conduct the test at each site so that, to the extent possible, samples were collected from the the same water volume with minimal mixing between this recharge water and the surrounding groundwater (or recharge water remaining from previous ASR cycles), during recharge, storage, and the first recovery sample. The second and third recovery samples were more likely to show mixing with surrounding water in the aquifer, as determined by evaluation of tracer concentrations such as chlorides. At each ASR site, volumes between 34 and 140 megaliters (9 and 37 MG) were recharged and then allowed to stay in the storage zone for periods that ranged from 48 to 191 days (mid-point of recharge to mid-point of recovery), following which all or a portion of the water was recovered. Samples were collected three times during recharge, five times during the storage period, and three times during recovery, plus one background sample was collected at each site. The total volume of water pumped out for sample collection was negligible compared to the volume initially stored. Samples were analyzed for a wide variety of water quality constituents, including total organic carbon (TOC) and ultra-violet radiation, both of which are indicative of THM and HAA precursors. In addition, calculations were performed to check the expected movement of the stored water during the storage period under the influence of the hydraulic gradient reported at each site. Natural tracers were utilized to estimate mixing between stored and native water at each site, although this was complicated by the small differences at some longer-term operational sites.

Figure 4.9 shows DBP concentrations, while **Figure 4.10** shows THM constituents and their differential rates of decline during ASR storage for the Thames Water Utilities site. **Figure 4.11** shows DBP formation potential decline during ASR storage at this site. Within 111 days from the end of recharge, THM concentrations had disappeared, while THM formation potential had declined by half from the end of recharge. Similarly, HAA concentrations disappeared within 3 days after the end of recharge, while HAA Formation Potential had declined by 76% within 111 days. **Figure 4.12** shows the reduction in dissolved oxygen and nitrate concentrations at this site during the sampling period. Denitrification is suggested by the reduction in nitrate concentrations soon after the end of recharge and roughly coincident with the loss of dissolved oxygen in the aquifer. Similar results are available for the other four sites.

Figure 4.13 shows a comparison of the THM data for the five sites, while **Figure 4.14** shows a comparison of the HAA data. **Table 4.8** includes a summary of the results, including the reduction in DBPs that might have been expected based upon mixing between recharge and native waters, as judged from natural tracers.

In general, it was concluded that

- THMs and HAAs are removed from chlorinated drinking water during aquifer storage.
- HAA removal precedes THM removal; the more brominated species tend to be eliminated earliest.

- In most cases, removal of these halogenated DBPs does not appear to occur until anoxic conditions develop, and frequently follows the onset of denitrification. A biological mechanism is suggested. Additional work must be conducted to establish the mechanism(s) responsible for removing these DBPs, and the conditions under which they occur.
- THM and HAA precursors are also removed to a significant degree during aquifer storage.

For the five sites investigated, THM and HAA reduction occurred during a few weeks of storage, as opposed to days or months. Site-specific testing is needed at other sites to establish reductions that may be achieved; however, it appears that seasonal ASR storage can provide useful water quality improvement benefits for many utilities. Whether or not THM formation potential is reduced at any particular site during ASR storage, the reduction in instantaneous THM concentrations may provide a significant reduction in total THM concentrations (instantaneous THM plus THM formation potential) once the recovered water is re-chlorinated prior to distribution.

This is a significant benefit attributable to ASR systems. For some ASR sites, it may be possible to recover water with low DBP concentrations, blend it with higher DBP water from the treatment plant, and thereby meet

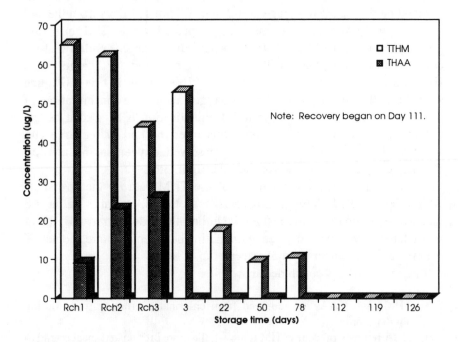

Figure 4.9 Disinfection byproducts during ASR storage, Thames Water Utilities, England.

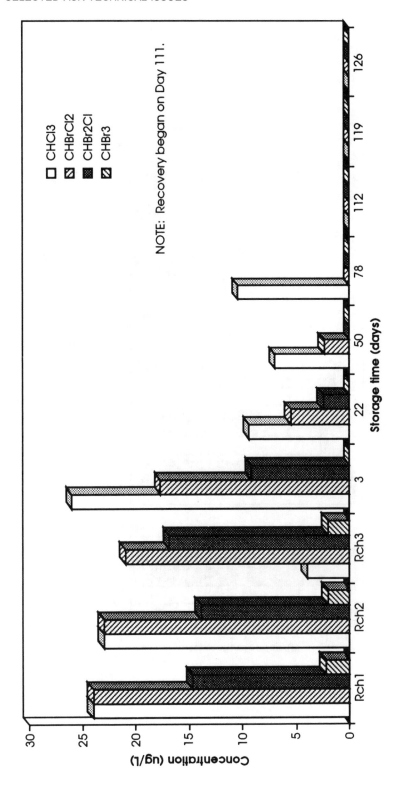

Figure 4.10 Trihalomethane constituent distribution during ASR storage, Thames Water Utilities, England.

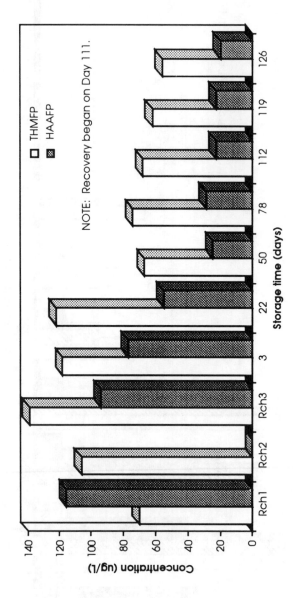

Figure 4.11 Disinfection byproduct formation potential during ASR storage, Thames Water Utilities, England.

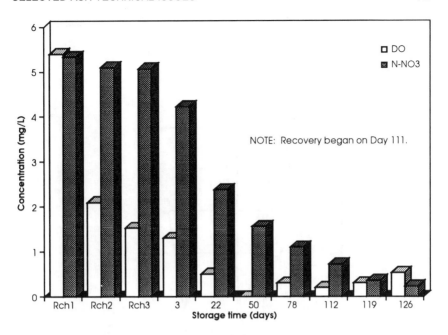

Figure 4.12 Dissolved oxygen and nitrate concentrations during ASR storage, Thames Water Utilities, England.

DBP standards that are expected to be lowered within the next few years. Other potential benefits are the opportunity to recharge water with a chlorine residual, avoid the addition of ammonia, and count on biological processes in the aquifer to reduce resulting DBP concentrations prior to recovery. As discussed in Section 4.2, Well Plugging and Redevelopment, changing from chlorine to chloramine disinfection in Los Angeles, CA

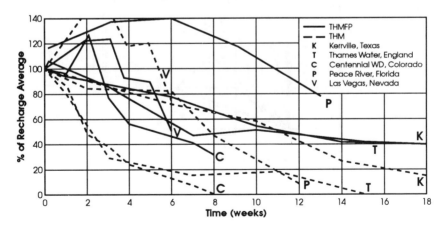

Figure 4.13 Trihalomethane reduction during ASR storage at five sites.

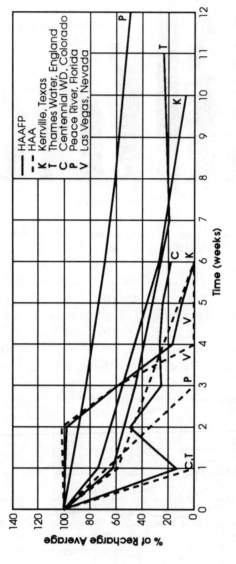

Figure 4.14 Haloacetic acid reduction during ASR storage at five sites.

TABLE 4.8 DISINFECTION BYPRODUCT REDUCTION DURING ASR STORAGE

Site	Storage Duration (days)	Expected Reduction Due to Dilution (%)	Disinfection Byproduct Reduction (%)			
			TTHM	TTHMFP	HAA	HAAFP
Thames Water Utilities, England	111	64	100	>51	100	>76
Upper Guadalupe River Authority, Kerrville, Texas	127	50	25	24	100	(4)
Centennial Water and Sanitation District, Highlands Ranch, Colorado	49	0	92	(<80)	100	12
Peace River/Manasota Regional Water, Supply Authority, Arcadia, Florida	90	0	100	(−11)	100	100
Las Vegas Valley Water District, Las Vegas, Nevada	36	25	33	44	100	90

Note: ( ) means concentration increase, not decrease.

injection wells has been shown to encourage bacterial activity around the well, increasing both plugging rates and redevelopment difficulty.

Following completion of the field testing for the AWWA Research Foundation program discussed above, preliminary laboratory investigations were conducted to evaluate the mechanisms governing the DBP reactions that had been observed. Results were inconclusive, reflecting the complexity of the laboratory procedures; the volatile nature of some of the compounds; and the need to maintain anoxic, aerobic, or sterile conditions in the sample vials. Further investigation is required to establish the mechanisms responsible for the observed DBP reactions occurring in the field.

4.6 PRE- AND POST-TREATMENT

As a general rule, treated drinking water can be stored and recovered from ASR wells without any need for further pre- or post-treatment other than disinfection following recovery. Occasionally a situation will arise where further treatment is necessary, usually limited to pH adjustment. In a few situations, chemical feed may also be required for other constituents such as calcium chloride or sodium bisulfite.

ASR experience to date with storage of treated drinking water has included only a few sites for which the recharge water requires pretreatment, and only one site for which the storage zone was pretreated. At these few sites, however, further pretreatment was required due to the following reasons:

- post-precipitation of alum floc in the distribution system between the water treatment plant and the ASR well
- presence of solids in the recharge water, other than alum (e.g., sand, rust, biomass, shrimp, etc.), generally due to local flow reversal in the distribution system
- occasional low pH of recharge water quality (below 8.0), which would mobilize manganese present in the storage zone
- occurrence of high concentrations of iron (13 mg/L) in the storage zone due to the presence of siderite (ferrous carbonate) and pyrite (ferrous sulfide)

Pretreatment to remove solids has been discussed in Section 4.3, Wellhead Filtration.

Pretreatment to elevate the pH to about 8.5 has been implemented at two sites to reduce the potential for ferric hydroxide precipitation and to ensure that manganese is not mobilized in the recovered water. At one site this was achieved with addition of sodium bicarbonate, while at the other site sodium hydroxide was used. In general, the latter approach is favored, primarily for operational reasons. Pretreatment at these two sites is required for all recharge flows.

Following is a brief discussion of pre- and post-treatment issues related to several water quality constituents or processes pertinent to ASR operations. Particular emphasis is included for iron and manganese issues, due to their prevalence in potential ASR storage zones.

Disinfection

During recovery, disinfection tends to reduce the pH of the water. Where recovered ASR water is blended with a larger flow of treated water, the pH of the blend may be only slightly affected. However, if the blend ratio is high and the alkalinity of the water is low, the pH reduction in the blended flow may be sufficient to make the water aggressive. pH adjustment is then required to restore equilibrium.

Ammonia addition may be required on the recovered water in order to control disinfection byproduct formation. As discussed previously in Section 4.5, DBP concentrations are reduced substantially during several weeks of ASR storage; however, some formation potential will undoubtedly remain, particularly in situations where storage time between recharge and recovery is brief. Testing is appropriate to determine whether ammonia is present in the recovered water, remaining from any chloramine residual present in the recharge water. Supplemental ammonia is then provided to achieve the desired concentration. **Figure 4.15** shows the ammonia present in the recharge and recovered water during test-cycle 5 for the Port Malabar ASR facility. Approximately half of the recharge ammonia was present in the recovered water after a storage period of 29 days between the end of recharge and the beginning of recovery.

Iron

Iron is present in the groundwater at many ASR sites, sometimes at concentrations that can cause problems either with ferric hydroxide precipitate plugging during recharge or with meeting drinking water standards during recovery. Several different approaches have been tried to control iron problems in ASR systems. A preferred approach is one in which a single aquifer pretreatment exercise eliminates subsequent iron problems. However, to date, this goal has proven elusive. An alternate approach involves continuous pretreatment of recharge flows for the life of the

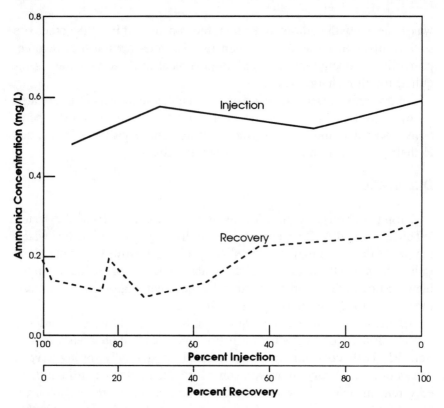

Figure 4.15 Ammonia concentrations observed during Cycle No. 5, Well R-1, Port Malabar, Florida.

ASR facility. Both approaches are discussed below, along with other measures for controlling iron concentrations and potential plugging with ferric hydroxide.

Aquifer Pretreatment

At Swimming River, NJ, iron concentration in the groundwater samples pumped from the proposed storage zone was about 13 mg/L. Cores from the storage zone indicated the presence of siderite, and occasionally, pyrite, both of which are iron-bearing minerals. Siderite is composed of ferrous carbonate and is the primary cause of the high iron concentrations, while pyrite, which is composed primarily of ferrous sulfide, contributes to the problem, particularly at lower concentration levels.

Following laboratory column tests to evaluate alternative procedures (see Section 5.6, Column Testing), it was decided to treat the aquifer around the ASR well for a radius of about 70 ft (21 m) with a solution of sodium bisulfite at a pH between 2 and 4. This was intended to strip dissolved oxygen from the recharge water while at the same time oxidizing the ferrous carbonate. About 35 MG (132 megaliters) of drinking water treated in this manner were recharged over a period of 93 days. Due to the slow rate of the deoxygenation reaction, recharge flow rate initially had to be very low to ensure adequate travel time before the recharge water became deoxygenated and then reached the aquifer. Initial flow rates of 70 and 150 G/min (0.38 and 0.81 megaliters/day) for two weeks each were followed by increasing rates in stages to 400 G/min (2.2 megaliters/day). As the buffer zone volume steadily increased, the rate of recharge also could be increased without risking plugging.

Testing indicated that the aquifer and well hydraulics were improved by this procedure since both transmissivity and specific capacity increased slightly. Sampling in a nearby storage zone observation well confirmed that ferrous carbonate dissolution products were moving away from the well under the influence of continued recharge of this low pH, deoxygenating solution, as shown previously in the column tests. A 35 MG (132 megaliters) buffer zone of deoxygenated water at normal pH was then recharged, followed by treated drinking water from the plant at a pH of 8.5 and dissolved oxygen concentration in the range of 7.5 to 9.0 mg/L.

There was no doubt from the results that the siderite in the formation around the well was eliminated by the pretreatment exercise, although the pyrite remained. However, the process also removed any oxidation coating on the pyrite grains, thereby enhancing the capacity of the pyrite to form ferric hydroxide when it came into contact with oxygen-bearing water. As a result, iron content of the recovered water was initially much higher than expected, reaching a concentration of 8 mg/L at the end of the first cycle. Continued cycles had the effect of forming a hydroxide coating on the pyrite, thereby reducing its reactivity with oxygen-bearing water. With successive cycles, the iron content of the recovered water steadily declined. After the seventh cycle, the iron content at 90% recovery was 0.4 mg/L. This may be compared with concentrations of 13 mg/L in the native water from this ASR well prior to pretreatment.

Recharge Water Pretreatment

After the first six test cycles at Swimming River, the pH of the recharge water was further increased to 9.0 with addition of sodium hydroxide, to address concerns regarding low residual concentrations of iron in the recovered water. This change also addressed new concerns regarding manganese concentrations in the recovered water, resulting from the acid treatment of the formation around the well. pH adjustment appears to have resolved both iron and manganese problems at this site. **Figure 4.16** shows the improvement in iron concentrations with successive cycles.

This is a more simple approach than aquifer pretreatment; however, it requires continuous pretreatment of the recharge flows. Due to the residual acidity in the aquifer around the ASR well, resulting from the initial aquifer pretreatment efforts, several ASR cycles may be required to overcome the residual tendency for low levels of iron and manganese in the recovered water. Initial results suggest that this approach will be successful.

pH adjustment is not expensive, particularly when considered in the context of overall savings associated with ASR implementation compared to other water supply alternatives. The relative simplicity and low cost of

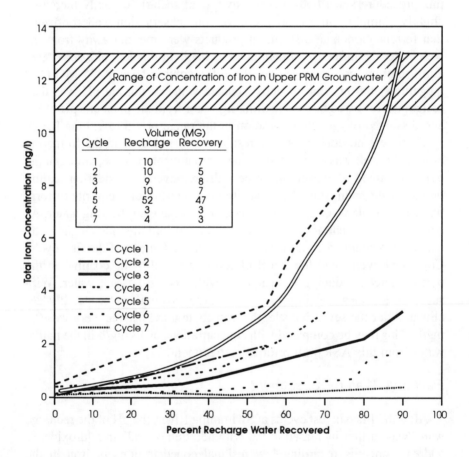

Figure 4.16 Cyclic improvement in iron concentrations during recovery, Swimming River, New Jersey.

recharge water pretreatment suggest that continuous pretreatment is probably a better approach for most situations where iron and/or manganese may be present in the storage zone. Nevertheless, the goal of a one-time aquifer pretreatment step is a worthy goal that should be considered at future ASR sites.

With blending between the flow from the water treatment plant, which is usually in excess of 30 MG/day (114 megaliters/day) during recovery months, and flow from one existing and two proposed ASR wells which should total between 2 and 3 MG/day (8 and 11 megaliters/day), the blend going to the consumers at Swimming River will meet the potable drinking water standard of 0.3 mg/L.

Vacuum Degassification

For the Swimming River example discussed above, deoxygenation was achieved through chemical addition. However, consideration was given to vacuum degassification to achieve the same objective. The depth to water level in the ASR well at planned recharge rates was such that cascading of recharge water into the well casing was expected to create a strong vacuum in the casing, which was sealed against entry of air. Installing a small vacuum pump attached to the casing would have extracted gases pulled out of solution during the recharge process, including dissolved oxygen.

This approach was not selected; however, it may be useful at some future ASR sites with similar technical challenges. The primary reason that this approach was not selected is that vacuum degassification was perceived as a continuing operating requirement, whereas the ASR test program at the Swimming River site was seeking a different approach that would accomplish the desired objective of iron removal in a single pretreatment operation, following which no further pretreatment would be needed. For the limited purpose of iron removal, the approach taken proved reasonably successful. However, pH adjustment of recharge flows for manganese control is now required at this site as a long-term operating requirement.

Buffer Zone Formation

Where the possibility exists that precipitates such as ferric hydroxide may occur in the storage zone, it is advisable to plan ASR testing and operations so that no reaction products are allowed close to the well during recovery. This is achieved by consistently leaving a small percentage of the recharge water in the aquifer. The minimum buffer zone radius around the well should be about 10 m (30 ft) or more during the first cycle, and should be augmented during each subsequent ASR cycle in order to provide a driving force that will slowly push any precipitates and other geochemical reaction products away from the ASR well.

In practice, this is best accomplished by leaving a specified volume underground during the first cycle, regardless of ultimate recovery efficiency expectations. For example, if the first cycle includes storage of 5 MG (19 megaliters), recovery could be to a volume of about 4 MG (15 megaliters) or to a target water quality criterion, whichever is less. In later cycles with presumably larger volumes stored, the volume left underground in each cycle may be smaller, so long as the tendency is for the buffer zone volume to continue expanding slowly.

Precipitation of ferric hydroxide floc in the storage zone should not inhibit ASR well hydraulic performance so long as it does not occur close to the well. In this regard, accurate measurement of ASR flows and cumulative volumes is particularly important in situations where ferric hydroxide precipitation close to the well is to be avoided. Precipitation of ferric hydroxide in the aquifer at a sufficient distance away from the well is unlikely to affect operational performance of the well, whereas the same process adjacent to the well screen can quickly plug the well.

Post-Treatment

Ferric hydroxide precipitation close to the well screen or borehole can theoretically be removed by frequent backflushing of the well in order to maintain its recharge capacity. Wildwood, NJ, has an ASR system that has been operational since 1968, backflushing each well on a daily basis for about 10 min. No testing has been performed to characterize the water quality of the backflush water; however, visual appearance, combined with experience at other ASR sites in the New Jersey coastal plain, suggests that the backflush water contains high concentrations of either rust or, more likely, ferric hydroxide precipitate. The backflush water is discharged to waste.

Manganese

Geochemical assessment may indicate the presence of manganese-bearing minerals in the storage zone. Recharge water with a pH below about 8.0 may mobilize the manganese so that after a few weeks of storage, recovered water may have unacceptably high manganese concentrations. Manganese reactions tend to be slow. One solution is to adjust the pH of the recharge water so that it is always above 8.0 and preferably about 8.5. This is usually most easily accomplished with addition of sodium hydroxide.

At Chesapeake, ASR testing was performed during a period of several months when pH of the recharge water was sufficiently high (7.4 to 9.4) that no manganese problems were apparent in the recovered water. Manganese concentrations during recovery were less than 0.01 mg/L. Subsequent process changes at the water treatment plant reduced the pH (6.2 to 7.0), with the result that manganese concentrations in the recovered water exceeded drinking water standards, reaching levels of 0.25 to 1.28 mg/L. Remedial measures included pumping out most of the stored water at the low pH and recharging water treated with sodium carbonate to raise the pH to within a range of 8.0 to 8.2, or higher. The sodium carbonate helps to buffer the acidity remaining in the aquifer from the large volume of water recharged at low pH. Once this acidity has been overcome, pH of the recharge water will be more likely to stay above 8.0 underground. A change to pH adjustment with sodium hydroxide will then be appropriate to facilitate operations. Operational testing conducted during 1993 indicates that the residual acidity in the aquifer is being overcome, as evidenced by increasing recovery efficiency without excessive manganese levels.

Figure 4.17 shows manganese concentrations for the Swimming River ASR site. As described above for Chesapeake, pretreatment with low pH water to control high iron concentrations was reasonably successful; however, it had the undesired side-effect of dissolving manganese present in the minerals of the storage zone. High manganese then became a problem requiring solution. As a result, the pH of the recharge water was adjusted with sodium hydroxide to about 9.0 during the seventh and subsequent cycles. As shown on Figure 4.17, the manganese concentrations in the recovered water subsequently declined to acceptable levels.

Arsenic

No field work has yet been conducted regarding the use of ASR wells for arsenic reduction. However, much attention has been directed in recent years to the public health significance of arsenic in drinking water, and new regulations are being considered by the U.S. Environmental Protection Agency that may reduce drinking water standards for arsenic.

In the near future it is anticipated that work will commence to establish the effectiveness of ASR wells in removing arsenic during aquifer storage. The reaction mechanism is expected to involve co-precipitation of arsenic along with ferric hydroxide in aquifers containing low concentrations of iron-bearing minerals and recharged with water containing dissolved oxy-

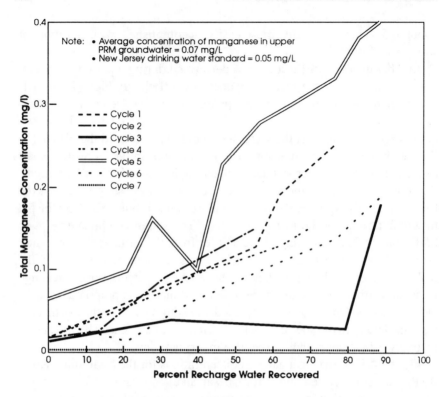

Figure 4.17 Cyclic improvement in manganese concentrations during recovery, Swimming River, New Jersey.

gen. If successful, this approach may provide a low-cost solution to many water utilities faced with the prospect of expensive, above-ground treatment processes for arsenic removal to meet new standards.

Radon

ASR wells appear to have little or no beneficial effect upon radon concentrations in recovered water. Experience with the Seattle ASR system suggests that if treated surface water containing no radon is stored in an aquifer containing minerals that release radon, the recovered water will contain radon at close to native water concentrations within a few days. Post-treatment alternatives include blending, storage time, and also aeration. Pumping the recovered water to a small storage tank for aeration and disinfection may be sufficient to provide the time required for radon reduction to within drinking water standards. The additional cost for pumping the water twice (once from the well to the tank and once from the tank to the distribution system) and the cost of the associated storage/

aeration facilities is a disadvantage of this approach. Until such time as EPA establishes the revised radon standard, the urgency of a solution to this water treatment problem will remain unclear.

Ion Exchange

None of the currently operational ASR systems in the U.S. have experienced plugging or water quality problems due to ion exchange between the recharged water and the native water in the aquifer. Geochemical modeling has been performed for almost every site, sometimes indicating the potential for ion exchange to occur. Water quality data collected at several ASR sites have confirmed the occurrence of ion exchange reactions, but not to an extent that would threaten ASR operations.

Perhaps the best-documented example of ion exchange plugging was at Norfolk, VA, during a U.S. Geological Survey injection well recharge investigation conducted in 1971–1972 [10]. The storage zone was a clayey sand aquifer containing brackish water. Recharge was with treated drinking water. During the first three test cycles, clay dispersion was shown to cause uniform plugging of the storage zone throughout the whole length of the screen interval, with 94% reduction in specific capacity of the well. Subsequent redevelopment and treatment with calcium chloride as a preflush prior to recharge Cycle 4 was successful in stabilizing specific capacity of the well at about one third of the initial value prior to testing. The authors concluded that "treatment of the injection well with a clay stabilizer prior to injection of any freshwater will minimize clogging and increase recovery of freshwater."

The pre-flush solution selected at Norfolk was calcium chloride, providing a divalent ion, calcium, to exchange with monovalent sodium ions in the formation clays. A trivalent ion such as aluminum would also achieve the same purpose.

Analysis of cores obtained from potential ASR storage zones, particularly in unconsolidated, brackish aquifers, can provide an indication of the potential severity of the clay dispersion problem. If smectite (montmorillonite) clays are present, even at concentrations of 1% or less, and the aquifer contains brackish water, the need for pretreatment may be indicated.

One element of the solution to clay dispersion problems may be to allow time for a gradual adjustment in salinity of the aquifer around the well. A slow transition may yield fewer plugging problems than a shock treatment. Since the greatest opportunity for rapid salinity change occurs next to the well screen and gravel pack, efforts to ensure a gradual adjustment in

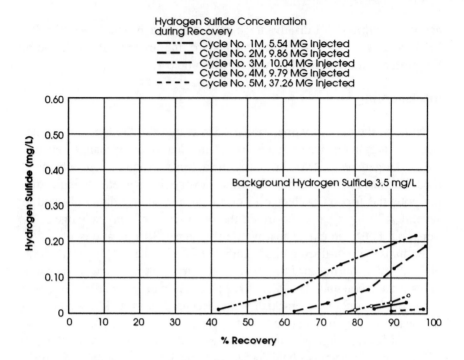

Figure 4.18 Hydrogen sulfide during ASR cycles, Cocoa, Florida.

salinity change should focus on initial recharge flows into the well. Low initial flows, or alternate periods of no flow, may help to achieve this objective.

While ion exchange may be reversible, the plugging associated with ion exchange is frequently not fully reversible. This is because clay particles that are dispersed through contact with water of low ionic strength tend to move with the water until such time as they become trapped in the aquifer matrix. Redevelopment may dislodge some of these particles and possibly remove a few of them from the well; however, it will not restore the particles to their original location.

Hydrogen Sulfide

Experience at several ASR sites has confirmed the rapid reduction in hydrogen sulfide that occurs during ASR storage. This is due to oxidation by chlorine and also by dissolved oxygen present in the recharge water. **Figure 4.18** is representative, showing the reduction at Cocoa, FL, from a concentration of about 3.5 mg/L in the native groundwater to concentrations of about 0.2 mg/L at the end of the first ASR cycle and decreasing to lower concentrations during subsequent cycles.

4.7 SIMULATION MODELING

Modeling is an important part of ASR system development, providing a cost-effective tool for system planning, design, and operation. However, experience to date suggests that the greatest value of the modeling tasks tends to occur following site-specific collection of data from the first ASR well and associated observation wells. Efforts to utilize modeling prior to construction and testing of the first ASR well are often frustrated by the lack of data with which to calibrate and verify the model, resulting in some loss of confidence in the results. Once the field data is obtained, models can be used to help resolve several important questions regarding ASR applications:

- How can the ASR component of a water system be best utilized to ensure the least-cost expansion path for water supply facilities to meet projected demands?
- How should the ASR wellfield be operated to meet projected seasonal, long-term, emergency or other needs with adequate reliability?
- How far will the storage bubble extend around the ASR well? How much of the stored water can be recovered? What, if any, will be the effect of biological reactions occurring underground upon recovered water quality?
- What geochemical reactions are expected to occur due to mixing between stored and native water in the presence of aquifer clays and minerals? Are these reactions significant and is the reaction rate significant to proposed ASR operations?

Each one of these sets of questions entails application of a different simulation model, as discussed in the following paragraphs.

ASR Water Supply System Model

This model has been developed by CH2M HILL, to simulate the operation of a water system comprising a river source of supply, intake structure, offstream reservoir, water treatment plant, and ASR system. It has been applied to the Peace River ASR system, to determine the most cost-effective facilities expansion path to meet projected demands with acceptable reliability. The number of variables, and the interrelationship between some of the variables, complicates any simple analysis of this system.

River flow and quality (as defined by TDS concentration) is simulated on a monthly time step, based upon the period of record. Simulation can utilize the historic record or an equally likely series of synthetic streamflow and quality records. Algal blooms in the river have historically caused time periods when no diversions were possible, even though adequate flow was available in the river. These are also simulated on a monthly basis.

A regulatory diversion schedule is incorporated in the model, along with alternate schedules that may be considered. In each month, diversion is defined as a percentage of streamflow above some minimum regulatory flow rate, up to a maximum defined by the capacity of the intake structure.

Raw water flows are routed to the water treatment plant, which treats water at its full design rate, or as limited by the sum of system demand and ASR recharge capacity. Any additional raw water is routed to the offstream storage reservoir where it is subject to mixing, evaporation, seepage, and rainfall augmentation. Any raw water exceeding the capacity of the offstream reservoir and also treated water needs is routed back to the river.

Treated water is used to meet system demands, with any remaining treated water stored in the ASR wells. During storage, the water mixes with surrounding brackish water in the aquifer, according to relationships established during initial testing and confirmed after several years of recharge-recovery operations. TDS of the recovered water is then blended with TDS of the water from the treatment plant, which also adds a small increment of TDS during the treatment process.

When no flow is available from the river due to flow or water quality constraints, the first source of water is recovery from ASR wells. To the extent that additional water is needed to meet demands or to reduce TDS concentrations going to the distribution system, water is pumped from the offstream reservoir, treated, and blended. By providing sufficient ASR wells to meet system peak demands, 100% flow reliability can be assured. By properly storing water in the ASR wells during periods when it is available up to a target storage capacity established for the wellfield, acceptable water quality reliability can also be assured.

The model is a very powerful tool to demonstrate how a complex water system incorporating ASR would best operate to meet demands, and the considerable cost savings associated with effective use of the ASR system. In particular, it shows the value of water storage during early years of an expansion phase, to help meet demands toward the end of that phase and thereby defer as long as possible the need for the next expansion. This is in addition to the more obvious role of ASR to help smooth out seasonal variations in both supply and demand.

Figure 4.19 shows a schematic of the model components while **Table 4.9** shows the least-cost facilities expansion path to meet projected water

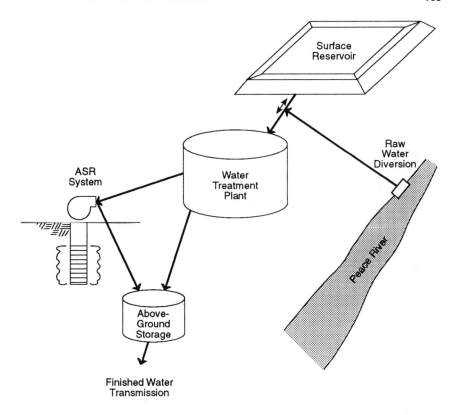

Figure 4.19 Peace River Water Supply System Model.

demands during the next planning period at Peace River. By deferring or eliminating the need for reservoir expansion and relying instead upon underground storage, the Peace River water system is expected to meet regional water demands at less than half the capital cost of other water supply alternatives.

Although applied initially at Peace River, the ASR model is adaptable to other sites with different water system components, hydrogeology, hydraulic, and water quality issues. Incorporation of the ASR/groundwater component in the overall system model provides the key to its value and success.

ASR Wellfield Operations Model

Currently the best available model for ASR wellfield operations is MODFLOW, which simulates water level response to recharge and recovery. This model is widely used in the hydrogeologic field and is certainly not uniquely applied to ASR systems. Hence, little further discussion of the model is pertinent to this endeavor.

TABLE 4.9 LEAST-COST FACILITIES EXPANSION PATH: PEACE RIVER, FLORIDA

Year	ADD (MG/day)	MDD (MG/day)	Peak Month (MG/day)	Alternative 1: Capacity ASR (MG/day)	WTP (MG/day)	Diversion (MG/day)	Reservoir (MG)	Alternative 2: Capacity ASR (MG/day)	WTP (MG/day)	Diversion (MG/day)
1991	5.7	7.8	6.7	4.4	12	22	625	4.4	12	22
1992	7.9	11.2	9.2	8				8		
1993	8.5	12.4	10.0							
1994	9.1	13.7	10.7							
1995	9.8	14.9	11.5							
1996	10.4	16.1	12.2							
1997	13.7	22.1	16.1	20	18			20	18	
1998	14.6	23.3	17.1							
1999	15.4	24.5	18.1							
2000	16.3	25.7	19.1							
2001	17.3	27.3	20.4							
2002	18.4	28.9	21.6	25	24	44		25	24	44
2003	19.4	30.5	22.8							
2004	20.5	32.1	24.1							
2005	21.6	33.7	25.3	28				28		
2006	22.6	35.3	26.6							
2007	23.7	36.9	27.8							
2008	24.7	38.5	29.1	32	30			32	30	
2009	25.8	40.1	30.3							
2010	26.9	41.7	31.6							
2011	30.8	47.4	36.2	40	45	66		40	45	66
2012	31.9	49.0	37.5							
2013	33.0	50.7	38.8				1500	48		
2014	37.7	57.3	44.2							
2015	38.7	59.0	45.5							
2016	39.8	60.6	46.8		60			52	60	
2017	40.9	62.2	48.1							
2018	42.0	63.9	49.3							
2019	43.1	65.5	50.6							
2020	44.2	67.2	51.9							

Note: Alternative 1 assumes 40 MG/day ASR maximum; alternative 2 has no ASR maximum.

This model has been applied to the ASR system at Kerrville, to evaluate whether use of ASR wells to maintain groundwater levels at a target elevation above mean sea level would enable the Upper Guadelupe River Authority to meet a repeat of the worst drought on record if it were to occur in the future. Further discussion of the Kerrville ASR system is included in Chapter 9, Selected Case Studies. The alternative water supply option for this site was the construction of a new offstream reservoir at a cost at least five times that of the ASR system. The operating rule for this site is to recharge whenever water is available from the currently 19 megaliters/ day (5 MG/day) treatment plant after meeting system demands. Recharge ceases when the elevation reaches 1500 ft in the ASR well. Water is recovered to meet seasonal peak demands each year, leaving some water underground as carryover storage to meet the drought demand at such time as it may arrive.

The same model has also been applied to the Peace River ASR system to estimate drawdown effects upon adjacent well owners due to seasonal recharge and recovery operations. The ASR recovery capacity at this site needs to equal maximum day demands so that these demands can be met during times when there is no flow available from the river to treat at the plant. Flow distribution to each well has to be balanced during both recharge and recovery so that well interference does not push stored water away from individual wells.

Solute Transport Models

Movement of the stored water around the well or wellfield is always of considerable interest in ASR systems, particularly where the storage zone contains water of inferior quality. Evaluation of water movement is best accomplished with a solute transport model, at least three of which have been applied on ASR projects.

Upon completion of field investigations, sufficient data may be available to consider whether or not a solute transport model of the ASR site would provide useful guidance regarding the extent and movement of the stored water bubble. This has not been typically required, reflecting the small radius of the bubble around the well under operational conditions and the limited movement of this bubble during storage periods under the influence of the regional hydraulic gradient or other nearby production wells. Usually the data required to calibrate the model are not available until after the first well has been constructed and tested. This reduces the potential value of the model as a site selection tool, but enhances its potential value as a guide for ASR expansion and optimization.

These models are useful tools to gain understanding of undergroundwater movement; however, they are frequently constrained by reliance upon assumptions regarding aquifer dispersion characteristics that are usually not available from test program data. The results, therefore, are not necessarily accurate representations of detailed system response to ASR operations. They do, however, provide a general response that is frequently useful.

The model HST3D, developed by the U.S. Geological Survey, was initially utilized for ASR purposes to gain insight regarding the potential for storing drinking water in a thin, confined, unconsolidated, seawater aquifer at Marathon. Input parameters were obtained from literature sources and also from data obtained during construction of an observation well at the site. Results suggested that the proposed ASR storage program might be successful, despite the large difference in density between the stored and the native water. The model was utilized as a tool to guide construction and test program design. Subsequently, the ASR facilities were constructed and tested, demonstrating the feasibility of emergency drinking water storage in a seawater aquifer for durations of at least 60 days. **Figure 4.20** shows the simulated relationship between recovery efficiency on the initial ASR cycle and storage volume, while **Figure 4.21** shows the expected improvement in recovery efficiency

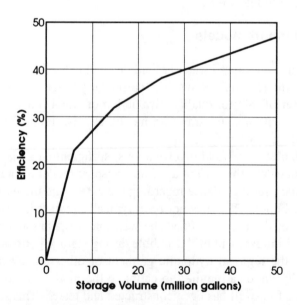

Figure 4.20 HST3D simulated relationship between recovery efficiency and initial storage volume, Marathon, Florida.

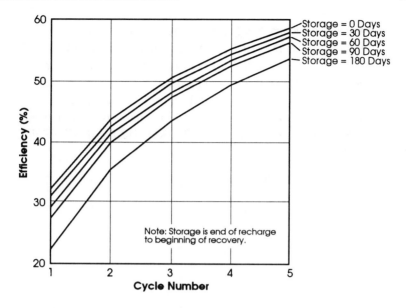

Figure 4.21 HST3D simulated recovery efficiency vs. number of cycles
for a given storage duration, Marathon, Florida.

associated with successive ASR cycles and increasing storage period durations, other parameters being equal.

This simulation was performed prior to ASR well construction and testing, based upon data available from an observation well at the site. It is pertinent to compare the results with those collected during subsequent cycle testing, as presented in Figure 4.3. The simulation modeling assumed that each cycle involved storage and recovery at a rate of 1.9 megaliters/day (0.5 MG/day) and a volume recharged of 24 megaliters (6.35 MG). Actual cycle testing was conducted at lower flow rates around 1.1 megaliters/day (0.3 MG/day) and with volumes of about 57 megaliters (15 MG). After adjusting for storage volume difference based upon Figure 4.20, and assuming that the predicted Cycle 5 results from Figure 4.21 approximate a fully developed storage zone, the predicted recovery efficiencies were very close to those observed, falling within a range of about 64 to 67%. Actual results showed greater sensitivity to storage time than did the predicted results.

Possibly the most applicable model at this time is CFEST, which has recently become available for use on a PC (486/33) computer. In addition to showing the direction and rate of movement of the stored water bubbles surrounding one or more ASR wells in fresh or brackish aquifers, the model output also includes estimated TDS concentration at the ASR well during successive recharge, storage, and recovery cycles. This model has been upgraded by CH2M HILL to make it more user-friendly and WINDOWS-driven. Once calibrated against existing data sets from operational ASR sites, it should prove useful as a predictive tool at proposed ASR sites to estimate recovery water quality and recovery efficiency.

Several other solute transport models are available that could be applied to ASR systems. Among these are RESSQU, KONIKOW-BREDERHOEFT, and SUTRA.

CHAPTER 5

Geochemistry

The world appears as you perceive it. It is not that your perceptions are wholly shaped by a so-called objective world. The habit of interpretation is interactive; we do things to test our hypotheses until we have created a complex web of sensory input and centrifugal manipulation. By the time we are "mature," we have created innumerable layers of interpretation and biased perception that become the templates for living. Of course, we could have fun with this situation. We could change the templates that we use to interact with the world.

Deng Ming-Dao, 1992, *365 TAO:* Harper San Francisco

5.1 INTRODUCTION

Our understanding of the geochemistry of ASR systems continues to evolve and become more clear as we learn from experience at different sites. Geochemical measurements, concepts, and issues that have proved particularly helpful in gaining an understanding of underground processes at these sites are discussed in this chapter. This is not intended to be a complete presentation on geochemistry, which is a broad and complex technical subject that extends far beyond ASR issues. However, a basic understanding of the processes, investigation approaches, potential fatal flaws, the approximate severity and impact of such fatal flaws, and the potential solutions can provide a useful guide to those considering ASR projects.

In areas with proven, long-term satisfactory performance of ASR systems and with no signs of geochemical problems, it is probably reasonable to assume that such problems should not occur in future projects. For example, this is the case in Florida ASR systems storing treated drinking water in brackish limestone artesian aquifers. It is also believed to be the case in Las Vegas, NV, where the storage zone is a fresh, unconfined alluvial aquifer.

In areas with little or no prior ASR or recharge well performance to use as a guide, it is wise to consider potential geochemical issues carefully and, based upon geochemical and other test results, adjust program development to reflect needed changes. It is important that test program development should reflect an awareness that ASR geochemistry is not yet an exact science. A potential problem that is indicated by initial laboratory analysis, core testing, and geochemical modeling may not materialize, or may occur so slowly as to be of no real significance. On the other hand, an unexpected geochemical problem may arise, requiring procedural changes, despite reasonable efforts to identify such problems in advance. With steadily increasing experience, the occurrence and frequency of unexpected problems should decline. This chapter presents experience through 1993, recognizing that the ASR geochemistry field is advancing fairly rapidly.

The outline of this chapter generally follows the geochemical steps that are considered appropriate for ASR feasibility investigations, starting with sampling of recharge water and native groundwater, laboratory analysis, and a preliminary geochemical assessment. This is frequently followed by observation, well construction and coring, core analysis, and more detailed geochemical evaluation based upon computer simulation of the mixing of recharge water and groundwater in the presence of the aquifer clays and minerals. This completes the procedure for sites that are considered geochemically benign. However, where uncertainty still exists regarding geochemical effects, column testing or laboratory batch testing may be performed to simulate in the laboratory the operations planned in the wellfield. It is less expensive to destroy a core than it is to irreversibly plug a well through geochemical reactions. This laboratory testing provides the basis for pretreatment or post-treatment plans that may be required during ASR well testing and subsequent operations.

5.2 WATER CHEMISTRY

Parameters

At least one sample of the recharge water is required for the geochemical analysis, taken at a time of the year that is representative of expected recharge water quality. In addition, a representative sample of the groundwater from the proposed ASR storage zone at or adjacent to the selected test well site is required prior to any recharge. These samples are analyzed for parameters listed in **Table 5.1.** Some of the analyses are performed in the field and the remainder in the laboratory.

Temperature, pH, and specific conductance are easily measured and need to be determined in the field. The pH of groundwater, and to a lesser extent surface water, typically changes as much as a full pH unit between the collection time of a sample in the field and the time it is measured in

TABLE 5.1   WATER QUALITY PARAMETERS FOR GEOCHEMICAL ANALYSIS

Parameter

Total alkalinity	Copper
Total dissolved solids	Manganese
Total suspended solids	Zinc
Turbidity	Cadmium
Color	Selenium
Specific conductance	Total hardness
pH (field)	Non-carbonate hardness
Temperature (field)	Calcium hardness
Dissolved oxygen (field)	Nitrate
Eh (field)	Phosphate
Chloride	Ammonia
Fluoride	Hydrogen sulfide
Sulfate	Total organic carbon
Carbonate alkalinity	Total halogenated hydrocarbons
Bicarbonate alkalinity	Specific gravity or fluid density
Total silica	Total coliform
Calcium	Chloroform
Magnesium	Bromodichloromethane
Sodium	Dibromochloromethane
Potassium	Bromoform
Iron	Total trihalomethane
Aluminum	

Note: For each parameter, estimate the expected value. If a seasonal variation occurs, estimate the annual range.

the laboratory. Specific conductance can be compared with the laboratory specific conductance to determine if some of the major ions have precipitated between sampling and analysis. It is also used to determine if there are dissolved ionic constituents that are not analyzed but may be as important as common major ions.

Although not commonly determined, the oxidation-reduction potential, or redox potential, commonly called either ORP (oxidation reduction potential), pE, or Eh (hydrogen electrode), is one of the most important physical measurements on *in situ* groundwater. This measurement is difficult both because it must be measured before the groundwater becomes exposed to the atmospheric oxygen (closed cell) and because it commonly takes time for the reading to stabilize. The Eh of the bulk groundwater, however, is particularly important where iron and manganese concentrations are above approximately 0.1 mg/L. This measurement allows the inorganic vs. the microbiota reaction paths to be ascertained. The Eh measurement is preferable to dissolved oxygen for groundwater, but for surface water recharge, particularly treated surface water, dissolved oxygen measurement is preferable. The residual chlorine in treated recharge water causes an inordinately high Eh value, which is meaningless.

Likewise, dissolved oxygen (DO) is an important field measurement that can be determined on both surface water (recharge water) and *in situ* groundwater to describe the oxidation-reduction condition of the water. Oxidized systems, including surface water and shallow groundwater systems in contact with atmospheric oxygen and with minimal organics, typically have 8 to 12 mg/L DO and an Eh of +300 to +500 mV. Increasingly reduced systems have lower DO concentrations. A reduced groundwater may indicate approximately 1 mg/L oxygen because the kinetics of oxygen reaction with water exposed to atmospheric oxygen is so rapid it is difficult to obtain a measurement below 1 mg/L DO even though the groundwater may be under extremely reducing conditions (an Eh of –100 to –400 mV).

If possible, the groundwater should have an oxidation-reduction-potential determination instead of the DO because it is more definitive of the oxidation-reduction conditions within the aquifer. However, a closed cell with no exposure to atmospheric oxygen is required. This may not be available or readily constructable and the measurement may take more time than samplers have the patience to endure. Relatively oxidized, low total dissolved solids (TDS) groundwater is poorly "poised" and the Eh meter may wander erratically, but reduced groundwater is typically poised and stable. It is important to take a series of Eh measurements with time to establish the asymptotic trend curve. Such a curve can be established by five to ten measurements over no more than 30 min.

The degree of reduction will distinguish between inorganic-dominated (abiotic) vs. organic-dominated (biotic or microbiotic) controlled groundwater systems. An inorganic-dominated system will generally give a plus mV Eh measurement, typically ranging from +100 to +500 mV. A water exposed to surface oxygen typically has an Eh ranging from 400 to 500 mV. A chlorinated water will have an Eh above 800 mV and, therefore, provides little real information on the Eh of the water itself. Microorganisms drive the Eh into the negative millivolt measurements ranging from approximately 0 to −500 mV. Sulfate-reducing bacteria active in a groundwater system generally have an Eh of approximately −100 to −300 mV. Methane-generating bacteria cannot compete with sulfate-reducing bacteria and, therefore, are generally under even more reducing conditions.

The ORP measurement requires a closed cell (no exposure to the atmosphere), two electrodes, and frequently considerable patience to get an accurate reading. The mV reading has to be corrected for the reference electrode and also the reference point used, the Eh. The reference electrode is commonly either a mercuric chloride or silver chloride electrode requiring a temperature correction also. In the absence of information regarding the reference electrode used in the field, an approximate value to be added to the meter reading from the platinum hydrogen electrode and the reference electrode would be +200 mV to give an approximate Eh value.

The water quality parameters listed in Table 5.1 represent a minimum required for the geochemical analysis. Other parameters may also be appropriate on a site-specific basis to meet local needs. Regulatory agencies are periodically adding parameters for compliance monitoring, some of which may need to be added to this list. It is a good idea to analyze the *in situ* groundwater for both priority pollutant organics (particularly volatiles) and metals before the aquifer is recharged. If these constituents are found after recharge has occurred, it is a difficult, costly, and frustrating process to prove it was not introduced with the recharge water.

During the evaluation of water quality data, it is important to recognize the potential variability in the precision and accuracy of results associated with different laboratories and different water quality constituents. A recent study of results from several highly reputable laboratories [1] showed substantial variability for many parameters important to ASR investigations, such as calcium, magnesium, silica, aluminum, iron, alkalinity, and total organic carbon. For all of these parameters, results within an acceptable range (±10%) around a known concentration were obtained from fewer than 57% of the reported values.

The relative quality of analytical data can be determined through use of properly calibrated and functioning field meters. pH, specific conductance, and temperature all change to some degree from the time the sample is collected to the time it is analyzed in the laboratory. Therefore, a field determination is important. Quality assurance checks include calculated vs. laboratory measurements of TDS; mass balance between cations and anions; field vs. laboratory measurements of pH and specific conductance, and the relationship between specific conductance and TDS. These checks and balances, along with commonly determined concentrations of major and trace ions, can be found in a report by Hem [2].

pH

Field pH values will generally be higher or lower than laboratory pH values by as much as ±1 pH unit or more. They are higher or lower for a large number of reasons.

Temperature may be a factor, with lower pH associated with higher temperatures, and higher pH associated with lower temperatures than when the pH was measured in the field. However, field values are commonly lower than laboratory values because collection, transport, or storage causes a release of dissolved carbon dioxide from the water into the head space of the sample and a release to the atmosphere when the sample is opened. Lower laboratory pH values than field values can result from the natural oxidation of dissolved ferrous ion to ferric iron forming hydroxide precipitates that can adsorb other metals. Oxidation typically lowers the pH through the formation and precipitation of hydroxides, thereby releasing hydronium ion into the water.

A neutral pH, at which the hydronium ion concentration equals the hydroxyl ion concentration, is defined as a pH of 7.00 only at 25°C. An increase of 5°C decreases the neutral pH approximately 0.08 pH units. Reducing the temperature to almost freezing results in a neutral pH increase of 0.48 units.

Table 5.2 [2] shows the change in pH caused by the oxidation of 135 mg/L dissolved ferrous iron (FeII or Fe+2) to ferric iron (FeIII or Fe+3) measured repeatedly in a single, unacidified sample over a period of approximately 3 months. Notice that the pH decreased 1 pH unit over the 2-week period between collection of the sample and analysis of the iron. The iron concentration was probably higher at the point of collection. More importantly, the iron continued to precipitate with a chemically-linked, continuing decline in the pH value for the sample. This illustrates

TABLE 5.2   pH CHANGE DUE TO IRON OXIDATION

Time (days)	pH	Iron (mg/l)
0	4.98	
14	3.98	135
35	3.05	87
70	2.81	41
105	2.69	2.2

Source: HEM, 1985.

TABLE 5.3
pH VALUES FOR NATURAL WATERS

Common	6.0–8.5
High	11.0–12.0
Low	0.0–2.0

Source: HEM, 1985.

the need to filter and acidify the sample in the field if reliably significant geochemical interpretations or decisions are to be made with the data.

Table 5.3 [2] shows the common, high, and low ranges of pH values for natural waters. The high range generally requires a significant bacteriological activity to achieve under natural conditions but alkaline, magnesium-silicate rocks can reach these pH values. Grout failure should also be considered since it can also cause high pH values. Low pH ranges are probably bacterially-catalyzed oxidation of primarily iron sulfides (pyrite, pyrrhotite, and marcasite). A pH of –2 has been reported in a mine but should be very rare under natural groundwater conditions since rocks surrounding the iron sulfides buffer the pH. Furthermore, the iron sulfides build up an oxidized iron hydroxide coating as a result of continued exposure to groundwater. This coating protects the sulfide, minimizing oxidation.

Specific Conductance

Specific conductance, whether defined in μmhos per centimeter or in microsiemens (μS), is defined or temperature-corrected to 25°C. The range in specific conductance varies from about 1 μS for distilled water to between 2 and 200 μS for rainwater and about 50,000 μS for seawater. Drinking water supplies rarely exceed a specific conductance of approximately 750 μS.

Specific conductance values measured in the field and in the laboratory should agree within approximately 5 to 10% unless there has been a chemical reaction following the collection of the water sample and prior to its analysis. The dissolution of the residue from a sample container that is not clean typically increases the specific conductance unless the residue causes a precipitation reaction, decreasing specific conductance. Precipitation may occur after sample collection, during shipment, or during storage prior to analysis. The lower specific conductance can be related to the amount of material precipitated, if this is the only reaction.

The ratio between TDS and the specific conductance should be between approximately 0.55 and 0.75 for most natural waters. A value of 0.67 is a common ratio for natural waters with less than about 1,000 mg/L TDS. This is the first check on the TDS concentration.

Total Dissolved Solids

The TDS concentration is approximately equal to the sum of the concentrations of all dissolved ions. The TDS should be both measured and calculated, and the difference should be within approximately 5%. For this calculation, the concentration of alkalinity should be multiplied by 0.5 (actually 0.4917), assuming half is driven off during the drying process as carbon dioxide and half precipitates as a carbonate.

If all of the water quality constituents listed on Table 5.1 are not determined, the difference can be used to check the gross values for those parameters not determined or determined by calculated difference. The measured value is typically larger for surface water, which commonly contains suspended organic and inorganic matter. If the measured groundwater value is higher than the calculated sum, the sample may contain either significant quantities of unanalyzed parameters, or an unfiltered sample may contain rust or other suspended solids from the casing, which are not present in the aquifer. The laboratory will acidify a sample for iron and other metals analysis if this has not been done in the field, putting these undissolved metals into solution. If metals are to be analyzed, the sample should always be filtered through at least a 0.45 μm (preferably a 0.1 μm) filter and acidified in the field. This minimizes the potential contamination of the samples with iron and other metals from the casing, suspended particulates, and other sources.

As discussed above, the alkalinity as mg/L bicarbonate should be multiplied by 0.5 to equal the mg/L carbonate, the solid form. Other constituents can also be driven off during the drying process such as nitrate, arsenic, and mercury, potentially reducing the real TDS. However, the calculated TDS is usually higher than the measured TDS. Gypsum (a calcium-sulfate mineral) is one of several common solid phases which can form in the dried sample and adsorb water from the air. A sample containing sulfuric acid cannot be brought to complete dryness under normal drying procedures used in the laboratory and it, too, will adsorb water from the air.

Suspended solids and organic compounds are usually the cause of higher calculated than measured TDS. The suspended solids may be coming from casing oxidation products in groundwater and particulates

TABLE 5.4   MASS BALANCE: CATIONS vs. ANIONS

Calcium (Ca^{2+}) mg/L × 0.04990 = meq/L
Magnesium (Mg^{2+}) mg/L × 0.08229 = meq/L
Sodium (Na^+) mg/L × 0.04350 = meq/L
Potassium (K^+) mg/L × 0.02558 = meq/L

Σ Cations

Bicarbonate ($HCO_3{}^-$) mg/L × 0.01639 = meq/L
Carbonate ($CO_3{}^{2-}$) mg/L × 0.03333 = meq/L
Sulfate ($SO_4{}^{2-}$) mg/L × 0.02082 = meq/L
Chloride (Cl^-) mg/L × 0.02821 = meq/L
Nitrate ($NO_3{}^-$) mg/L × 0.01613 = meq/L

Σ Anions

$$\text{Mass balance} = \frac{(\Sigma\ \text{Cations} - \Sigma\ \text{Anions})}{\dfrac{(\Sigma\ \text{Cations} + \Sigma\ \text{Anions})}{2}} \times 100 = \text{Percent difference}$$

Percent difference   ≤5%   Excellent
                         5–10% Good
                        10–20% Fair
                         >20%   Poor

(Specific conductance/100) ~ Σ Cations ~ Σ Anions

(natural and/or treatment products) in surface water. The organics can be natural organics which may be precursors to trihalomethanes and haloacetic acids when the water is chlorinated, or may be halogenated hydrocarbon contamination. It is extremely important to differentiate between natural organics and organic contamination before any recharge is allowed to occur.

Mass Balance

A mass balance can be calculated for both the cations (positively charged ions) and the anions (negatively charged ions) in units of milliequivalents per liter (meq/L). This unit of measurement takes into account both the differences in gram atomic weight (molar) units of each ion and the valence or charge units for each ion. Calcium, magnesium, sodium, and potassium are the minimum parameters for cation balance calculation and alkalinity, sulfate, and chloride for the anion balance calculation. Conversion factors from mg/L to meq/L are included in **Table 5.4** [2].

The mass balance, therefore, is both a weight and charge balance. The meq/L difference should be compared using the following equation:

$$Error[\%] = \left[\frac{cation\ sum\ meq/L\ -\ anion\ sum\ meq/L}{(cation\ sum\ +\ anion\ sum)/2} \right] \times 100$$

The error should be within 5% for a high quality analysis. A 10% error is allowable if TDS is less than approximately 100 or more than 5000 mg/L. An error equal to or greater than 20% is not acceptable for meaningful interpretation of aquifer geochemistry. However, water containing a high iron concentration can give a significant cation error, and a high nitrate concentration can cause a significant anion error if not included in the ionic balance. A very high pH water will not balance unless the hydroxide is added.

A comparison between the calculated meq/L of either the cation sum or anion sum should be approximately equal to the specific conductance (μmhos/cm at 25°C) divided by 100. These are typically within 5%, assuming that the major cations and anions have been analyzed and the TDS is approximately 1,000 mg/L or less. The cation or anion sum closer to the value of the specific conductance divided by 100 is probably correct, and the sum farthest from the value probably contains any error. The laboratory should be consulted and questioned about any potential error.

Water Chemistry Diagrams

Stiff and Collins Diagrams

These are two of several methods for showing water chemistry types graphically. They show patterns, allowing the eye to discriminate water chemistry types. The size of the individual diagrams indicates the relative amount of TDS because the diagrams plot each cation and anion in milliequivalents. **Figures 5.1** and **5.2** illustrate these two types of diagrams. These diagrams are best used for illustration of water chemistry on maps.

Trilinear (Piper) Diagrams

As illustrated in **Figure 5.3,** a trilinear diagram should be plotted for any area involving more than one aquifer or more than one well for which common ion chemistry data exists. For each sample, cations and anions are plotted as a percent in their respective triangle and then extrapolated up into the diamond. The list of sample codes is typically placed on the right hand side of the page listing temperature, specific conductance, TDS, pH and Eh or DO as well as other ions of interest such as metals or organic compounds. The cations typically indicate mixing, solubility, and ion

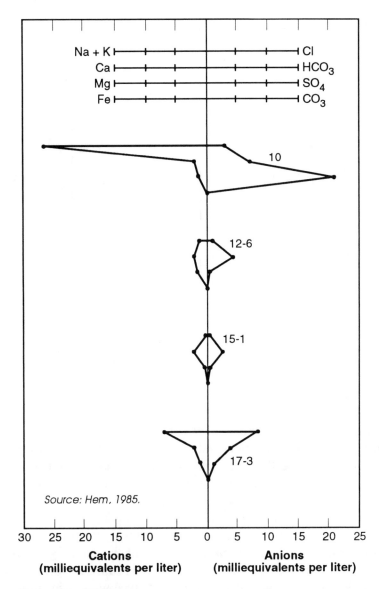

Figure 5.1 Stiff diagram.

exchange processes but can also include precipitation reactions. Cation chemistry typically indicates the degree of ion exchange. Anion chemistry, compared with cation chemistry, typically indicates solubility and precipitation reactions. This comparison occurs in the diamond field of the diagram.

The diamond field is used to estimate the percent mixing of waters of two or more chemistries. A single line of water samples on the diamond indicates a simple mixture or increasing dissolution/ion exchange/precipi-

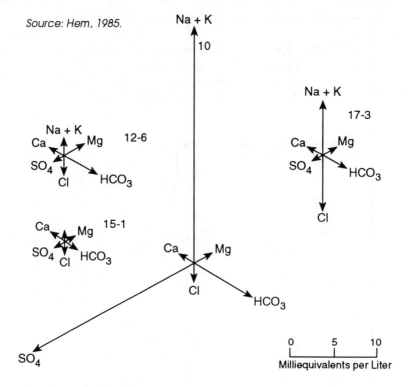

Figure 5.2 Collins diagram.

tation of a simple ion mineralogy. Conversely, a scattering of water samples in the diamond field indicates diverse water chemistry types resulting from more than one aquifer, complex aquifer mineralogy, or mixing between waters of diverse origins and chemistry. Groundwater chemistry can be followed along a flow path to determine what changes and their causes have occurred along the flow path.

Several typical areas for specific water chemistry types by source have been added to the diagram in **Figure 5.4.** For example, shallow groundwaters in alluvial systems are typically calcium bicarbonate water chemistry types, commonly referred to as calcium-bicarbonate type water. This means that the cation (calcium) and anion (bicarbonate) both constitute more than 50% of the cations and anions, respectively. If calcium and magnesium (the next most abundant cation) are both required to exceed 50% the water would be referred to as a calcium-magnesium-bicarbonate water.

Hydrogeochemical Processes

Rainfall has a typical pH of about 4.3 in a temperate climate, which is associated with a balance of anions and cations that is characterized by a relatively high concentration of ammonia (typically ammonium when dissolved), created by lightning. Ammonia is one of the best ion exchangers, displacing other cations from clays and providing a nutrient to both soil microbiota and plant roots. This is why plants grow better when watered by natural rainfall instead of irrigation.

Water flowing through the soil environment undergoes major changes in ionic composition and pH due to several processes [3]. Among these, the nitrification process is conversion of ammonium to nitrite, which then typically converts rapidly to nitrate. Nitrification is an aerobic process occurring under oxidizing conditions. Most oxidation processes are acid-producing processes. In this case, nitric acid is one product, and the pH decreases from 4.3 to 3.95. pH is a logarithmic scale so the change in ionic balance is substantial. Weathering, or the chemical attack on soils and sediments, produces most of the inorganic ion load and buffers the pH to

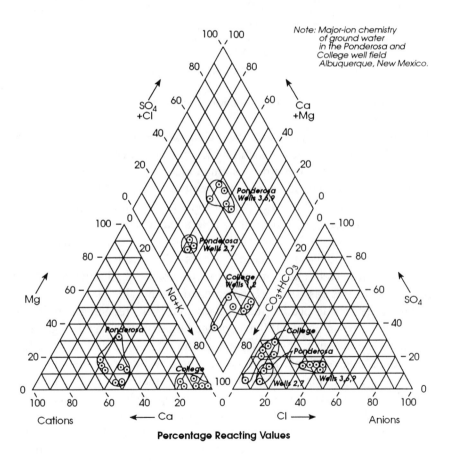

Figure 5.3 Trilinear diagram.

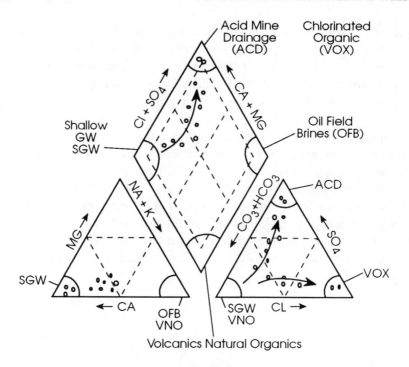

Figure 5.4 Trilinear diagram — environmental interpretation.

4.9 in this case. Weathering alkaline rocks, like volcanics, will increase the pH to above 8.0 and yield a sodium-bicarbonate water type.

Biomass production assimilates ammonium, sulfate, calcium, magnesium, potassium, iron, and aluminum into the bodies and fluids of microorganisms, thereby converting part of the ammonium to nitrate and producing organic acids that lower the pH less than occurs with nitrification.

Denitrification, on the other hand, is conversion of nitrate and nitrite to nitrogen gas, and is an anaerobic process that occurs under reducing conditions. Most reduction processes are hydroxide, or alkaline, producing processes. In this case, the pH increases to about 4.7.

Sulfate reduction occurs under relatively severe reducing conditions, typically below an Eh of -100 mV, created mostly by the microorganisms responsible for converting dissolved sulfate into either hydrogen sulfide gas (rotten egg odor, pH less than 7.0) or bisulfide ion (HS^-, no odor, potential precipitation of iron, pH equal to or greater than 7.0). Actually, the microorganisms produce organic-sulfur compounds, thio-species, which can break down into one of the above inorganic species depending upon transport into an oxidizing environment (acid-producing, hydrogen sulfide gas) or reducing environment (alkaline-producing, bisulfide ion). If the microorganisms responsible for sulfate reduction are still active, the pH will remain between 6.5 and 8.0 because hydrogen sulfide gas is also toxic to the sulfate-reducing bacteria. Methane generating bacteria cannot compete with sulfate-reducing bacteria but are present in the most reducing environments (Eh typically less than –300 mV), usually beneath the sulfate reducing bacteria.

An awareness of the fact that hydrogeochemical processes occurring in ASR wells will tend to alter the pH and oxidation-reduction potential is fundamental to the successful design and operation of ASR systems, particularly in geochemically sensitive aquifers. Both changes can significantly affect solubility and precipitation reactions, which have a direct effect upon plugging rates, recovered water quality, and possible pre- or post-treatment requirements.

Changes in Groundwater Chemistry with Depth

As groundwater flows through an aquifer from the point of recharge to a point of discharge, it undergoes changes in quality, as shown in **Figure 5.5.** The evolution of water chemistry in a groundwater basin proceeds from the near surface oxidizing conditions typically producing calcium-bicarbonate water chemistry to deep, reducing conditions producing sodium-chloride water chemistry. The evolution involves cation exchange, such as between dissolved calcium and dissolved sodium; oxidation, in which organic material is oxidized to carbon dioxide and water; and chemical weathering (diagenesis), a dissolution and leaching of inorganics from soil and sediment particles creating a relatively stable mineralogy compatible with the water chemistry. Acid created by the oxidation process increases TDS. The water chemistry acquires the imprint of the particle mineralogies. For example, if the area contains gypsum, a hydrated-calcium-sulfate mineral, present either through oxidation of sulfides from a mineralized area or an area with high surface evaporation rates, the groundwater acquires a calcium-sulfate water chemistry. Metals concentrations are typically below detection near recharge areas lacking sulfide minerals.

Oxidation of the organic matter creating the acid that leaches the soil and sediment particle mineralogies also consumes oxygen, and the groundwater becomes more reducing with depth. In marshes and peat-producing wetlands, reducing conditions occur within inches of the surface. In temperate climates with high concentrations of organic matter, groundwater can become reducing within moderate depths of about 30 m (approximately 100 ft). In arid climates, with no significant sulfide mineralization, groundwater may still be oxidizing at depths that can exceed 300 m (1000 ft), as found in the Tucson Basin, AZ. Sulfide oxidation consumes large

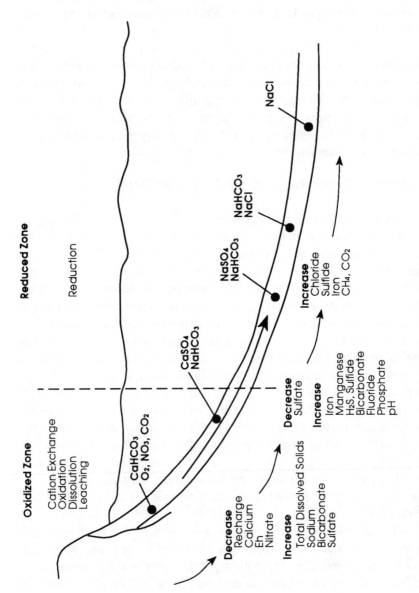

Figure 5.5 Changes in groundwater chemistry with depth.

amounts of oxygen, creating sulfuric acid. As groundwater moves to deeper parts of a groundwater basin, it loses oxygen and becomes more reducing.

As the groundwater loses oxygen, sulfate-reducing bacteria become viable and can reduce the sulfate to bisulfide, which then precipitates metals, particularly iron. This is a classic environment for uranium deposition. The groundwater becomes more alkaline with a dominant sodium-bicarbonate composition, which in turn allows dissolved fluoride and phosphate to increase in concentration. Calcium is the major controlling cation for fluoride and phosphate. The subsurface microorganisms, including heterotrophic bacteria near the surface and sulfate-reducing and methane-producing bacteria at depth, produce carbon dioxide. This contributes bicarbonate and carbonate to the groundwater and the water achieves a sodium bicarbonate composition. An ammonium ion increases the cation exchange and release of sodium and calcium, but calcium will be precipitated as calcite, whereas sodium remains in a dissolved form and increases in concentration.

Metals solubilities, particularly iron and manganese, increase with increasing degree of reduction. Metals concentrations are indirectly related to the amount of available bisulfide. Low bisulfide allows high iron and manganese concentrations. High bisulfide precipitates and thereby decreases the iron and potentially the manganese concentrations. The iron concentration of most natural groundwater systems is commonly approximately one order of magnitude higher than the manganese concentration. If manganese concentrations are equal to or higher than iron, the well is within the transition zone between oxidizing and reducing conditions.

Dissolved manganese concentrations increase faster than does dissolved iron as sulfides are oxidized. The initial oxidation usually causes iron oxyhydroxides to precipitate at or near the oxidizing sulfide because this reaction occurs rapidly, even at the relatively low pH created by the sulfuric acid. This reaction occurs within a few minutes as the bulk groundwater chemistry increases the pH. Manganese oxidation and precipitation reactions are very slow at low pH, typically requiring months at a pH of 6.0. If the aquifer mineralogy contains siderite (iron carbonate), the iron concentration can be several orders of magnitude higher than manganese concentrations.

Finally, TDS increases and the water becomes more reducing. Chloride from fluid trapped in aquifer pores and from anionic positions on aquifer minerals becomes a dominant anion. Under natural conditions, this water chemistry type is nearer the surface if clayey sediments of marine origin dominate the lithology and deeper if sands and gravels dominate the section. However, this water chemistry type is also characteristic of septic systems, landfills, and heavily industrialized areas.

Analysis of water chemistry from a well in a proposed ASR storage zone can suggest the dominant processes and mineralogy potentially affecting the quality of water stored at that site and the associated geochemical plugging characteristics.

Eh–pH Diagrams

Figure 5.6 [4] shows the approximate position of common natural environments as characterized by Eh and pH. Groundwater can range from an Eh of about +500 to –200 mV.

Figure 5.7 [2] shows an Eh–pH diagram for iron. The shaded area in the figure corresponds to combinations of Eh and pH that would cause iron to precipitate (ferric state). The kinetics of this reaction are typically very rapid, occurring within minutes in near neutral water. ASR operations will frequently cause changes in Eh–pH relationships during recharge, storage, and recovery, potentially causing iron precipitation. The dominant change is oxidation of the previously moderately-reduced groundwater system.

Dissolved iron in groundwater systems is frequently at higher concentrations than would be predicted by using the Eh–pH diagram. Iron forms strong complexes with a variety of other inorganics, such as silica and hydroxide, and also with natural organics. Eh–pH diagrams assume equilibrium conditions between the element and selected solid phases, including minerals and compounds. This assumption ignores the role that other dissolved species and organics may have in affecting the equilibrium estimation.

Equilibrium conditions as displayed on an Eh–pH diagram are idealized and overly simplified when compared with natural groundwater. However, these diagrams will generally show the major controlling conditions in dilute, simple water chemistries usually available in recharge situations.

Iron oxyhydroxide forms very rapidly when a sample containing dissolved iron at near neutral to alkaline water is exposed to air, but the resulting precipitate may not coagulate in a water sample fast enough to be filtered. Iron oxyhydroxide precipitates as a colloid small enough to go through a 0.45-μm filter and fresh precipitates frequently go through a 0.1-μm filter.

Samples shown in Figure 5.7 were collected by filtering through a 0.45-μm filter after a suitable well volume of groundwater had been pumped. However, waiting until the water appears clear is not a suitable operating procedure for collecting and filtering a water sample. The water can contain enough iron oxyhydroxide that went through a filter to give an

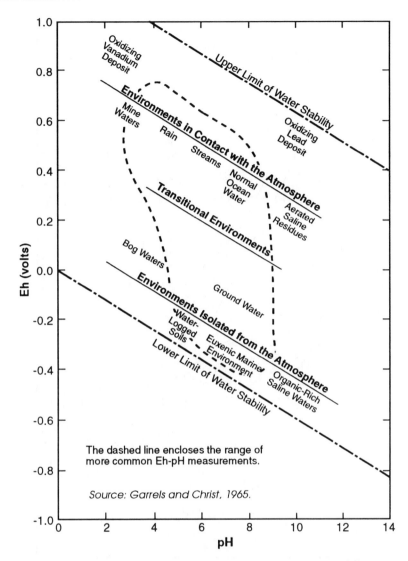

Figure 5.6 Approximate position of some natural environments as characterized by Eh and pH.

erroneously high iron concentration. In addition, iron oxyhydroxide adsorbs other metals and will exaggerate their dissolved concentration as well.

It is important to pump at least three well volumes and document that field water quality parameters have stabilized before collecting a water sample for metals determination. The best groundwater sample is one collected from a pumping well after at least 24 hours of continuous pumping and both filtered through a 0.1-μm filter and acidified to a pH of less than 2 immediately after collection.

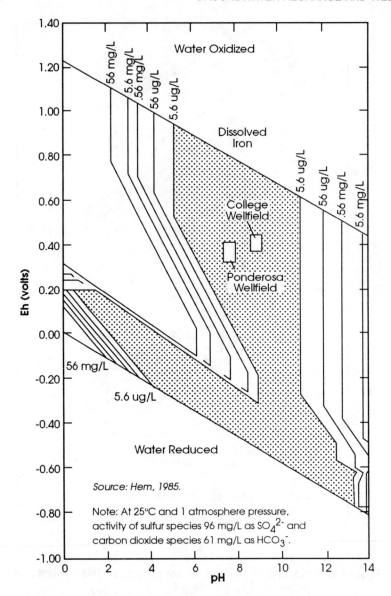

Figure 5.7 Equilibrium activity of dissolved iron as a function of Eh and pH.

Unlike iron, the kinetics for manganese are extremely slow. **Figure 5.8** shows an Eh–pH diagram for manganese for the same wells shown in Figure 5.7 for iron. The Eh and pH relationships that appeared out of equilibrium in the iron Eh–pH diagram indicate equilibrium with manganese. This is a common relationship. Dissolved manganese from a filtered sample is usually a more reliable indicator of Eh–pH conditions in the

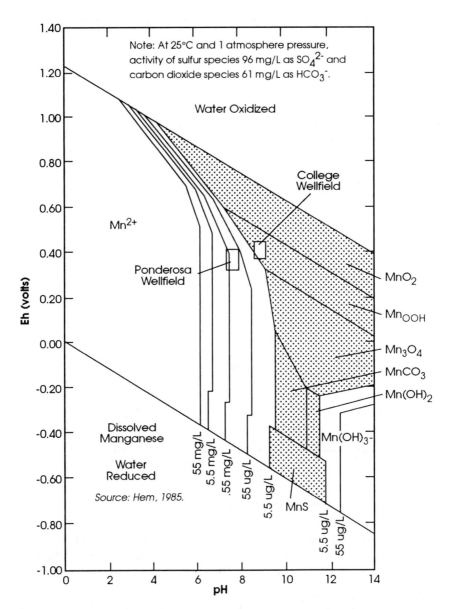

Figure 5.8 Manganese stability and equilibrium as a function of Eh and pH.

aquifer than is dissolved iron, particularly under moderately oxidizing conditions. Additional analysis of the groundwater at this site indicated a probable organic complex that caused the iron to appear out of equilibrium.

Figure 5.9 shows the Eh–pH diagram for sulfur species [5], showing why sulfate is the dominant dissolved species for groundwater. Notice that the Eh is at least −200 mV (pH of 7) before either bisulfide ion or hydrogen

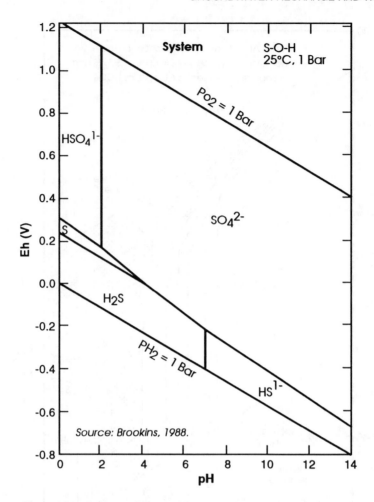

Figure 5.9 Sulfur — Eh vs. pH.

sulfide become significant. Although accurate for groundwater containing dissolved sulfate, this diagram is not accurate if sulfides are being dissolved, because sulfite and thio-sulfate ions complicate the dissolved-sulfur species.

Dissolved nitrogen species are more complicated than sulfur species, as shown in **Figure 5.10** [5]. Nitrogen gas is the most stable species over most of the natural water Eh–pH conditions. The formation of nitrate from nitrogen gas takes very oxidizing conditions, and ammonium takes very reducing conditions. However, once in solution, the nitrate and ammonium equilibrium is shown by the dashed line on the diagram. Natural groundwaters indicate that this is a relatively reliable equilibrium relationship.

5.3 AQUIFER CHARACTERISTICS

Once water samples for the potential ASR storage zone and for the recharge water source have been obtained and analyzed, the next logical step in the geochemical assessment is to use a geochemical simulation model to evaluate in a preliminary manner whether mixing these two waters in various proportions can potentially result in precipitation reactions that may plug the aquifer. Such models are discussed subsequently in Section 5.4, Geochemical Models.

When combined with local ASR experience at other operational sites, the results of this assessment may indicate that geochemical issues are unlikely. In this case it is appropriate to proceed directly to construction

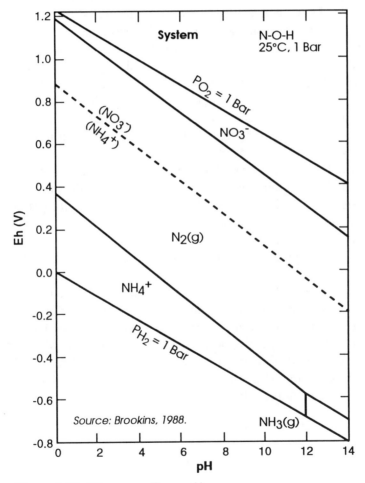

Figure 5.10 Nitrogen — Eh vs. pH.

and testing of ASR facilities. In many cases, however, the results are not conclusive, reflecting lack of data regarding aquifer mineralogy, and lack of local ASR experience demonstrating that geochemical problems do not occur. It is usually appropriate in such cases to obtain core samples of the aquifer under consideration for ASR storage, in order to characterize the physical, chemical, and hydraulic characteristics of the environment in which the recharge water will be stored.

Continuous wireline cores of unconsolidated sediments or consolidated rock, through the full thickness of the storage zone, provide the opportunity for detailed lithologic analysis and comparison to geophysical logs obtained after completion of the initial test hole at any proposed ASR site. Core intervals may then be selected for analysis, usually one each in the overlying and underlying confining layers, plus at least three in the storage zone, depending upon aquifer heterogeneity as judged from the lithologic analysis and geophysical logs. In the event of uncertainty, it is usually wiser to select a greater number of cores for detailed analysis, ensuring that they represent a reasonable balance of different lithologies in the storage zone.

Selected core intervals, usually about 30 cm (12 inches) in length, are then prepared for shipping to a core laboratory. This may involve rapid freezing in dry ice, encasing in wax, or other approaches. If coring occurs over an extended period, cores will need to be wrapped in clear plastic sheeting; marked (top and bottom); labeled and stored appropriately, such as in core boxes, in sealed PVC tubes, or frozen. In this case, core selection may entail cutting previously frozen cores prior to shipping. Most of the operational ASR sites in the U.S. for which core testing has been performed, utilized frozen cores.

Cores obtained from the test well but not sent to the core laboratory should be stored for the duration of the ASR test program and, preferably, for a longer period that would include the first three years of operation. If subtle geochemical shifts or changes in recovered water quality occur later during ASR operation, supplemental core tests can be performed to verify reactions and evaluate well response to potential remedial measures.

Core laboratories are typically capable of running a wide variety of analyses required for petroleum industry operations. Analyses that have proven useful for ASR projects are discussed in the remainder of this section, including the following:

- physical characteristics
- colored pictures of cores
- mineralogy (X-ray diffraction, XRD)
- cation and base exchange capacity
- scanning electron microscopy (SEM)
- thin section petrography

There is some overlap in the information provided by these analyses. The overlap provides support for conclusions based upon the overall core analysis. For example, cation exchange capacity should confirm the presence of smectite indicated by the mineralogy analysis, while the SEM analysis shows how the smectite occurs in the formation, whether pore-bridging, pore-plugging, or grain-coating placement forms are present.

Sidewall cores were substituted for wireline cores at one ASR site for the Centennial Water and Sanitation District, Highlands Ranch, CO. These cores are obtained using small explosive charges attached to a mast that is lowered into the well opposite the interval to be cored. It is not unusual for about 40% of these small cores to be lost during the coring and core recovery process. The cores that are recovered are not representative of formation structure, due to the manner in which they were collected. However, they do provide some indication of particle mineralogy and gross grain-size distribution. Clay mineralogy, cation exchange capacity, and placement forms are generally not possible.

Physical Characteristics

Typically, this includes vertical and horizontal permeability, grain-size distribution, porosity, and specific gravity. Permeability estimates are useful for comparative purposes between different core sections but are not useful for prediction of bulk permeability of the aquifer as determined from pumping tests on the completed ASR well. Grain-size distribution is useful for screen design and also provides a rough indicator of potential geochemical problems, as determined by the percentage of silts, clays, and other fine materials present in the section.

Colored Pictures of Cores

A picture of a longitudinal section of the core provides a useful, quick reference point for judging the homogeneity of the core and the likelihood that core properties determined from test results are representative of adjacent portions of the aquifer. Organic matter in the form of lignite bands or discrete nodules within the sand matrix are important because these usually contain iron sulfide minerals.

The core color can also be an indicator of whether oxidizing or reducing conditions occur in the aquifer. Color may change between when the core is first collected and when it is photographed in the laboratory, so drillers' logs can provide a useful source of information. Reduced iron and manganese and high organic content sediments tend to be blue or grey. Blue clays tend to occur within marine or lacustrine deposits. Oxidized iron is associated with red, yellow, and white formations, indicating abiotic oxidation.

Brown colors usually indicate biotic oxidation, although in some cases brown can indicate reducing conditions.

Mineralogy

X-ray diffraction (XRD) analysis provides a description and percentage of the clays and minerals present in the core. The relative abundance and types of clays provides an important indicator of potential geochemical or physical plugging. Smectite (particularly montmorillonitic) clays are most sensitive to changes in TDS potentially occurring during ASR operations. Kaolinite clays are more likely to cause plugging due to physical movement of the clays in the moving water during recharge and recovery operations, causing bridging of pore spaces. Such plugging is difficult to reverse.

Table 5.5 shows a representative XRD analysis for an ASR test well at Myrtle Beach, SC.

Cation and Base Exchange Capacity

A representative sample from the core is flushed with ammonium acetate solution to determine the concentrations of various cations in exchange positions of the aquifer mineralogy. Cation exchange capacity (CEC), expressed in meq/L, is an indicator of the potential for ion exchange to occur in the aquifer during ASR operations. Higher CEC values above about 5 meq/L are generally indicative of the presence of smectite clays while lower values are associated with other clays. The leachate from this test is then analyzed to determine the base exchange capacity.

Clay mineralogy and placement is particularly important if the recharge water has a considerably different pH, cation chemistry, or TDS than the native groundwater. These conditions can cause clays to become unstable enough to detach from the aquifer particles and move into the pore throats, causing irreparable decrease in the permeability of the aquifer.

Table 5.6 shows the cation exchange capacities of several clays and minerals. Also shown in this table is their surface area, expressed as m^2/g. Those with higher surface areas tend to be more geochemically reactive.

Scanning Electron Microscopy (SEM)

SEM photographs of the cores can be very useful in providing a direct assessment of potential geochemical issues, if the photograph is reasonably representative of the remainder of the core. Typically, the photographs are provided at a series of magnifications, such as 50X, 200X, 500X, and 2000X. It is possible to identify clays and other minerals, and also to determine how the clays occur in the formation pores.

TABLE 5.5 X-RAY DIFFRACTION ANALYSIS FOR ASR TEST WELL: MYRTLE BEACH, SOUTH CAROLINA

Sample Identification Number	Quartz	Feldspar	Calcite	Dolomite	Siderite	Pyrite	Kaolinite	Illite/Mica	Smectite	Plagioclase Feldspar	Feldspar K
1012'10"	76	16	trc	trc			3	2	3	7	9
1086'7½"	82	11	trc	trc			4	1	2	2	9
1120'5"	85	8	trc		trc		2	trc	4	1	7
1222'1"	77	16		1			3	2	2	5	11
1297'10"	76	18	trc	1		1	1	2	1	4	14
1356'4"	84	8	1	trc			4	1	2	2	6

Note: trc = trace.

Figure 5.11 shows a typical SEM photomicrograph for a relatively clean sand interval, as indicated by the 30X magnification. Drilling mud is visible on the grain surfaces. With higher magnification, partially disaggregated thin-bedded shale is shown infilling an intergranular pore.

Thin Section Petrography

Thin section petrography shows a larger-scale picture of the grain and mineralogical characteristics of the cored interval. The roundness and grain-size distribution is shown. These data are important to ascertain the susceptibility for plugging. The distribution of pyrite grains and coatings are also shown. These data are important to ascertain the amount of oxidation and potential for plugging that may occur in the aquifer with the recharge water.

TABLE 5.6   CATION EXCHANGE CAPACITIES OF SELECTED CLAYS AND MINERALS

Constituent	CEC (meq/L)	SurfaceArea (m²/g)
Zeolites	230–620	
Organic matter	150–500	>1000
MnO_2 (pH 8)	260	
Soil organic matter	>200	
Vermiculite	100–150	750–800
Smectite	80–150	600–800
Montmorillonite	80–150	600–800
Saponite	80–120	700–800
Soil micas	20–60	
Soil clay	5–60	150–200
Halloysite ($4H_2O$)	40–50	21–43
Chlorite	4–47	25–40
Illite	10–40	100–200
Soil clay loam	4–32	150–200
Palygorskite	20–30	
Soil silt loam	9–27	50–150
Various soils	4–25	
Soil loam	8–22	50–100
Glauconite	10–20	
Soil sandy loam	2–18	10–40
Kaolinite	3–15	10–20
Halloysite ($2H_2O$)	5–10	
Soil sand	2–7	
Oxides and hydroxides	2–6	14–90
Pyrophyllite	4	
Basalt	0.5–3	
Quartz, feldspars	1–2	1–3

Note: Multiple sources, notably Dragun, 1985.

A.   50X   ⌐200 ⴴ⌐

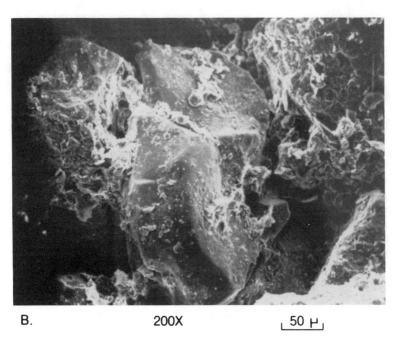

B.   200X   ⌐50 ⴴ⌐

Figure 5.11 Scanning electron microscope (SEM) photomicrograph *(continued on next page)*.

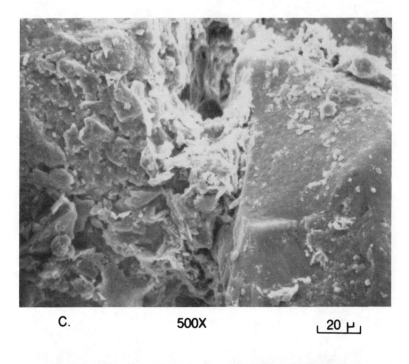

C.   500X   ⌊ 20 μ ⌋

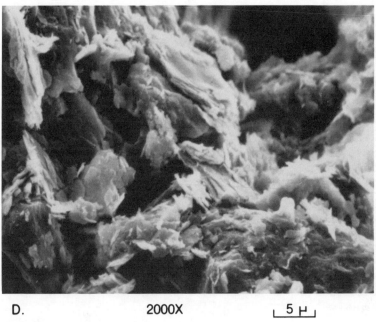

D.   2000X   ⌊ 5 μ ⌋

Figure 5.11 *(Continued.)*

5.4 GEOCHEMICAL PROCESSES

A brief discussion of the geochemical processes that are more likely to occur during ASR storage is presented in the following approximate sequence in which their effects may occur. Physical clogging, bacterial activity, ion exchange, and adsorption processes are presented first while more subtle reactions that may not become apparent for several months (dissolution processes) are presented last.

Other processes also occur underground but are less significant in ASR projects. These include complexing and retardation, both of which are important in hazardous waste investigations and with shallow soils but usually not significant in aquifer storage and recovery. The exception to this is the reduction in trihalomethanes and haloacetic acids, present in the treated recharge water and removal of precursors for their subsequent formation prior to recovery.

Suspended Solids Clogging

Physical plugging by suspended solids is perhaps the most important technical fatal flaw for ASR projects. In aquifers with low hydraulic conductivity, TSS concentrations in waters meeting turbidity standards for drinking water can have significant adverse effects upon hydraulic conductivity of the filter pack and aquifer sediments near the well. Turbidity is not a good indicator of TSS.

As little as 2 mg/L TSS can significantly reduce the recharge rate in most wells. The operational response to the reduction in recharge rate may be to increase the head (pressure) on the aquifer. This response is the most dangerous to the ASR well because it forces the TSS into the pore throats of the aquifer sediments. As the pressure builds up, it becomes increasingly difficult to remove these solids by periodically backflushing the well. A schedule of backflushing cycles needs to be established through testing at each well site to remove the TSS particles from the ASR well with a frequency that avoids residual clogging. If the required backflushing frequency causes operational problems, such as excessive on-off cycling of large electric motors, greater attention to well and wellhead design is needed to reduce solids buildup in the ASR well.

Since the required backflushing frequency cannot usually be confirmed until after the first ASR well in an area is constructed and tested, initial care with well and wellhead design is appropriate. As discussed in Section 4.2, Well Plugging and Redevelopment, some progress has been made in the development of a predictive tool to estimate probable well-plugging rates and redevelopment frequency prior to well construction.

In **Figure 5.12** [6] the units of TSS are expressed in terms of milligrams per liter TSS per square centimeter area of sand face. As little as 0.1 mg/L TSS per square centimeter of sand face within the ASR well can cause significant reduction in recharge rates at the well.

It is frequently useful to obtain a 0.45-μm filtered sample during the backflushing operation at an ASR well to determine the mineralogy of the TSS. This, in turn, identifies sources of the solids clogging the well. Freshwater diatoms were the most abundant at one ASR site in Seattle, retained in the water from a lake which was used for recharge. Another ASR site at Swimming River, NJ, had an aluminum-sulfate composition resulting from operation of the water treatment plant. Others involved reversing flow in old, unflushed municipal water lines. Each of these problems can be addressed satisfactorily once the origin is known.

Biofouling

Bacteria are present in aquifers to depths of at least 500 m (1,500 ft) and probably deeper. Many years ago it was believed that bacteria were only present in shallow aquifers as a result of septic systems. Bacteria were sampled by collecting filtered water samples. We now know that most of the bacteria in the groundwater are attached to aquifer particle surfaces and, therefore, are poorly sampled using groundwater as the sampling media. Cores are required. Even with cores, little is known about the role these bacteria perform in the groundwater system. Extrapolation from the reasonably well-known role in the soils indicates that the bacteria may be either directly or indirectly responsible for most of the changes in groundwater chemistry in ASR wells.

As shown by Dragun [7], bacteria are typically present in surface soils at populations of 0.1 to 1 billion cells per gram of soil and may be present at populations as high as 10 billion cells/gram. In groundwater, bacteria are typically present at concentrations of 100 to 200,000 cells/ml, and possibly as high as 1 million cells/ml. Biofilms form around various sizes of sand grains and glass spheres between 0.12 and 1 mm in diameter, reaching equilibrium thickness of about 10 to 60 μm between 8 and 10 days. Several studies have addressed the pH sensitivity of bacterial species and activity.

Figure 5.13 [8] shows the major parameters and their correlations that affect the growth of microorganisms at or near the wellbore, another potential fatal flaw for ASR projects. A disinfectant residual should be maintained in the recharge water and also in the wellbore during storage periods, to control and preferably eliminate the growth of microbiota at and near the wellbore. Letting a well sit idle when not in use can result in the growth of microbiota at and near the wellbore as a result of no

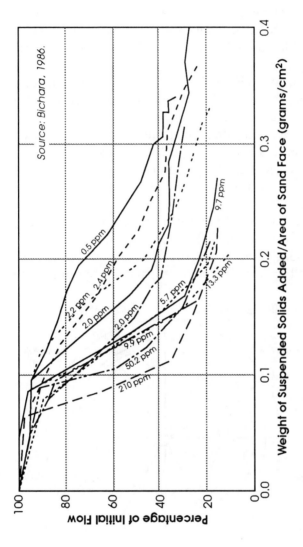

Figure 5.12 Change in flowrate during recharge with different concentrations of suspended solids.

disinfectant and a combination of these parameters. A potable recharge water quality not only meets all regulatory criteria but also avoids all or most of the parameters leading to biofouling, which is as effective as TSS in reducing permeability around the ASR well.

Temperature between 20 and 40°C, pH between 7.8 and 8.6, total phosphorus exceeding 0.1 mg/L, nitrate exceeding 1 mg/L as N, dissolved organic carbon exceeding 5 mg/L, total iron exceeding 1 mg/L, dissolved oxygen exceeding 3 mg/L, and a slow flow sequence strongly enhance biofouling potential.

The semi-circles on Figure 5.13 indicate the significance of the associations between parameters. For example, total nitrogen, nitrate, total phosphorus, pH, and temperature are significantly correlated with biological load. This is the only significant correlation between total nitrogen and the other parameters. Similarly, particulate organic carbon is significantly correlated with biological load, particulate volume, dissolved oxygen, total iron, flow sequence, and dissolved organic carbon, but has few other associations.

Wells equipped with oil-lubricated pumps commonly contain a layer of oil that can be several feet thick on top of the water column in the well as a result of leaking lubrication oil from the pump motor and/or filling of the oil reservoir. This layer excludes atmospheric oxygen from the ground-

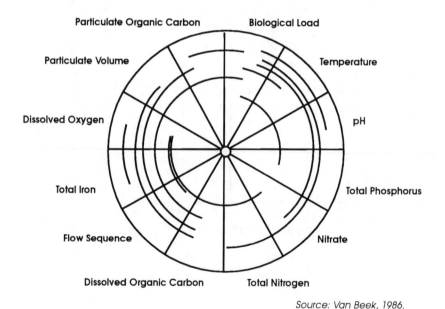

Source: Van Beek, 1986.

Figure 5.13 Factors affecting growth of microorganisms and associated well plugging.

water in the well, resulting in reducing conditions and a very high dissolved carbon concentration from the degradation of the oil. These conditions can result in an excellent environment for sulfate-reducing bacteria, particularly if the groundwater is also under reducing conditions, or if there is little movement between aquifers with different heads through the wellbore. Sulfate-reducing bacteria can effectively clog a well screen.

Sulfate-reducing bacteria are only one of hundreds of different types of bacteria that can cause biofouling. Most of the bacteria can only be described by the methods used to determine their numbers. Their functions, roles, and methods to control their numbers are virtually unknown.

Adsorption

Adsorption processes can occur at several levels, from precipitated flocs to individual ion adsorption. Iron oxyhydroxide and organic flocs are particularly sticky substances and usually become adsorbed to aquifer particles at the most constricted but important part of the tortuous flow path through the aquifer — the pore throat. This form of adsorption is the most important because this permanently reduces the amount of water that can move through the aquifer. Once this occurs it irreversibly destroys aquifer permeability. Acidification can be attempted and will recover some permeability but, if significant, this process will not allow the acid to flow into the affected area.

Natural iron and other oxyhydroxides attached to the aquifer particles adsorb the low concentrations of other metals present in the recharge water. Adsorption may also be partially responsible for the removal of trihalomethanes and haloacetic acids. Iron and other oxyhydroxides precipitated from the recharge water that are not significant in concentration, or occur slowly as the recharge water moves through the aquifer, will enhance this removal process.

Ion Exchange

Most aquifers contain some ions in exchangeable amounts. Both the amount and the cation in the exchangeable position is important. If the groundwater is brackish to saline, the clay typically has sodium in exchange positions and is stable in this high TDS condition. Recharge with a dilute calcium-bicarbonate water will impact the clays in two ways. The reduction in TDS increases the reaction between groundwater and dissolved ions. Calcium will replace the sodium in the exchange positions, converting the clay to a calcium clay. Calcium montmorillonites have a plate-like morphology and sodium montmorillonites have a ribbon-like

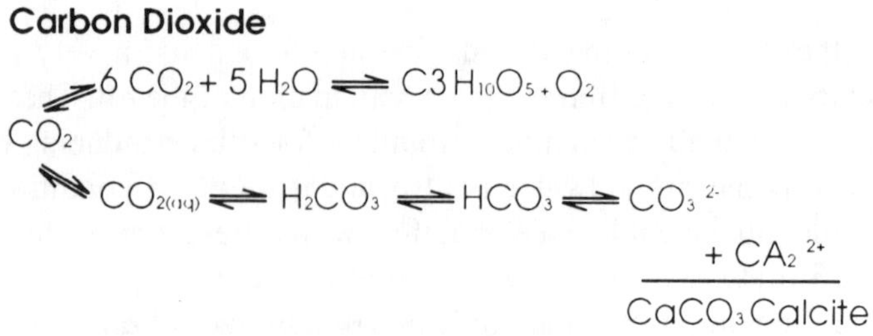

Carbon Dioxide

$$6\ CO_2 + 5\ H_2O \rightleftharpoons C3\ H_{10}O_5 \cdot O_2$$

$$CO_2$$

$$CO_{2(aq)} \rightleftharpoons H_2CO_3 \rightleftharpoons HCO_3 \rightleftharpoons CO_3{}^{2-}$$

$$+\ CA_2{}^{2+}$$

$$\overline{}$$

$$CaCO_3\ Calcite$$

Figure 5.14 Solubility equilibria for calcite.

morphology. Either or both of these conditions can potentially destabilize the attached clay particles and allow them to move into the pore-throat of the aquifer, causing plugging that is difficult, if not impossible, to reverse.

Oxidation

Most aquifers are under either moderately oxidizing or moderately reducing environments. Recharge water will be more oxidizing than the native groundwater and will cause chemical reactions to occur with aquifer mineralogy in equilibrium with the native groundwater oxidation reduction conditions. Iron carbonate (siderite) and iron sulfides (pyrite, marcasite, etc.) are the most susceptible to oxidation by recharge water. These minerals will immediately begin to oxidize, creating iron oxyhydroxide floc. Depending upon how much and where these minerals are in the aquifer and the physical characteristics of the aquifer, these flocs can destroy significant proportions of the hydraulic conductivity as they plug pore throats.

Dissolution

Solubility is a complex process that can involve several phases. Consider the common chemical reaction in which calcite is formed. Carbon dioxide participates in microbiotic reactions to form organics and in inorganic reactions by first becoming an aqueous species. Then, depending upon pH, carbonic acid, bicarbonate, and carbonate form in sequence with increasing pH. Finally, with sufficiently high calcium and carbonate concentrations, calcite precipitates. This sequence is shown in **Figure 5.14.**

The relative solubility of a series of common, naturally occurring minerals is ranked in **Table 5.7** [9] along with the major controlling mechanisms for solubility. The most soluble minerals are controlled by transport of the cation to the anion in the solution, resulting in sufficient concentration to precipitate the mineral.

Dissolution is strongly dependent on the velocity of transport away from the mineral. For example, calcite, generally considered a soluble mineral, is less soluble than is opaline silica, a common mineral in volcanic ash. For less-soluble minerals, dissolution is increasingly controlled by the reactions at the surface of the mineral. Their solubility depends upon the defects resulting from substitution of "alien" cations or anions and structural defects at the surface. Alien ions substituted into the mineral structure increase the solubility at the point of substitution. For example, strontium for calcium, or sulfate for carbonate in calcite, would increase the solubility of the calcite. Cracks in a mineral increase the solubility of the mineral by increasing the surface area exposed to the groundwater.

As shown in **Figure 5.15** [2], dissolved carbon dioxide converts to carbonic acid at a relatively slow rate beginning at an approximate pH of 4.5. Carbonic acid begins to lose one hydronium ion and becomes 50% bicarbonate at a pH of approximately 6. Bicarbonate ion is the most common form of dissolved carbon in natural waters. Bicarbonate begins to lose the final hydronium ion at an approximate pH of 8.2, becoming 50% bicarbonate and 50% carbonate at a pH of approximately 10. Most natural groundwater has a pH of less than approximately 9.5 but carbonate springs from alkaline rocks are known to achieve a pH of as much as 11.

Most groundwater chemistry is controlled by a calcium ion (actually calcite solubility) to a pH of less than approximately 8.2. If a groundwater has a pH higher than 8.5, sodium is usually the dominant cation. A pH higher than 11 is usually considered to result from an anthropogenic (man-induced) source. Temperature does affect the carbonate speciation but generally not more than half a pH unit, between 0 and 50°C.

TABLE 5.7
MINERAL SOLUBILITIES AND CONTROLLING MECHANISMS

Salt NaCl	5 mol/l	
KCl	4	Transport control
$NaCO_3$ $10H_2O$	3	
Na_2SO_4 $10H_2O$	0.2	
Gypsum $CaSO_4$ $2H_2O$	0.005	Mixed control
Anglesite $PbSO_4$	0.0004	
Opaline silica	0.002	
Calcite $CaCO_3$	0.00006	
Strontianite $SrSO_4$	0.00003	Surface control
Barite $BaSO_4$	0.00001	
Albite $NaAlSi_3O_8$	0.0000006	
K-feldspar $KA1Si_3O_8$	0.0000003	

Source: Berner, 1978.

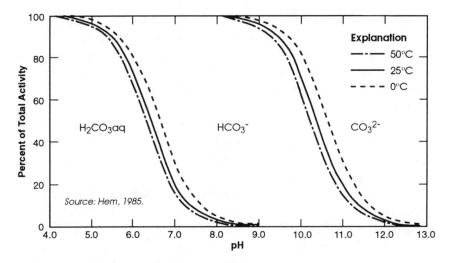

Figure 5.15 Alkalinity — carbonate species.

Particle size can have a dramatic effect upon solubility. As shown in **Figure 5.16** [7], dolomitic limestone (magnesium-rich calcium carbonate) was screened to discrete grain-size samples, and a constant weight of each sample was placed into an equally constant volume of water. One sample of water was used as a control. The pH was measured at regular intervals for 18 months to determine the rate at which the limestone dissolved.

The pH of the water declined to approximately 4.5, indicating equilibrium with the carbon dioxide in the air. Gravel-sized particles, 4 to 8 mesh, reacted very slowly and barely attained a pH of 5. Coarse sands, 8 to 20 mesh, did little better, achieving a pH of approximately 5.2. A pH of near 7 was not achieved after 18 months until the particle size was in the 40- to 50-mesh range, which is medium sand. Fine sand (80 to 100 mesh) dissolved more rapidly but both fine and medium sand-sized dolomitic limestone achieved a pH of approximately 7 only after 18 months in the water. Dissolution is not a linear function of particle size but smaller particle sizes with their larger surface areas dissolve more rapidly.

This data suggests that the initial ASR testing cycles may not achieve equilibrium with the aquifer mineralogy. The initial cycles are more likely to demonstrate the surface reactions, such as ion exchange and adsorption, and only begin to see the dissolution reactions. Longer term cycles over a period of several months are more likely to show the equilibrium between the aquifer mineralogy and the recharge water. Caution is therefore appropriate when faced with the natural tendency to conduct the testing program within a short time interval.

Similar to carbonates, iron complexes with hydroxide control the solubility of iron. Iron achieves a minimum solubility between a pH of approximately 6.5 and 8.0 for colloidal-sized, yellowish-brown iron hydroxide. The iron hydroxide typically becomes attached to aquifer particles
coating the area around the pore throats more than to the whole exposed
particle. Fresh iron oxyhydroxide has a very high adsorptive capacity for
other metals but loses capacity as it ages. As the iron oxyhydroxide ages,
it loses hydronium ions and becomes more of an oxide. With this loss of
hydronium ion the color changes from brown to red. Metallic iron is black
and reduced iron is generally some shade of gray.

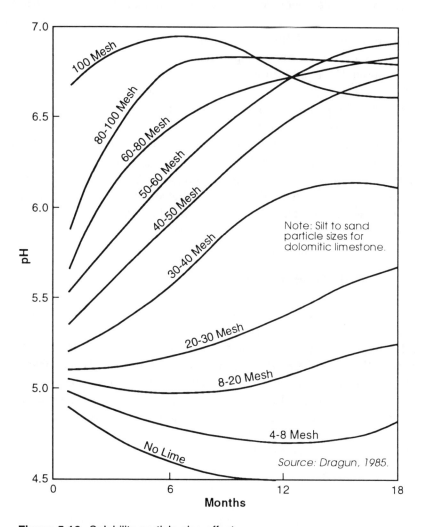

Figure 5.16 Solubility particle size effect.

Kinetics

It is becoming increasingly apparent that the kinetics of chemical reactions are important and, indeed, may control equilibrium reaction paths in the aquifer. Kinetics, the rate at which a chemical reaction occurs, has been described in the above description of iron and manganese precipitation rates. Iron precipitation reaction rates are much faster than manganese rates.

Reactions that occur rapidly change the eventual products and water chemistry suggested to potentially occur by equilibrium calculations using thermodynamic computer models. Equilibrium calculations do not indicate the rate at which the reaction occurs and the kinetics of most chemical reactions are not yet defined sufficiently to be generally applied. General gross rates are known. For example, clay minerals may be calculated to be supersaturated by equilibrium calculations but at common pH values (6 to 8) the reaction rates are too slow to be significant. However, if the pH is 9 or above, the precipitation of smectite clays (common swelling clays) becomes increasingly important.

Experience with many different ASR projects under widely different environmental conditions and geochemical results over increasing time intervals and cycles has clearly demonstrated that kinetics is both important and complex. As these data are assimilated and interpreted, the kinetics and, more importantly, their impact will become more apparent and predictable.

5.5 GEOCHEMICAL MODELS

Once pertinent data are available regarding water chemistry for both the recharge water and the native water in the proposed storage zone, a preliminary geochemical screening should be performed to identify potential fatal flaws for the ASR test program. Several relatively user-friendly computer simulation models are available for this purpose, each of which can be used to evaluate different proportions of recharge and native water in order to determine whether precipitation or solution reactions have a tendency to occur. Among these models are WATEQ, PHREEQE, and MINTEQ, as discussed briefly below.

Model results are no better than the available data. The poorest data are usually critical. For the equilibrium models mentioned above, simplifying assumptions include no biologic activity and no kinetics. Equilibrium constants are limited and of variable quality. The models should be used to try to understand primary existing conditions and to make predictions of the effects of changing these conditions.

If model results indicate no potential problems, and if ASR experience locally suggests that geochemical issues should not be a concern, then it may be appropriate to proceed with ASR well construction and testing without further significant attention to geochemical issues.

If local experience, model results, or other information sources suggest that geochemical issues may be significant, then it is appropriate to obtain cores of the storage zone and conduct testing to determine aquifer characteristics, as discussed in Section 5.3 above. Once such data are available, it is possible to conduct a more thorough geochemical assessment to evaluate possible reactions between recharge and native waters in the presence of aquifer minerals. Models such as EQ3NR/EQ6 can be used to determine not only the reaction products, but also the rates at which these reactions occur and the geochemical paths that the reactions may follow. For some reactions, as discussed above under kinetics, the rate is sufficiently slow that they are of no real significance to ASR operations.

When evaluating the model results, it is important to bear in mind that many areas of uncertainty underlie geochemical analysis. Some of the reaction mechanisms occurring underground are not well understood, particularly bacterial activity. Information on aquifer characteristics based upon core analysis is partially dependent upon the number and location of core intervals selected for analysis; core preservation, shipping, and handling; and the skill of the core laboratory analyst. It is important to perform a geochemical analysis for ASR projects, as outlined above. The results provide a valuable tool for understanding possible reactions occurring underground and for planning subsequent ASR well construction and investigations. However, they are not necessarily conclusive. The need for field investigations remains.

Following is a brief synopsis of the models discussed above, along with pertinent references. These models, and those yet to be created, will continue to improve geochemical understanding of rock-water interactions involving both inorganic and organic (including microbiota) reactions and processes that will lead to more accurate and reliable predictions of ultimate water quality and reaction products resulting from ASR operations.

WATEQ

This name stems from the words "WATer EQuilibrium." The model was developed by the U.S. Geological Survey and is the most user-friendly of the equilibrium models. It has been thoroughly documented and evaluated. Equilibrium equations include 30 chemical elements, 227 aqueous species, 12 organics, 3 gases (carbon dioxide, oxygen, and methane), 3 redox elements, and 309 minerals and compounds. All calculations are based upon 25°C and 1 atmosphere pressure. The model does not consider either adsorption reactions or reaction paths [10].

PHREEQE

The name stems from "pH REdox EQuilibrium Equations." This model was also developed by the U.S. Geological Survey and is fairly user friendly, although most versions are user-modified. It is well documented and evaluated. It includes aqueous speciation, solubility equilibrium and also the reaction path, including the dissolution and precipitation of common minerals. The basic model includes 19 elements, 120 aqueous species, rudimentary organics, 3 gases (carbon dioxide, oxygen and methane), 3 redox elements, ion exchange, and 21 minerals and compounds. It is based upon temperatures between 0 and 100°C and 1 atmosphere pressure [11, 12].

MINTEQ

Developed by the U.S. Environmental Protection Agency, this model is relatively user-friendly and is well-documented; however, some training in its use is advisable. The model addresses aqueous speciation, dissolution/precipitation equilibria, and adsorption. However, it does not address reaction path calculations or solid/liquid phase equilibria. It includes 31 chemical elements, 373 aqueous species, organic acids, 5 gases, 20 redox elements, 450 minerals and compounds, and 6 different adsorption models. The model considers temperatures between 0 and 100°C and 1 atmosphere pressure [13].

EQ3NR/EQ6

This is a highly sophisticated model developed at the Lawrence Livermore National Laboratory in California. EQ3NR addresses aqueous speciation and solubility equilibria, while EQ6 addresses adsorption as well as the reaction path and kinetics of the geochemical reactions. Training is necessary to use this model, although it is well documented. The model includes 47 chemical elements, 650 aqueous species, no organics, 15 gases, 16 redox elements, and 716 minerals and compounds. Temperatures may range from 0 to 300°C and pressures may vary from 1 atmosphere to saturation pressure at temperatures exceeding 100°C. The model includes both vapor and aqueous phase chemistry [14].

5.6 LABORATORY TESTING

It can be quite expensive to lose a significant portion of ASR well capacity due to irreversible geochemical plugging. Where geochemical model results are inconclusive, or local experience suggests that geochemical issues may be very significant, consideration should be given to simulating ASR operations in the laboratory, using core material from a test well or observation well constructed at the ASR site. Such core testing has been performed at a few ASR sites in the Atlantic coastal plain, which is characterized by the presence of freshwater-sensitive clayey sands in brackish aquifers, confined by clay formations. These core tests have included column tests using whole cores and also batch tests using material from the cores.

Column Testing

These tests have been performed at three ASR sites: Myrtle Beach, Chesapeake, and Swimming River. For each site, several frozen core sections with diameters ranging from 2.5 to 4 inches (60 to 100 mm) and lengths in the range of 12 to 24 inches (300 to 600 mm) were wrapped and sealed with adhesive in clear plastic sheeting, with wire mesh on each end. Steel end caps were machined to include tubing for movement of water across the entire end face of the core, and also to include axial threaded rods to restrain each end of the core. The wrapped core, screens, and end caps were then supported on a small tray constructed of circular lexan pipe and inserted into a 200 mm (8-inch) lexan cylinder with a length of 30 inches (830 mm). Cylinder end plates, seals, and four tie rods were machined so as to provide a pressure vessel capable of being pressurized to at least 30 psi. The wrapped core was centralized along the axis of the pressure vessel, and thrust restraint was then provided by the axial threaded rods.

Three cylinders were constructed of 8-inch (200 mm) PVC pipe and end caps, each with a capacity of about 15 G (57 L). One was provided with a glycerine-water mixture. The second was provided with recharge water, and the third with native water from the ASR storage zone. A nitrogen cylinder at 2000 psi with a regulator was also provided, along with accurate flowmeters, pressure gauges, and sampling ports for water at each end of the cylinder.

Once the pressure vessel was sealed with the core inside, it was filled with the glycerine-water mixture and pressurized, allowing sufficient time for the core to thaw and reach thermal equilibrium. Recharge water was then passed through the core at an inlet pressure below the confining pressure surrounding the core. Samples were collected of the recharge water and also the water leaving the core. The sampling frequency was established so that a relationship could be determined between water quality and the number of estimated pore volumes. Pressures and flow rates were monitored, and observations were made regarding the appearance of the product water, any apparent leakage through the core wrap, temperature, and cumulative volume.

Some of the cores required a few hours to simulate a series of ASR cycles, in each of which recharge water and native water were passed through the core in opposite directions. Other cores required several days to complete one simulated cycle. A few cores indicated leakage through the end caps or other logistic problems, and had to be discarded. In general, the equipment worked well but was expensive to construct to the required tolerances and also expensive to operate due to the long duration of the core testing and the skill level required of the operators.

Laboratory water quality data was then combined with flow and pressure measurements and other observations to calculate trends in hydraulic conductivity of the core material, and water quality of the flow leaving the cores. pH adjustment of the recharge flows was tested on several cores, to estimate the aquifer response to pretreatment measures under consideration. Ion exchange, oxidation, precipitation, dissolution, and possible bacterial reactions were evident in the quality of water leaving the cores.

Despite the expense and complexity of this effort, the results proved useful in evaluating potential geochemical reactions in subsequent ASR well construction and testing. Water quality data collected from ASR wells and nearby observation wells tended to match results predicted from the column tests. The column tests were less useful in predicting long-term well hydraulic response to ASR operations since it was concluded that bacterial activity, particle rearrangement or some other comparable mechanism adversely affected hydraulic conductivity following a few days of testing for almost all of the cores. The equipment may be useful for evaluation of core and aquifer response to pre-flushing operations designed to control potential clay dispersion in brackish aquifers; however, this has not been investigated to date.

Figure 5.17 shows the column testing apparatus while **Table 5.8** shows typical results for a core obtained from an observation well at Swimming River. The core was one of several subjected to column testing. It contained a significant concentration of siderite (ferrous carbonate) and was

Figure 5.17 Column testing apparatus.

flushed with oxygenated water with a pH of 8.27 to evaluate the effect upon iron concentrations and plugging associated with recharge of treated drinking water adjusted to this pH. In another core test from this site, acidification of the aquifer around the well with deoxygenated, low pH water (2.79) was simulated, demonstrating an increase in permeability and

TABLE 5.8 ASR COLUMN TEST RESULTS: SWIMMING RIVER, NEW JERSEY 03/01/90

Time	Total Volume Passed (mL)	Incremental Volume Passed (mL)	Flow Rate (mL/sec)	Injection Pressure (psi)	Confining Pressure (psi)	Temp. (°C)	Hydraulic Gradient	Permeability (cm/sec)	Pore Volume	Volume Passed (mL)	pH	Cond. (µmhos/cm)	Turb. (NTU)
15 24	0	—	—	4.80	9.30	25.0	11.1983	Effluent slightly milky					
15 35	240	—	—	4.80	9.30	25.0	11.1983	1.04E-03	1	0–240	6.54	463	97
15 36	—	50	0.3704	4.80	9.30	25.0	11.1983	—					
15 47	480	—	—	4.80	9.30	25.0	11.1983	9.19E-04	2	240–480	6.73	462	92
15 48	—	50	0.3226	4.75	9.70	25.0	11.0816	—					
15 59	720	—	—	4.75	9.70	25.0	11.0816	8.90E-04	3	480–720	6.75	—	—
16 5	—	50	0.3125	4.75	9.70	25.0	11.0816	—					
16 13	960	—	—	4.75	9.70	25.0	11.0816	8.48E-04	4	720–960	6.56	366	28
16 15	—	50	0.2976	4.75	9.60	25.0	11.0816	—					
16 28	1200	—	—	4.70	9.60	25.0	10.9650	8.47E-04	5	960–1200	6.52	360	19
16 30	—	50	0.2941	4.70	9.60	25.0	10.9650	—					
16 42	1440	—	—	4.70	9.55	25.0	10.9650	8.42E-04	6	1200–1440	6.23	345	21
16 44	—	50	0.2924	4.70	9.55	25.0	10.9650	—					
16 56	1680	—	—	4.70	9.55	25.0	10.9650	8.32E-04					
16 59	—	50	0.2890	4.70	9.55	25.0	10.9650	—					
17 11	1920	—	—	4.70	9.50	25.0	10.9650	8.18E-04	8	1680–1920	6.21	329	12
17 13	—	50	0.2841	4.70	9.50	25.0	10.9650	—					
17 15	End test	—	—	4.85	9.95	24.1	11.3149	—					

Note: Effluent samples collected as composites of the 2nd and 3rd pore volume (A), and sixth and seventh pore volume (B). Results: calcium declined from 22 mg/L in the recharge water to between 3 and 5 mg/L in the effluent; iron was about 3 to 4 mg/L in the effluent, sodium increased from 36 mg/L in the recharge water to 87 mg/L in the effluent after three pore volumes and 63 mg/L after seven volumes.

TDS, rapid release of iron in the effluent, and an increase in manganese. After a few pore volumes, high concentrations of these parameters in the effluent declined rapidly. Subsequently, the core was tested with oxygenated water at the same pH, showing little further water quality response. The very high concentrations of iron flushed from this core within a few pore volumes matched the results in the observation well when this procedure was subsequently implemented at the ASR well.

Batch Testing

This testing approach has been conducted for the Chesapeake ASR site. Compared to column testing, this approach is relatively inexpensive. Samples of core material are thoroughly mixed and are then subjected to a progressive series of leaching tests at steadily increasing pH values to determine dissolution, ion exchange, and other reactions occurring during each step in the series. Results provide considerable insight into the expected water quality and geochemical response to ASR operations. However, they are not useful for directly evaluating aquifer hydraulic response, except as may be inferred from the geochemical potential for plugging or dissolution.

For the Chesapeake ASR site, the issue was manganese concentrations in the recovered water which exceeded drinking water standards. No manganese problems had been detected during previous ASR cycle testing at pH values ranging from 7.4 to 9.4. However, subsequent ASR operations had been conducted at a lower pH (5.9 to 7.4) relative to cycle testing, causing manganese to become solubilized. Manganese concentrations in the recovered water ranged from 0.25 to 1.28 mg/L.

The following sequence of batch-testing operations was conducted on core samples selected from a repository created during the coring to be representative of the storage zone, based upon geophysical logs, drillers logs, and core laboratory data. During the first test, core samples from four selected depths were sequentially analyzed to determine the manganese concentrations associated with leaching, first in water at low pH and then in ammonium acetate. From this test, the core with the greatest potential for manganese release was identified. Analysis showed it contained manganese oxide, organic manganese and 137 mg/kg of manganese associated with iron oxyhydroxides. Another sample from this core was then split and each portion tested, one with pH 6.5 water and one with pH 8.0 recharge water for a period of 33 days, to determine the released manganese concentrations.

Based upon batch testing results, pH adjustment of the recharge water was implemented to control manganese reactions in the aquifer. An optimum pH of 8.3 was selected. This followed recovery to waste of a significant portion of the water stored at low pH. Subsequent recharge water is steadily overcoming the residual acidity in the aquifer and is expected to restore geochemical conditions to those that occurred during initial ASR cycle testing.

5.7 FIELD INVESTIGATIONS

For ASR site investigations that incorporate significant geochemical concerns, activities during the test program include monitoring of recharge and recovery water quality to determine if geochemical reactions are occurring and the resultant water quality effects. Field investigation activities may include pretreatment of either the recharge water, the storage zone, or both. They may also include post-treatment of the recovered water. Following the test program, geochemical activities may include periodic monitoring of water quality to confirm that no adverse effects are occurring, and to address and resolve any adverse effects that may arise.

Pre- and post-treatment processes are usually limited to disinfection and, occasionally, pH adjustment but may include other processes such as filtration, deoxygenation, and aeration. Disinfection of recharge and recovery flows is assumed to be required for all ASR systems storing potable water. Wellhead filtration to remove solids has been discussed previously in Section 4.3, Wellhead Filtration, and is not repeated here. Pre- and post-treatment issues associated with iron and manganese removal have been addressed previously in Section 4.6, Pre- and Post-Treatment.

Other geochemical issues will undoubtedly arise as the number and variety of ASR applications continues to grow. The role of the geochemist on an ASR project team will remain challenging as new regulatory issues such as arsenic, radon, and disinfection byproduct reduction are addressed from an ASR perspective to capitalize on the ability of aquifer systems at many sites to improve the quality of stored water.

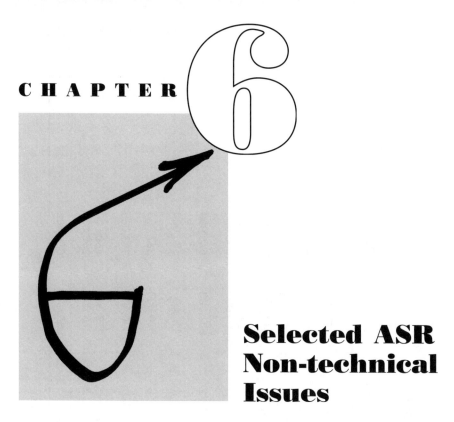

CHAPTER

Selected ASR Non-technical Issues

Many centuries ago in a desert that we now know as Arizona, a native American carved a sign in a rock face, to guide his companions to the nearest waterhole. The sands of time have since obscured the ancient trail, but the sign and the waterhole remain.

6.1 ECONOMICS

ASR feasibility has been demonstrated at a growing number of operational sites in the U.S. It is a practical, cost-effective, and environmentally acceptable water management alternative for water utilities and other water users. When compared to surface storage reservoirs, aquifer storage is very low cost, since land requirements are minimal and the storage capacity is provided by nature for the relatively low cost of a few ASR wells. In addition, water transmission and treatment facilities can be operated more efficiently with ASR systems, often requiring less capacity and associated construction costs.

Most utilities can use ASR to meet water system expansion goals while achieving significant cost savings. However, feasibility must be confirmed

TABLE 6.1   CAPITAL AND OPERATING COSTS FOR ASR SYSTEMS

Site	Year	Yield (MG/day)	Cum. No. Wells	Capital Cost	$/MG/day	Operating Cost $/MG/day
Peace River, Florida	1984	1.5	2	702,000	468,000	
	1988	3.4	6	1,342,000	395,000	
	Est. 1994	3	9	1,300,000	433,000	20,000
	Est. 1997	14	23	8,200,000	586,000	
Cocoa, Florida	1987	1.5	1	444,000	296,000	
	1992	6.5	6	1,314,000	202,000	6,000
Marathon, Florida	1993	0.5	1	827,000	1,654,000	
	Est. 1995	3.0	8	3,000,000	1,000,000	40,000
Kerrville, Texas	1991	1.8	2	987,000	548,000	
Centennial, Colorado	1992	0.7	1	410,000	586,000	
Seattle, Washington	1993	5.1	3	1,670,000	327,000	
Swimming River, New Jersey	1993	1.7	1	600,000	353,000	
Calleguas, California	1991	1	1	459,000	459,000	
	Est. 1994	6.5	6	1,278,000	256,000	
Murray Avenue, New Jersey	Est. 1994	1.5	1	950,000	633,000	

through satisfactory completion of an ASR test program, with all associated permitting, legal, economics, water rights, environmental, and other issues resolved. This typically requires 2 to 3 years to complete, after which it is appropriate to begin adjusting water system expansion plans to accommodate the new ASR technology and begin to realize the associated cost savings. Until an ASR test program is completed, it is recommended that ongoing expansion plans be continued. Therefore, maximum savings from application of ASR technology can best be achieved by starting ASR feasibility investigations at least 3, and preferably 5 years before any major decision regarding investment of capital and sizing or location of facilities.

To obtain an understanding of ASR economics, construction and engineering cost data were obtained for nine ASR sites in six states that are operational or about to begin long-term operation. Cost data ranged from projects begun in 1984 to expansion projects expected to start operation in 1994, with all but two of the projects coming online after 1988. Some of the sites used retrofitting of existing wells to achieve ASR objectives while others used new wells. Several sites provided data on successive phases of ASR expansion. Individual well yields ranged from 1.9 to 5.7 megaliters/day (0.5 to 1.5 MG/day). Three sites had information on operation and maintenance costs and a few provided information on costs of water supply alternatives to ASR. All costs are expressed in 1993 U.S. dollars. **Table 6.1** presents the resulting cost data, including the site location and the year in which the ASR facilities were, or are expected, to go into operation.

Based upon this analysis, the following conclusions may be drawn:

1. Unit costs for ASR facilities generally range from about $50,000 to $160,000/megaliters/day ($200,000 to $600,000/MG/day) of recovery capacity, with an overall average of about $100,000/megaliters/day ($400,000/MG/day). Higher unit costs are typically associated with the first new ASR well at any site, sites requiring extensive piping to tie them in to the existing water system, and sites with low recovery capacity per well. Lower unit costs are associated with retrofitting existing wells at sites close to existing piping facilities, higher yield wells, and also with multiple-well ASR expansion projects.

2. Unit costs for the second and subsequent ASR wells at any site are typically lower than for the first well, reflecting generally reduced efforts to obtain regulatory approval. The first ASR well incurs additional cost in order to demonstrate ASR feasibility. The reduction in unit cost is typically in the range of $26,000 to $53000/megaliters/day ($100,000 to $200,000/MG/day).

3. When comparing capital cost per unit of new capacity, ASR typically is less than half the cost of other water supply and treatment alternatives. In some

cases, the cost savings approach 90%. This savings reflects the efficient use of major facilities such as pipelines, pumping stations, and treatment plants, and the relatively low cost of using underground storage capacity when compared with a similar storage volume provided in surface reservoirs.

4. Annual operating cost ranges from about $1,600 to $10,600/megaliters/day ($6,000 to $40,000/MG/day) of recovery capacity, although data availability is sparse. This includes the marginal cost for power and chemicals during recharge and recovery, plus an allowance for operation and maintenance. One of the three sites is somewhat atypical in that treated drinking water is stored in a seawater aquifer at Marathon, FL. Consequently typical annual operation costs based upon this data are probably closer to about $15,000/ MG/day recovery capacity.

When comparing ASR to other water management alternatives, it is important to compare them on the same basis. When comparing capital costs, an appropriate comparison is usually the cost/unit production capacity since ASR increases system peak capacity even though it may only recover water for a few months each year. It is usually inappropriate to compare capital costs on the basis of dollars per m³ or dollars per thousand gallons recovered, since total annual production from ASR facilities may be small, depending upon the duration and extent of the peak demand period.

Similarly, when comparing operating costs, it is more appropriate to compare the annual costs/unit production capacity rather than dollars per unit volume, since ASR wells typically are not in operation all year.

In some areas of the U.S., ASR facilities are used primarily for long-term storage, or "water banking," without significant consideration of the time or location of eventual withdrawal. The recharge well has a pump that is used to redevelop the well by backflushing periodically, thus maintaining the well's injection capacity. The objective of these systems is recharge, not recovery. For such projects, the unit cost of ASR systems (dollars per acre ft, or dollars per m³ recharged) may be higher than comparable surface recharge systems because ASR wells typically require a higher level of treatment prior to recharge. However, if and when this stored water is recovered at the same site, the overall unit cost for both recharge and recovery is usually less for ASR systems than for surface recharge systems because the same facilities are used for both recharge and recovery and because the recovered water usually does not need retreatment, other than disinfection.

If recharge is the sole objective, surface recharge, if feasible, is the lowest cost approach to getting water into the ground. If recharge is the

**TABLE 6.2   COMPARISON OF CAPITAL COSTS,
WITH AND WITHOUT ASR**

Location	Expansion Cost ($Million)	
	With ASR	Without ASR
Wyoming, MI	9	31
Peace River, FL	46	108
Manatee County, FL	2	38
Florida Keys, FL	3	38
Kerrville, TX	3	30

sole objective but surface recharge is not feasible because of hydrogeologic constraints, high land costs or other issues, then ASR can also achieve this objective at a higher unit cost due to the probable need for a higher level of water treatment prior to recharge. If the objective is to recover and use the stored water within a few years at or near the same site, then ASR systems will probably be more cost-effective than surface recharge systems because no additional facilities will be required for recovery, treatment, and use.

Table 6.2 shows a comparison of capital costs for five water utility systems, with and without ASR, in order to meet comparable levels of service. In some cases, the savings with ASR is due to reduction or elimination of surface reservoir storage. In other cases, the savings is in terms of a major pipeline or treatment plant. In all cases, the savings is in excess of half. This appears to be typical for most ASR systems placed in operation to date. With these kinds of savings, the ASR system can experience and respond to a variety of technical, regulatory, and other challenges while still providing cost-effective service to the owner and the consumer.

6.2 WATER RATE IMPACTS

The savings in capital costs provided through ASR create the opportunity for a corresponding reduction in water costs and associated rates. A 50% reduction in the capital cost of facilities expansion can lead to a corresponding reduction in water rate increases required to finance the expansion. In some cases, the need for an expansion program and associated rate increase can be deferred for several years, reduced or eliminated through more efficient use of existing facilities.

Where rate increases are required for systems utilizing ASR wells, it is pertinent to consider certain fiscal questions that are unique to an ASR mode of operation. Two key questions are as follows:

- If a water utility system supplies several water users, meeting their demands with a system that incorporates ASR, who should pay for water stored but not recovered within the same fiscal year and when should they pay?
- If there are multiple users, some of whom wish to recover water to meet short-term seasonal needs, and others who have longer-term needs with varying storage periods, or emergency needs, how should the associated costs be apportioned among the users?

There is probably no single, best answer to these questions, since the ultimate balance between simplicity and equity at each site is usually a political decision that can easily vary between sites. At the risk of venturing onto new ground once again, some observations may be helpful in guiding resolution of these issues at any particular site.

Water customers are willing to pay for certainty in their water supplies. Payment should occur when the costs are being incurred. A fair and efficient water allocation process can lead to sound economic development and social stability, and is best achieved through market pricing, over and above a monthly sustenance volume for each household.

Where the primary purpose of an ASR system is to provide seasonal storage, the annual costs for debt service, operation, and maintenance will all tend to occur in the same year. Therefore, they may be distributed among all users in a manner similar to other utility costs, with no particular consideration of the marginal cost of the ASR component of the system. Capital costs would be amortized evenly over the expected life of the facilities or the bond repayment period. Operation and maintenance costs will be incurred each year and the full benefit of the water stored will be realized by the consumers in the same year. Not only will they utilize the stored water, but they will also benefit from lower water rates due to use of ASR wells for seasonal storage.

Where water is stored by a utility in one year for recovery in a later year in order to supply a single water user or wholesale customer, the utility may build the cost into the rate base for each year in which some cost is incurred. This is a "pay as you go" philosophy. During the year in which water is treated and stored, the associated marginal cost for chemicals and power to supply off-peak water for ASR recharge may be reflected in the rate base. During the year in which the water is recovered, the associated marginal cost for pumping and disinfection may be reflected in the rate base. The water user benefits from the economies associated with both seasonal and long-term water storage, which ultimately tend to reduce water rates, and also benefits from utilization of the water when it is recovered. A "pay as you go" philosophy distributes the costs and benefits in a reasonably fair manner that is consistent with normal utility system planning and budgeting.

In the more complicated situation with one utility but multiple major water users or wholesale customers, each with separate needs for recovery of stored ASR water in different years and under different situations (different peaking factors, emergency demands, etc.), water rates will tend to be negotiated separately with each user, depending upon their unique circumstances. However, the same philosophy seems appropriate. Costs may be built into the rate base as they are incurred and may be apportioned among the users according to a formula that reflects some consideration of each user's reliance upon ASR storage as opposed to providing peaking capacity with other, more costly facilities.

If it is a particularly wet year and water demands are consequently depressed, considerable ASR storage can be accomplished at relatively low marginal costs. During drought years, demands will tend to be higher and the opportunity for ASR storage will be reduced. Costs may be incurred either for limited ASR storage or for recovery, or possibly both. Assuming the same number of service connections in each of these cases, the water user would pay a slightly higher rate in wet years so that excess water can be stored. The same customer will still benefit from lower unit cost for service, higher reliability during drought years when he utilizes the stored water, and possibly a lower rate during later years in which water is primarily recovered instead of recharged.

Practical considerations will probably tend to smooth out the rate variability and associated utility revenues from year to year. During relatively wet years when demand is low, the ASR component of the rate base will tend to sustain revenues. During drought years when demand is high, ASR will tend to ensure that the system can meet demands without service restrictions. This should tend to sustain utility revenues even though ASR costs would be minimal. As discussed previously, all water rates will tend to be reduced due to incorporation of ASR into utility operations.

This wet year–dry year variability in costs and rates suggests the possibility of a rate structure that incorporates a seasonal and long-term incentive for aquifer storage. Such a program has been implemented in recent years by the Metropolitan Water District of Southern California in order to encourage member agencies to buy water from the regional system during months when it is readily available, and store it locally in aquifers. This helps to reduce peak demands on the regional system at times when the water is least available, thereby providing the opportunity for avoiding significant costs for peaking facilities. Peaks can then be met from local aquifers at much lower costs. To ensure the long-term viability of such a program, it is important that the seasonal cost incentive matches, or does not exceed, the avoided costs for regional peaking facilities.

Where the water utility provides both the water and the ASR storage, the concept of a seasonal rate incentive is internalized within the utility and its rate base. Awareness of the approximate marginal cost for water stored and recovered, and the more costly alternatives available for meeting peak demands, can probably guide ratemaking in an appropriate direction.

For some utilities, ASR storage within the service area provides the opportunity for negotiation of favorable terms in wholesale purchase agreements for imported water. Water can be purchased during periods of low demand and/or high supply, thereby achieving low marginal costs. Transmission facilities may possibly be downsized, reflecting planned conveyance of water during off-peak months at lower flow rates, so that peak demands can be met locally from ASR storage.

A significant challenge in establishing rates with ASR is to overcome the traditional utility rate-making process thinking in which a major water user perceives that he is buying "system capacity" on a "take-or-pay" basis. With ASR, it is not as simple as paying some money and having access, whenever needed, to a fixed amount of facilities capacity. Instead, the water user is buying a commitment to supply water at some time in the future, whether months or years away. This requires reasonable planning and forecasting of water needs so that adequate facilities can be built and operated in time to store sufficient water so that it will be available when needed. The payback for this planning is the reduction in water costs, since facilities for treatment, storage, and conveyance can be substantially downsized if ASR wells are used to help meet future peak demands. This is where the 50% or greater reduction in capital cost amortization is achieved. Conversely, if a water user insists on a conventional type of rate agreement, the facilities will probably have to be built up front to meet that full demand without ASR storage and the cost will therefore be much higher.

The role of ASR in the overall water supply plan is a concept that is not always easy to understand. When elected officials are faced with the need to establish rates and make commitments based upon water demand projections, it is frequently easier to defer construction of ASR and other facilities and thereby hold down short-term costs. This can reduce that water storage volume that can be achieved. It can also limit the opportunity to reduce the size of the associated treatment facilities, thereby increasing long-term costs to the water users.

The impact of ASR upon water rates is perhaps less significant than other factors potentially affecting water rates. By making more efficient use of existing facilities and aquifer storage, ASR helps to hold down overall costs, but may not substantially affect the rate structure or its underlying philosophy. Where ASR is implemented to augment water supplies to meet increasing projected demands, the same demands can perhaps be met through a sharply inclined rate structure over and above a base monthly volume per household. The resulting water conservation effect can also extend the service life of existing facilities.

6.3 LEGAL AND REGULATORY ISSUES

The principal frontier for ASR activity in the next few years is probably associated with legal and regulatory issues at the federal and state levels. How these issues are understood, framed, and resolved will probably determine the extent to which ASR technology can ultimately benefit water users throughout the country.

As an aid to those who are considering how to regulate ASR activities, and the appropriate water quality standards to follow, Appendix A contains the current (1993) standards adopted by the U.S. Environmental Protection Agency, the European Community, and the World Health Organization to govern drinking water quality.

EPA Surface Water Treatment Rule

EPA implemented the final Surface Water Treatment Rule (SWTR) on June 29, 1989. Under the terms of this rule, utility systems dependent upon surface water sources are required to provide filtration treatment and adequate disinfection to inactivate viruses and *Giardia lamblia* cysts, deleterious constituents sometimes found in surface water supplies. The rule became effective in June 1993.

This rule will generally improve the quality of water available for ASR recharge in some areas by reducing the organic and solids content of recharge waters and by providing an adequate disinfectant residual. For most utility systems impacted by the rule, a part of the solution may be to provide increased contact time with free chlorine or chloramine residuals. This may entail construction of additional above-ground storage reservoirs in order to provide the required contact time. The cost of this storage may be substantial.

Some utility systems may be able to provide the requisite contact time by using ASR storage instead of aboveground reservoir storage. Certainly, data collection at several ASR sites has confirmed that disinfectant residuals tend to continue underground for one or more days, providing extended contact time. This may be a cost-effective alternative, particularly where

ASR storage also achieves other system objectives, such as diurnal or seasonal storage, maintenance of distribution system pressure, emergency storage, or other benefits.

EPA Disinfection Byproducts Rule

Another important element that could affect utility system response to the Surface Water Treatment Rule is the Disinfection Byproduct (DBP) Rule, which was issued by EPA during 1993 after a negotiated rulemaking process. This rule provides more stringent regulation of trihalomethanes and regulate other carcinogenic organic compounds in drinking water supplies, including haloacetic acids. The concentrations of such contaminants depend on the concentration of organics in the source water and the free chlorine residual and contact time during the treatment process. High DBP formation potentials are typically associated with highly colored source waters and disinfection with free chlorine. As a result, many utilities are changing their treatment processes to provide secondary, and in some cases primary disinfection with chloramines. These are less powerful as disinfectants, however, they ensure acceptably low concentrations of DBPs.

When used with ASR storage, it is quite possible that continued use of a free chlorine residual may be a worthwhile and cost-effective approach to meeting the DBP and SWTR requirements, since data collected at several ASR sites during recent investigations have shown that DBP concentrations are typically reduced significantly during seasonal ASR storage and the associated DBP precursors are also reduced. In other words, incorporation of ASR facilities into water treatment operations may provide a cost-effective seasonal or long-term storage reservoir and also provide adequate chlorine contact (CT) times, while at the same time meeting DBP requirements, as a result of further treatment occurring during several weeks of aquifer storage. Further development of this concept would probably entail site-specific evaluation of its potential application at a few representative locations to determine whether the idea is viable and cost-effective and also to better define reaction mechanisms for the benefit of regulatory agencies. This is discussed in greater detail in Section 4.6, Disinfection Byproducts.

EPA Groundwater Rule

The draft EPA groundwater rule was announced in July 1992. Formal proposal of the rule is currently scheduled for August 1994, with its promulgation 24 months later. EPA proposes that all public water systems using groundwater disinfect the source water from its wells unless (1) "natural disinfection" requirements are met, or (2) the system qualifies for a variance. The two exceptions are intended to include only those systems where wells are not vulnerable to viral contamination. All sources would have to provide a disinfectant residual during distribution.

The regulatory position related to recovery of treated drinking water from ASR systems has not been determined. Most ASR sites are in deep, confined aquifers and are not likely to be affected by overlying land-use activities. Consequently, it should be possible to demonstrate the absence of viruses and pathogenic bacteria in ASR recovered water, thereby meeting the "natural disinfection" requirements when they are implemented.

EPA Underground Injection Control Program

An awkward legal and regulatory framework currently exists to guide ASR activities in the U.S. Other countries interested in applying ASR technology may benefit from consideration of the U.S. experience, drawing from the beneficial aspects and avoiding those that are inappropriate. In time it is hoped that an improved legal and regulatory framework can be established.

In the U.S., groundwater management and development is a right reserved by the individual states that is not subject to direct control by the federal government. In different states, groundwater laws range from complex to almost non-existent. However, the federal government has enacted legislation protecting the rights of all citizens to a safe drinking water supply. This reflects the rapid movement of people between states, the movement of water supplies between states, and also the movement of potential sources of contamination between states. This interstate movement provides a legal basis for federal involvement in state groundwater issues.

The 1974 Safe Drinking Water Act was designed to address growing concerns at that time regarding the safety of drinking water supplies and vulnerability to contamination. Part C of this act addresses underground injection control, commonly known as "UIC." The text of Part C is relatively brief and the key wording is contained in paragraph 300h(day)(2), as follows:

> Underground injection control endangers drinking water sources if such injection may result in the presence in undergroundwater which supplies or can reasonably be expected to supply any public water system of any contaminant, and if the presence of such contaminant may result in such system's not complying with any national primary drinking water regulation or may otherwise adversely affect the health of persons.

It is quite clear from this wording that the intent of the act is to ensure that no injection practice should occur if it would jeopardize the ability of a water treatment plant treating water from the same aquifer to remove any contaminants prior to distributing treated water for public consumption.

In the process of developing regulations to implement the act, the U.S. Environmental Protection Agency (EPA) interpreted this language as meaning that water should meet primary drinking water standards prior to injection into the well. These standards are generally becoming more restrictive. The UIC regulations provided for individual states to accept responsibility for implementation of the UIC program so long as state standards are at least as strong as the federal standards. As a result, approximately 39 states have developed their own UIC regulations which are patterned after the federal regulations, and have therefore accepted responsibility for implementation of the act.

The federal UIC program includes classification of injection wells into five categories: Classes I through V. ASR wells are generally considered as Class V wells, which include a very wide variety of injection practices other than ASR. Some of these practices represent a significant threat to groundwater quality.

The UIC program also includes an aquifer exemption process that provides for those situations where the quality of the injected water does not meet primary drinking water standards. The process depends upon the quality of water in the receiving aquifer. If the TDS (total dissolved solids) concentration exceeds 3000 mg/L, a minor aquifer exemption is required that can be handled at the state and regional level (one of ten EPA regions around the country). However, if the TDS concentration in the aquifer is less than 3000 mg/L, this is considered a major aquifer exemption requiring action at the EPA headquarters level. To date, EPA has approved very few minor aquifer exemptions and no major exemptions for Class V wells. Minor exemptions have been generally issued for regional areas rather than for individual wells. Exemption criteria are listed in Section 146.4 of the 40 Code of Federal Regulations (40 CFR).

The issuance of such exemptions is understandably sensitive from a regulatory point of view. Once contaminated, potable aquifers require many years to restore their water quality, if full restoration of quality is achievable. Consequently, regulatory positions have generally been conservative and consistent in requiring treatment to potable standards prior to injection into Class V wells and potable aquifers.

Currently, ASR storage of treated drinking water in fresh or brackish aquifers is generally considered to be an acceptable practice, as evidenced by the growing number of operational ASR systems in the U.S. Ten states currently have operational ASR systems. State regulatory agencies have generally applied UIC requirements in a manner appropriate to each state's needs. The challenge for the future lies in the application of the UIC process for ASR storage of water not meeting all potable standards.

It is anticipated that the demonstrated ability to hydraulically control movement of stored water in a small radius around a well for a period of several months, combined with a growing body of investigations demonstrating water quality improvement during aquifer storage in both saturated and unsaturated portions of an aquifer, may provide a basis for regulatory approval of such ASR systems that do not meet all potable standards. The regional water management benefits would probably outweigh the environmental risks and costs of such a practice for many parts of the U.S. Resolution of the regulatory issues associated with such storage projects is most likely to originate in California and Florida, although a few other states are beginning to address these issues.

There are several significant drawbacks to the 1974 Act and the UIC regulations implemented pursuant to the Act. ASR was not anticipated or considered at the time the Act was passed or the regulations developed. Although the 1974 Act was designed to protect the public drinking water supplies from contamination so severe that it cannot be readily treated at water treatment plants, the UIC regulations seem overly restrictive when applied to many ASR applications.

For example, recharge of surface water from high quality water sources would require issuance of an aquifer exemption since the surface water would undoubtedly include constituents that exceed primary drinking water standards such as turbidity and coliforms. Once an aquifer exemption is issued, however, the aquifer loses all federal protection under the UIC program. Since the state programs generally follow the federal UIC regulations for those states that have accepted responsibility for implementing this program, it is probable that the ASR site would lose important protection from possible injection activities by other water users in the exempted area. This is particularly true for a regional aquifer exemption, which is the principal type of exemption issued to date for Class V wells. Such loss of regulatory protection may cause the quality of ASR stored water to be threatened by injection of water elsewhere in the exempted area that is much inferior in quality. In short, the aquifer exemption process is very arduous and, once issued, the aquifer exemption may create more problems for ASR than it solves.

A model state legal and regulatory process for ASR is needed. This process would provide for ready implementation of ASR programs where recharge water quality meets drinking water standards. It would also provide for ASR implementation where water quality meets these standards with the possible exception of certain constituents that do not threaten public health and the environment; are readily treated either in the aquifer or in public water treatment facilities; and meet the original intent of the 1974 Act. In this way, EPA can approve state-level ASR programs; continue with the UIC program implementation for non-ASR systems; and help to attain the broad national, state, and local environmental benefits associated with widespread ASR implementation. To date, such a model legal and regulatory process does not exist. However, it is the subject of considerable activity in several states and may emerge in a form suitable for consideration by other states and countries in the near future.

Some of the constituents that may not meet drinking water standards but may be quite suitable for ASR recharge purposes in some situations include the following:

- coliforms
- turbidity
- color
- sodium
- chloride
- total dissolved solids
- iron
- nitrate
- corrosivity

Each of these constituents would not represent a threat to public health if stored seasonally in a brackish aquifer and recovered to meet non-potable uses, such as irrigation. Water sources that would be defined by this list include high-quality sources of surface water, untreated groundwater from uncontaminated surficial aquifers, and high-quality reclaimed water. These potential ASR water sources are discussed in greater detail in Chapter 7, Alternative ASR Applications. With reasonable care, barriers in space, time, and required wellhead treatment processes would separate the ASR stored water from any potential domestic water supply wells close to the ASR well sites.

Seasonal storage of water from these sources in a freshwater aquifer may also be appropriate in some circumstances, depending upon site-specific consideration of water quality constituents, aquifer treatment mechanisms, and careful assessment of public health and environmental risks and benefits.

It is important that any model code that provides for easier implementation of ASR should not be interpreted as an opportunity to authorize water management practices that, in fact, contaminate the environment. For example, poor quality urban runoff may contain many constituents that violate drinking water standards, such as metals, oil and grease, and other parameters that may be carcinogenic, mutagenic, teratogenic, or otherwise deleterious. ASR storage of this water in a freshwater aquifer would be unacceptable. Seasonal storage of urban runoff in a brackish aquifer following pretreatment, such as detention/retention, may be a beneficial water management practice in some situations. The legal code and associated regulations adopted for any particular state or region may follow a model code or may follow the example established in another state; however, it will undoubtedly reflect local priorities for water use and water quality protection.

Suggested elements of such an ASR Model Code include the following:

1. Water should be recovered from the same well(s) in which it is recharged.
2. The cumulative volume recovered should not exceed the cumulative volume stored.
3. Annual restrictions on volumes stored and recovered should be avoided or minimized in order to achieve the full benefits of ASR technology.
4. Significant adverse impacts upon other existing well owners utilizing the same aquifer during recharge and recovery should be mitigated.
5. ASR is recognized as a storage measure rather than a new water supply source. Once water rights have been obtained to utilize the water for a beneficial purpose, underground storage of this water and subsequent recovery for the same purpose should not affect the rights to the water following recovery. No other entity or individual has the right to recover the water from ASR storage.
6. Recharge water quality criteria should be established by each state to meet its own unique needs, opportunities and circumstances.
7. Separate state-level permitting for recharge and recovery of stored water should be avoided. An ASR permitting process that integrates these two functions is desirable. Similarly, ASR permitting should be based upon a framework of existing water supply considerations, rather than existing wastewater disposal regulations.

Ownership of the Stored Water

As discussed previously, groundwater ownership in the U.S. is determined according to the laws of each state. Several legal doctrines are followed, with eastern states tending to follow a riparian rights, reasonable beneficial use legal doctrine based upon English common law. In the western states, many of which have fewer water resources, a prior appropriation water law doctrine is followed.

For ASR projects in both eastern and western states, the ownership of the stored water is an important issue. In general, experience is supporting the position that, if the water user has the right to the water prior to ASR storage, he also has the right to recover that water. In other words, the state provides some protection of his right to store the water. However, in some states groundwater law is not adequately defined. In these states it is theoretically possible for another water user to construct a well adjacent to the ASR facility and pump out the stored water. Alternatives available to a water user to protect his rights to the stored ASR water include location of the ASR facility a sufficient distance from property lines that the risk is minimized; municipal zoning or land-use control in the vicinity of the ASR site; local municipal ordinance, or changes in the state law to provide for ASR storage. Fortunately, it is reasonably common for the radius of the storage bubble around the ASR well to be quite small, frequently facilitating judicious siting of ASR facilities to minimize or eliminate this problem.

As a part of the regulations, states may impose constraints upon recharge and recovery operations that go beyond water ownership. These may include consideration of impacts upon other water users who may be in existence at the time the ASR system comes on line. Recharge and recovery rates, or selection of storage zones, may be regulated so that unacceptable adverse effects upon existing legal users of the water resource do not occur, or are mitigated.

There is a need to integrate groundwater, surface water, and ASR permitting so that the situation is avoided in which a water user obtains the right to divert, treat, and store surface water underground, but is unable to recover the stored water at desired recovery rates due to groundwater permitting restrictions. It is usually desirable to recover all of the stored water. In some cases, such as in areas of depleted groundwater resources, it may be desirable to leave a small percentage of the stored water in the ground to slowly replenish the resource until such time as target water levels are reached.

Non-Degradation of Groundwater Quality

Several states have laws against degradation of groundwater quality by proposed injection or other land-use practices. The interpretation of these laws is still developing; however, with only one exception, it has not inhibited ASR storage of treated drinking water.

The one exception is in Oregon, where current state policy precludes recharge of any water with constituent concentrations higher than native groundwater quality. This policy was enacted many years before ASR was considered as an alternative for meeting increasing utility water demands. Currently regulators and utilities are working together to amend the policy in such a way that ASR can be considered, while also protecting the quality of the state's groundwater resources. The environmental and public health risk is perceived to be very small, and the potential economic savings are substantial.

Aside from this one exception, state regulatory agencies have noted that recharge water concentrations for a few selected constituents are higher than for the native groundwater, although both are well within potable standards. However, this has not been viewed as potential degradation of groundwater quality. For example, seasonal ASR storage of water with a typical chloride concentration of 30 mg/L in an aquifer with a chloride concentration of 5 mg/L is usually viewed as being consistent with state non-degradation legal requirements since the drinking water standard for chloride is 250 mg/L.

Interpretation of state non-degradation laws for ASR storage of water that does not meet all drinking water standards is still in the development stage. In California, current draft-state regulations require treatment to potable standards, plus reduction of total organic carbon (TOC) concentrations prior to reclaimed water injection into a *potable* aquifer. ASR recharge into a *brackish* aquifer, with surface water or reclaimed water for seasonal irrigation purposes, is under consideration but has yet to be resolved. In Oregon, ASR has been proposed for high quality surface water from the Columbia River to be stored in a fresh, basalt aquifer to control regional water level declines in a large farming area. As discussed above, non-degradation issues there have not yet been resolved.

During the next few years it is probable that experience with an increasing number and variety of ASR applications in different states will shed some light upon how individual states wish to enact and interpret non-degradation laws. The ultimate benefit achievable from ASR will depend upon each state's assessment of the tradeoff between recharge water quality, risks to public health and the environment, and ASR benefits.

Seasonal vs. Long-Term Storage

Initial ASR permits are sometimes issued providing only for seasonal storage and recovery, so that annual volume recovered cannot exceed the volume stored in the same year. However, greater benefit can be achieved by writing the permit to provide for carryover storage from one year to another. In this way, excess water available during wet years can be carried over to meet needs during drought years. Similarly, water stored during early years following a water treatment plant expansion can be carried over to meet increasing peak demands in later years when treatment plant capacity may be insufficient. If long-term storage is feasible and permitted, it can defer and downsize the next facilities expansion phase, achieving considerable savings. A preferred approach is to regulate according to long-term cumulative storage rather than seasonal or annual variability.

A related issue is the desirability of providing for seasonal allocation of recharge water. It is quite common for recharge water to be available according to an annual volume allocation, based upon safe sustained yield of the water source during the dry season. An ASR allocation process could be seasonal, making more efficient use of water available during wet months.

Recovery Percentage

In most states, it is desirable for technical, economic, and regulatory reasons to achieve full recovery of the same volume of water stored, whether recovery is seasonal or long-term. Where less than 100% recovery efficiency is planned, regulatory support appears probable in situations where water management benefits are reasonably shown to exceed the costs, economic and otherwise, associated with less-than-full recovery.

In some areas where ASR is being implemented due to declining groundwater levels, a requirement is being included in permit provisions that allows some small percentage of the water stored to be left in the aquifer. This requirement helps to restore groundwater levels; however, it also increases the unit cost of ASR.

Water Level Impacts

The impact of ASR operations on water levels in surrounding areas is an issue that is addressed during the permit process in some states. However, the modeling approach selected can sometimes lead to incorrect assessment of these impacts. In particular, the modeling approach must incorporate the increase in regional water levels that typically occurs during long-term recharge operations, as well as the decline occurring during recovery. Unless both of these operations are simulated and calibrated against actual observation well records, the resulting predictions of water level impacts may be erroneously high.

Location for Recovery of Stored Water

ASR wells are used for both injection and recovery. In addition to enabling the well to be periodically backflushed, thereby maintaining its injection capacity, this approach also facilitates hydraulic control of the storage bubble around the well, thereby minimizing mixing with surrounding native groundwater. Furthermore, this approach is quite cost-effective in that the same facilities are used for both recharge and recovery. Regulatory approval of ASR facilities and operations has been facilitated by these inherent advantages.

However, some situations may arise where it is desired to recharge at one location and recover at another location. One possible reason for this is the need for blending two different water qualities to achieve a relatively uniform product water quality. This can be achieved by designing the location of injection and recovery wells and well-head facilities with enough spacing and capacity as to achieve the desired blending ratio. A more common reason is to use the aquifer as a means of conveyance and also long-term storage. Water injected at point A is recovered from the same aquifer at point B, even if the distance between the two points is such that the travel time may be hundreds of years or more. The net volume of water in the aquifer, and associated water levels, are maintained through artificial recharge practices.

This is a sound water management approach; however, it is not ASR. Two fundamental issues challenge the widespread application of this approach. First, injection wells tend to plug unless they are periodically redeveloped. The redevelopment frequency and associated cost may be such that an ASR design for the injection well would provide better service since the pump in the injection well would help to maintain injection capacity. Second, water rights laws in some western states that follow the prior appropriation legal doctrine can inhibit the ability of a water user to operate recharge facilities in this manner. For example, such a practice is legal in California, Arizona, Texas, and possibly other states but is not legal in Colorado.

Permit Timing Relative to ASR Feasibility Investigations

The 13 states in which ASR systems are either operational or under development, have used a variety of permitting approaches. In some states, virtually all permit issues must be addressed before the ASR facilities can be constructed and tested to confirm feasibility. In other states, two stage permitting is required. The first stage addresses well construction issues and formulation of ASR testing plans, while the second stage addresses more substantive ASR feasibility and viability issues and is initiated when field investigations are completed. In other states, ASR permitting is addressed after initial ASR facilities have been constructed and tested and feasibility has been confirmed.

Experience with all of these approaches suggests that each can work. However, the most cost-effective approach that still provides adequate environmental protection is to defer consideration of substantive permitting issues until after the ASR facilities have been constructed and tested. In this way, the data used to support projections of water level impacts, storage bubble movement, water quality changes, and potential contamination sources, are based on actual field data collected at the ASR site under full-scale testing conditions, rather than on literature values or intermediate testing on wells at nearby sites.

If the ASR program is properly planned and implemented, the risk of failure or of completing a system that cannot eventually be permitted appears to be very small, based upon experience to date. However, this approach increases the importance of completing a Phase 1 ASR feasibility assessment, before initiating Phase 2 field investigations and cycle testing. The Phase 1 effort is essential to properly locating and designing ASR facilities and developing the associated testing program.

ASR Education

A common theme among the various ASR legal and regulatory issues and processes is a lack of understanding of ASR technology, applications, and experience. In some cases, this extends to significant misconceptions that must be overcome before permitting can be completed successfully. This is not surprising in view of the recent development of this technology. In 1983, only three ASR systems were operational, whereas in 1994, at the time of writing, 20 such systems were operational and about 40 additional systems were under development in 13 states. Successful completion of ASR permitting requires an initial effort to educate regulatory personnel regarding the technology and its operational experience at comparable sites. This is best accomplished when the Phase 1 ASR feasibility report is completed; this provides the opportunity to present report findings and recommendations to regulatory staff, including well-thought-out plans for field investigations to be completed under Phase 2.

6.4 ENVIRONMENTAL IMPACTS

A distinct advantage of ASR is the associated environmental benefit. To date, there have been no instances of environmental opposition to an ASR project storing potable water.

Opposition was encountered in Kerrville, TX, where a whitewater kayaking organization opposed issuance of the final operating permit for ASR operations to store drinking water since the water diverted from the river during peak flows would reduce the river's value as a recreational amenity. The permit was ultimately issued after extensive hearings. Opposition was also encountered in eastern Oregon, in response to a proposal to store Columbia River water in a potable basalt aquifer, for irrigation water supply purposes. Quality of the recharge water was excellent; however, the proposal is on hold pending resolution of non-degradation issues, as discussed above.

By increased reliance upon water sources during times of high flow and low demand, and upon storage of the water primarily in confined aquifers with no impacts upon surficial aquifer water levels, the environmental effects of ASR operations are positive in that they help to sustain human welfare without adversely impacting aquatic and terrestrial ecosystems. This benefit has been widely recognized by environmental interest groups.

As the ASR technology is increasingly considered for application to other water sources that may not be as high quality as drinking water, it is anticipated that environmental issues associated with ASR will lead to some debate over the relative risks and benefits. Where the line is ultimately drawn may tend to be different in each state. As a starting point, it is suggested that seasonal storage of high quality surface water or reclaimed water in brackish aquifers for irrigation purposes does not threaten public health or the environment, and should receive environmental support since it will conserve water supplies without creating the need for new surface reservoirs. Extension of ASR technology to seasonal storage of water from the same sources in freshwater aquifers is more likely to precipitate environmental debate, the resolution of which will tend to depend upon site-specific needs, opportunities and alternatives.

6.5 PUBLIC INVOLVEMENT

Decision-making relative to ASR planning, feasibility investigations, and implementation is increasingly subject to public involvement, as with all issues facing water management agencies. Public meetings and formal hearings are frequently a very important part of ASR programs. Adequate preparation for these events is essential to the ultimate success of the program.

While the general concept of aquifer storage and its associated economic and environmental benefits are readily explainable to a public group, important details usually require careful explanation to avoid conveying or supporting misconceptions. Key issues that invariably have to be addressed include ownership and control of the stored water, how far the storage bubble extends from the well, how far the water level effects extend from the well during recharge and recovery, how much of the stored water will be recovered, how much the program will cost, and the associated potential costs and savings for the ratepayers. Attendees at these meetings are frequently well informed and can pose sophisticated questions.

Where storage is planned in brackish or non-potable aquifers, it is also necessary to discuss mixing between stored and native water, recovery efficiency, and disposal of brackish water to the environment during drilling and testing. Where sources other than drinking water are under consideration, water quality impacts will have to be addressed. These issues require great care and excellent graphics in order to avoid misunderstanding.

Public support for ASR programs has generally been strong. Opposition has generally been due to support for other water supply alternatives, such as surface reservoirs and major transmission pipelines, which may have considerable, long-standing momentum in the community. Sensitivity to these underlying issues can frequently provide guidance as to how to handle them.

It is quite common for serious consideration of ASR as a water management alternative to begin after years of effort to construct a surface reservoir or long transmission pipeline against increasingly strong environmental and other opposition. Polarization of supporters and opponents to the surface reservoir or pipeline project is frequently far advanced by the time that ASR enters the discussion. It is usually unwise to tackle this issue directly since surface reservoir or pipeline momentum is frequently strong and supported by the same individuals who would ultimately need to support a change in direction toward ASR. ASR is therefore in the position of presenting a viable and cost-effective alternative to their preferred long-standing project, at a time when this may be unwelcome for a variety of reasons.

In such situations, the only viable recourse is usually to emphasize the need for site-specific field investigations to confirm feasibility and economics before any serious consideration of ASR to meet local needs. This may enable the ASR program to move forward quietly, without reducing momentum for the alternative project. In some cases, it is also appropriate to point out the merits of combining both projects, assuming that the surface reservoir or pipeline will definitely be built. In many cases an excellent case can be made that the combination of ASR and surface storage can work together to achieve benefits not attainable with either system by itself.

This scenario is common to perhaps one-third of all ASR projects to date. That the ASR technology has moved forward as rapidly as it has during the past 10 years primarily indicates its cost-effectiveness relative to other alternatives, some of which continue to have strong support. Sensitivity to these important public involvement issues, and careful preparation for public and other ASR presentations, is most likely to achieve ultimate ASR success.

Alternative ASR Applications

"From water, God made every living thing."

Surah 21, Al Anbiya, The Koran

7.1 INTRODUCTION

With the growing success and acceptance of ASR for storage of drinking water supplies, attention has begun to focus on the potential application of this technology for storage of water from other sources, such as untreated or partially treated surface water from rivers, canals, lakes, and reservoirs; untreated groundwater; and reclaimed water from wastewater treatment plants. Common to each of these other sources is that they are generally of a quality that falls short of meeting drinking standards. This presents a challenge, not only in technical terms but also for regulatory reasons.

Due to the immense potential benefits associated with ASR storage of water from these sources, it is anticipated that considerable effort will be devoted in the future to the resolution of technical constraints and reevaluation of the existing regulatory framework governing such applications. Where such waters can be safely stored underground without significant risk to the environment, water quality or public health, these benefits can be realized so long as the associated projects can be permitted with reasonable effort and cost. On the other hand, there is no justification for underground storage of waters of such low quality that wells may clog or that aquifer contamination may occur. Much work remains to be done; however, progress is underway in several states. In the remainder of this chapter, the current status of ASR applications for non-potable water sources is discussed.

7.2 SURFACE WATER STORAGE

As discussed in Section 4.2, Well Plugging and Redevelopment, and elsewhere in this book, ASR wells tend to plug with particulate matter unless the recharge water quality is excellent or the storage zone is highly transmissive. It is no coincidence that ASR success to date has been associated with storage of water meeting drinking water quality standards. Consequently, it is reasonable to assume that well recharge of surface water containing particulate matter is more likely to rapidly plug a well, requiring frequent backflushing and redevelopment.

Technical Considerations

Four technical approaches may be considered to achieve successful ASR recharge with untreated or partially treated surface water:

- Select a storage zone with acceptably high transmissivity.
- Select a water source of high quality, approaching drinking water standards where possible, and containing low suspended solids (silt) content.
- Pretreat the water as necessary to reduce suspended solids concentrations, whether by wellhead filtration or by movement through surficial sandy soils.
- Implement a backflushing program to remove solids from the well with a frequency that maintains an acceptable recharge capacity.

These approaches are in the approximate order in which they should be implemented, each step being added if the previous steps prove inadequate or not feasible. For most surface water sources, all four approaches will probably be required.

Experience with well recharge of surface water is not widespread. Approximately 400 drainage wells in the vicinity of Orlando, FL, have operated for several decades, draining surface runoff into highly transmissive karst limestone aquifers with typical transmissivities exceeding 1 MG/day/ft^2 (12,600 m^2/day). The wells typically require cleaning every few years. The Lake Okeechobee, FL, ASR test well was similarly constructed into a limestone aquifer with a transmissivity estimated at 4.6 MG/day/ft^2 (58,000 m^2/day). Testing with recharge water from a regional drainage canal indicated no plugging problems at recharge rates of about 5 MG/day (19 megaliters/day) for periods of up to 60 days. Total suspended solids concentrations from this source ranged from less than 1.0 to 11.5 mg/L during the test period, but were generally under 5 mg/L. There are probably several other recharge well locations with karst aquifers and high transmissivities that have operated successfully in the U.S.

On the other end of the hydrogeologic spectrum, the Seattle, WA, ASR system receives water for recharge from a reservoir on the Cedar River. The water from the reservoir is of very high quality, requiring only disinfection to meet drinking water standards. ASR operations at this site have proven successful, although during recharge periods some well plugging has been noted, which is reversed by recovery and backflushing. Analysis of the solids in the backflush water has shown that it contains diatoms, or single cell algae present in the recharge water. The transmissivity of the aquifer in this area ranges from about 2175 m^2/day (175,000 G/day/ft) at the Boulevard Park site to 4350 m^2/day (350,000 G/day/ft^2) at the Riverton Heights site. Total suspended solids concentrations in the recharge water during the test program ranged from about 0.5 to 12 mg/L, averaging 4 mg/L. At the Riverton Heights site, backflushing the well every 2 weeks is sufficient to maintain specific capacity. At Boulevard Park, more frequent backflushing is required to prevent residual plugging.

At Ft. Myers, FL, an injection well test program utilized untreated water from the Caloosahatchee River as a recharge source. The well rapidly plugged, reflecting the low transmissivity of the aquifer (5600 G/day/ft^2, 70 m^2/day) and also substantial solids in the recharge water. Another recharge well program in St. Lucie County, FL, experienced similar plugging problems when recharging an aquifer with a transmissivity of about 570 m^2/day (46,000 G/day/ft^2). The water source was a shallow well adjacent to a drainage canal. Iron bacteria were suspected of contributing to the plugging. These examples suggest that recharge of surface water into low transmissivity aquifers may be infeasible without substantial wellhead treatment.

There seem to be few reference points in the U.S. to provide the opportunity for interpolation between the two ends of the transmissivity spectrum for recharge of most surface waters. A few long-established projects are operational in very high transmissivity aquifers. These were all "grandfathered" under existing Underground Injection Control legislation, and were therefore operational prior to about 1981. Probably many other projects prior to 1981 were tried and failed due to clogging in low transmissivity aquifers. The combination of surface water recharge and periodic backflushing and well redevelopment in intermediate transmissivity aquifers is not currently practiced in the U.S. This is partly due to technical constraints but primarily due to the regulatory framework, as discussed in Section 6.3, Legal and Regulatory Issues.

For karst limestones, transmissivity may not be a very good criterion since the flow usually follows one or more major fractures or solution channels. If these features are absent, transmissivity is substantially reduced and clogging may occur. For unconsolidated aquifers, hydraulic conductivity may be more important than transmissivity as an indicator of clogging characteristics. Experience with recharge of treated drinking water into unconsolidated aquifers, as discussed in Section 4.2, Well Plugging and Redevelopment, suggests that recharge of untreated surface water with higher suspended solids loadings will probably be infeasible except for waters of excellent quality. Some pretreatment such as wellhead filtration will probably be required in such cases. Wellhead filtration may be achieved with manufactured equipment or with surface recharge systems, possibly drained by shallow wells that could supply deeper ASR systems with filtered water.

In the absence of pertinent experience from other sites, caution is appropriate in considering the suitability of a proposed aquifer for storage of surface waters. A test program is required to gather the necessary operating experience so that such questions can be answered.

Until such time as this data becomes available, it seems appropriate to consider a transmissivity of about 0.25 MG/day/ft^2 (3,100 m^2/day) as a preliminary guide, above which well recharge with high quality surface water may be feasible and below which the chances of success diminish. Obviously, the quality of the recharge water source is critical. This guideline assumes that recharge water approaches potable quality, falling short on such parameters as coliforms, color, turbidity, corrosivity, and possibly sodium, chloride, and total dissolved solids. To the extent that solids such as silt and algae are present in the recharge water, clogging potential will increase.

It is important to emphasize that little operating experience with untreated surface water ASR systems is available to guide planning and decision-making. Questions regarding ASR feasibility in intermediate transmissivity areas will need to be determined through field investigations until such time as greater experience is available upon which to base judgments of preliminary feasibility.

Remedial measures may be implemented to reduce solids in the recharge water. Some of these have been discussed in Section 4.3, Wellhead Filtration. Other alternatives under consideration at potential surface water ASR sites include use of horizontal wells, shallow wells or Ranney wells adjacent to surface water bodies. In this way, surface water receives filtration as it moves from the source, through the shallow sediments to the supply well, and then to the ASR well for recharge.

New horizontal well technology enables laying a flexible, horizontal, high density polyethylene slotted pipe covered with a fabric sleeve in a backfilled trench at depths of up to about 25 ft. This is connected to a PVC riser pipe,in which a pump is placed. It is anticipated that this technology will evolve in the near future to permit installation of horizontal wells at greater depths. These wells have been in operation for up to 5 years without failure, relying upon the sand backfill and the fabric sleeve to reduce solids in the pumped water to small particle sizes. Selection of finer fabrics can reduce the particle size accordingly. Horizontal wells are constructed with a trenching machine modified to continuously lay the pipe, sleeve and backfill material. A distinct advantage of these wells is the diffuse impact upon the water table, in contrast with a network of shallow vertical wells producing at the same combined rate. In environmentally sensitive areas, this can be particularly important in order to avoid adverse water table impacts. This technology is available through Horizontal Wells, a division of HDSI, PO Box 150820, Cape Coral, FL 33915–0820 (Phone: 813–995–8777).

Regulatory Considerations

As discussed in Section 6.3, Legal and Regulatory Issues, the most critical issues in the U.S. pertaining to ASR storage of untreated surface water are probably in the regulatory arena. Whether or not the technical issues can be resolved economically at any particular site, the regulatory issues can be complex, expensive, time-consuming, and of uncertain outcome. Until the regulatory framework can be developed to facilitate such practices, ASR storage of untreated surface waters in fresh or brackish aquifers will tend to evolve slowly.

The present regulatory framework protects the nation's aquifers against contamination, but does it in such a way that benign ASR practices never envisioned in 1974 when the enabling legislation was passed by Congress, are difficult to implement. This is in spite of the fact that the potential adverse impact of such practices upon the environment, groundwater quality and public health is recognized to be minimal and their overall potential benefits are significant. Each of the states that have accepted responsibility for EPA Underground Injection Control (UIC) program implementation interprets the UIC regulations to meet its own needs and priorities, even though EPA retains the right to reject or approve state decisions under this regulation. The reader is referred to Section 6.3, Legal and Regulatory Issues, for a more complete discussion of this important issue.

Within the U.S., California and Florida are currently leading efforts to store non-potable waters in brackish aquifers. Storage of water from such sources in fresh aquifers has been proposed at one or more sites but is not under active consideration, except following treatment to potable standards.

Economics

The marginal value of water from untreated surface sources, which is stored in an ASR well, should be lower than for treated drinking water sources. This is because the associated costs include only the pumping costs to divert and recharge the water, plus the operating costs associated with periodic backflushing. If some filtration or other wellhead pre- or post-treatment is added, marginal costs will increase accordingly. In western states with prior appropriation water rights, the marginal cost would also include the cost of the water rights for those flows and months in which recharge may occur.

Where this water is stored in a highly transmissive fresh aquifer, the same volume stored may be recovered. However, where it is stored in a highly transmissive brackish aquifer, the high transmissivity may reduce recovery efficiency below 100%. The value of the water stored but not recovered should be evaluated to confirm that it does not exceed the value of the water recovered when needed. In some cases the water not recovered may also serve a useful purpose.

Efforts are underway in southeast Florida to store good quality surface water in a highly transmissive brackish aquifer containing water with total dissolved solids that may reach 7000 mg/L at some sites. Recovery efficiency under long-term operating conditions has yet to be confirmed, although early indications suggest that it may be in a probable range of 40 to 70%. The water stored but not recovered will serve as a source of recharge to the aquifer, which is increasingly being used for desalination water supplies in coastal areas. As overall water demands for agriculture, industry, and public water supplies increase, water that would otherwise leave the system rapidly as storm runoff can be stored in the brackish aquifer for seasonal withdrawals to meet peak demands. Water not recovered will recharge and also freshen the aquifer, improving its value for desalination supplies.

Agricultural Applications of Surface Water ASR

A potentially important application for ASR storage of untreated or partially treated surface water is for agriculture. This sector of the economy has much to gain from resolution of the technical and regulatory uncertainties associated with surface water ASR.

A typical example is for the potato farming area around Echo Junction in eastern Oregon. Groundwater supplies from a basalt aquifer have sustained agricultural operations for many years; however, groundwater levels have continued to decline and will soon force a curtailment of irrigation activities. Since there is little rainfall in this area, the economic and social impact of reduced irrigation would be substantial. Nearby is the Columbia River, a major potential source of irrigation water with excellent quality approaching and frequently meeting drinking standards. A plan has been proposed to irrigate with water from the river, seasonally storing excess water in the basalt aquifer using ASR wells to restore groundwater levels and also to augment seasonal irrigation supplies. Certain technical and economic issues remain to be resolved through testing. However, implementation of the testing plan is constrained by concerns regarding aquifer water quality impacts and the state's anti-degradation policy.

A second example is for the Indian River citrus area in southeast Florida. Climatological and soil conditions in this area have helped citrus producers to maintain high productivity and relatively high resistance to the effects of occasional freezes, such that Indian River citrus has a well-earned reputation worldwide for premium fruit. Water management is through network of canals that provide drainage and also a source of irrigation water. During about two months in the spring of each year, rainfall is frequently inadequate to maintain irrigation requirements. At such times several hundred irrigation wells are utilized to supplement water supplies. These wells generally discharge by artesian flow into the drainage network, from which the water is pumped for irrigation. Water quality from the wells is deteriorating, such that within 10 years about half of the wells will no longer be useful due to excessive chloride concentrations. A plan has been proposed to utilize ASR technology to store seasonally available water supplies from the drainage network. The water would be stored in the brackish aquifer underlying the area, and would be recovered when needed to meet irrigation requirements. Testing required to confirm the viability and the cost-effectiveness of this plan has yet to be initiated. Meanwhile work continues on an alternative plan to build a surface storage reservoir, which would be several times as expensive as ASR storage and will require taking several hundred acres of citrus out of production permanently.

7.3 UNTREATED GROUNDWATER STORAGE

Untreated groundwater is frequently of excellent quality, meeting potable standards except for a few parameters such as iron, color, fluoride, and hydrogen sulfide. Under pending standards, several existing groundwater supplies, particularly in western states, will fall short of meeting drinking water quality standards for arsenic and radon. In some areas, groundwater is treated to remove hardness. Other parameters may be of concern in selected areas of the country.

As with surface water sources, a distinct advantage of untreated groundwater supplies for ASR storage is the relatively low marginal cost of the water. This usually includes only electricity costs for pumping the water to the ASR site.

With groundwater, the source of supply is usually relatively steady throughout the year, drawing upon the immense storage capacity in the source aquifer. Where a variation in water demand occurs during the year, the wellfield may be pumped to capacity during off-peak months, particularly when this corresponds to extended periods of heavy rainfall. The water not required for treatment to meet system demands can then be stored, if a suitable storage zone is available. The storage zone could be in the same aquifer but at a different location, or it could be in a different aquifer, usually deeper. Potentially, it could be stored at the same site as the supply well, but in a shallower or deeper aquifer. The supply well is presumably fresh; however, the ASR storage zone can be fresh or brackish.

This approach is quite cost-effective as a short-term means of augmenting raw water supplies in areas where availability of wellfield sites is limited, such as by urban development or contamination. However, it may not be the most long-term, cost-effective ASR application for public water systems.

When recovered, the stored groundwater requires treatment to meet potable standards. Water treatment plant capacity then has to be sized to meet peak demands, treating water from the conventional production wells and also from the ASR wells. Where recovery efficiency is acceptable, it is probably more cost-effective to store and recover treated drinking water rather than untreated groundwater. This enables deferral or elimination of plant expansion requirements and downsizing of treatment facilities capacity, with associated cost savings. A site-specific feasibility investigation will indicate the most cost-effective plan for individual sites. This is addressed in Section 2.1, Phase 1, Feasibility Assessment.

No ASR systems currently store untreated groundwater in the U.S. However, such a system is being developed for Dade County in southeast

Florida to augment limited available wellfield capacity. A portion of the water produced from the county's wells during the wet season will be stored in a deeper brackish aquifer, for recovery to help meet dry season water demands.

Horizontal wells, discussed in Section 7.2, Surface Water, are considered to produce partially treated groundwater even though they may be constructed parallel and adjacent to a canal. Consequently, water produced from horizontal wells and recharged into an ASR well would fall into this category. To date, no such systems exist, although this should change in the near future.

7.4 RECLAIMED WATER STORAGE

As urban, agricultural, and industrial water demands continue to climb to sustain the needs of a growing population, the most reliable supply of water available to meet these needs is frequently reclaimed water. In contrast to many other sources, it is available throughout the year at a relatively steady rate. When treated to meet ever-tightening standards required for discharge into the environment, the quality of this water is steadily improving. In a few locations, nutrient and other surface discharge requirements are so restrictive that water quality approaches or exceeds drinking standards. For a growing number of water users, the reliability of this source, its high quality, rising competition for limited available water supplies and regulatory pressures to conserve water provide four good reasons to incorporate reclaimed water into long-term water supply plans.

Reclaimed water is generally utilized for irrigation purposes, including parks, golf courses, playing fields, commercial and residential property, and landscaping. It is also used for industrial cooling, agricultural irrigation, aquifer recharge, and maintenance of injection well salinity intrusion barriers. Its increasing acceptance by water users reflects many years of effort by the water and wastewater industry, regulators, and others to overcome aesthetic concerns associated with this water source by conducting comprehensive health and epidemiological studies and also by demonstrating the successful and cost-effective application of reuse technology.

By substituting reclaimed water, other water sources can be utilized more efficiently to meet higher value needs. Increasingly, the allocation of limited available water supplies to meet environmental priorities is causing more serious consideration of efficient use of water from other sources.

A principal constraint upon more widespread utilization of reclaimed water is the associated cost for pipelines, pumping stations, individual connections and other facilities to convey the water from the wastewater treatment plant to where it is needed, and to store the water at times when it is not needed. Storage requirements depend upon the variability in supply and demand for reclaimed water during the year, and also upon the viability of discharging the water to waste during periods of heavy rainfall or other times when it is not needed.

At some locations, ASR storage of reclaimed water in brackish aquifers may be viable and cost-effective, storing when excess reclaimed water is available and recovering when irrigation or other demands exceed available supplies. By providing inexpensive storage of reclaimed water, ASR may help to expedite funding and implementation of expensive regional plans to make better use of reclaimed water for beneficial purposes.

Reclaimed Water Injection Experience

No reclaimed water ASR systems are currently operational in the U.S. However, three systems inject reclaimed water into potable aquifers. These are at Water Factory 21, operated by the Orange County Water District in southern California; El Paso, TX; and Gainesville, FL. Orange County Water District has operated Water Factory 21 during the past 15 years to help maintain the Talbert Gap salinity intrusion barrier. Since 1986, El Paso has operated the Fred Hervey Water Reclamation Facility to recharge the Hueco Bolson aquifer, which supplies a portion of the potable water supply for the city; and for about 20 years Gainesville has injected highly treated wastewater into an aquifer in an area drained through sinkholes, with no surface outflow and few viable land application options.

Several other water systems are developing plans for well injection of reclaimed water in brackish aquifers, unlike the fresh aquifer examples mentioned above. These are located in California, Florida, and possibly other states. Examples include Orange County Water District, which is currently evaluating use of reclaimed water as a portion of the total injected flow at the Alamitos Gap salinity intrusion barrier. A project is under development at Monterey, CA, to reduce demand on potable groundwater supplies by seasonally storing reclaimed water for agricultural use in brackish portions of the aquifer system in the Salinas Valley area. Reclaimed water produced during winter months would be stored. During the peak irrigation season the reclaimed water would then be recovered. A similar project is being evaluated at Eastern Municipal Water District in southern California. West Basin Municipal Water District in southern California is evaluating the use of reclaimed water instead of potable water for their seawater intrusion barrier. Hilton Head, SC, is evaluating seasonal storage of reclaimed water in a deep brackish aquifer to meet peak irrigation demands and thereby reduce potable water demands upon their stressed shallow potable aquifer. Manatee County, Charlotte County and the City of St. Petersburg, all located in southwest Florida, are planning seasonal storage of reclaimed water in a brackish aquifer for irrigation purposes.

Assuming that the reclaimed water is of high quality, the principal challenge relating to ASR storage of this water is regulatory. Some of the regulatory aspects are discussed for the EPA Underground Injection Control program in Section 6.3, Legal and Regulatory Issues. Seasonal storage of reclaimed water in a brackish aquifer is not likely to be an environmental or water quality threat at most sites. However, the regulatory framework governing potential implementation is not well suited to ASR opportunities.

California Draft Regulations

California has led the way in the development of draft state regulations (July 1992 draft) governing the use of reclaimed water for recharge of potable aquifers, whether through wells or surface recharge facilities. These draft regulations have been developed over a period of several years and reflect considerable effort by many experts. As a result, they provide useful experience to guide those interested in developing comparable regulations for other areas.

A principal component of the California draft regulations is to ensure that water produced at nearby domestic production wells in the same aquifer does not exceed specified percentages of reclaimed water. Required blending ratios at the nearby production well depend upon the associated level of treatment for the reclaimed water. Under the draft regulations, all recharge waters would have to undergo biological oxidation and disinfection, with well injection also requiring filtration and organics removal. Surface recharge would require filtration when the water of reclaimed origin may exceed 20% of the total flow from any domestic water supply well in the vicinity, and would require filtration plus organics removal when this percentage exceeds 50%.

Oxidized wastewater must not exceed 20 mg/L total organic carbon (TOC), 30 mg/L suspended solids and 30 mg/L biochemical oxygen demand. Filtered wastewater is defined as not exceeding an average of 2 nephelometric turbidity units (NTU), and not exceeding 5 NTU more than 5% of the time during any 24-hour period. Disinfection requirements depend upon the level of prior treatment. Organics removal depends upon the method of recharge and also upon the reclaimed water contribution as a percent of total flow from any domestic water supply well in the vicinity.

For surface recharge, draft TOC limits vary from 20 to 6 mg/L, while for well injection they vary from 5 to 2 mg/L. A minimum spacing of 2000 ft between the injection well and the nearest domestic supply well is specified. Travel time between the two wells has to exceed 12 months.

The TOC requirements for reclaimed water that is injected directly into an aquifer are designed to prevent more than 1 mg/L of TOC from reaching domestic supply wells. If the TOC in the reclaimed water is 5 mg/L, 20% of the water pumped by a nearby domestic supply well can be reclaimed water. If the TOC of the injected water is 2 mg/L, then the nearest domestic supply well can get up to 50% of its water from the reclaimed water source. The regulations assume no TOC reduction in the aquifer.

The cost of reducing TOC to meet well injection criteria can be substantial, reflecting treatment by such organic removal processes as granular activated carbon adsorption and membranes. The difference between organic removal requirements for surface recharge and well injection is primarily based upon the demonstrated efficiency of aerobic biological processes occurring in the vadose zone to remove such organics, and the presumed absence of such processes in the saturated zone below the water table.

As discussed previously in Section 4.6, Disinfection Byproduct Reduction, organics removal does occur in the saturated zone based upon recent investigations. Disinfection byproducts are only a small component of the entire range of organic compounds in the aquatic environment. However, their significant reduction during aquifer storage under confined, saturated conditions suggests the possibility that other organics may also be removed in a similar manner, probably due to biological activity in the aquifer and possibly to adsorption processes. It is too soon to draw conclusions regarding the efficiency of a saturated aquifer system to reduce concentrations of a broad range of organics, however, the suggested direction for further research is clear. Similarly, the fate of organic compounds around an ASR well needs to be established. Breakdown into innocuous compounds such as carbon dioxide, nitrogen, and water would undoubtedly be satisfactory; however, breakdown into other organic compounds may be less desirable. Further understanding of organic removal processes and the fate of these compounds in a saturated environment will provide a basis for the design and operation of facilities to optimize organics removal during aquifer storage. It will also provide an improved basis for assessment of the risks and benefits associated with seasonal storage of reclaimed water in brackish aquifers.

Recent studies in St. Petersburg, FL, where reclaimed water is injected into a 213 to 338 m (700 to 1110 ft) deep saline injection zone, revealed that the hydrogeologic system within this and the overlying brackish aquifers is very reactive from a geochemical and biochemical viewpoint [1]. Results were similar to those for several case studies at other injection sites in the U.S., as determined from a literature search. In particular, metals, nutrients, and trace organics are removed by a variety of processes in these confined aquifers, including precipitation, ion exchange and bacterial activity. Generation of methane and nitrogen gas is a common metabolic byproduct of these processes. Particularly active areas in injection zones occur within the immediate vicinity of injection well boreholes and within the injection front of the fluids that were injected.

Suggested Regulatory Strategy

The cost of treating wastewater to potable standards for agricultural purposes is generally prohibitive. Consequently, it is appropriate to consider whether an alternative regulatory approach would achieve reclaimed water storage needs without risking adverse water quality, environmental, or public health effects.

With ASR technology, the opportunity exists to seasonally store reclaimed water at suitable locations such as the point of proposed irrigation use, and to minimize the degree of mixing with surrounding native waters. This would be an appropriate strategy if the goal were to separate waters of wastewater origin from those of non-wastewater origin, preserving the latter for higher quality uses such as domestic water supply. This approach may be contrasted with the current approach in the California draft regulations that are based upon the implicit assumption that widespread dispersion of waters of wastewater origin into the aquifer is unavoidable. The two strategies are both valid and feasible, but quite different.

The current California strategy is most appropriate for regulation of continuous recharge of potable aquifers with reclaimed water, which was its intended purpose. However, extension of this strategy to regulate seasonal recharge of, and recovery from, brackish aquifers with reclaimed water may not be the best approach. If efficient organics reduction in a saturated portion of an aquifer can be demonstrated, whether the aquifer is confined or unconfined, then it may be possible to use ASR technology to seasonally store and recover high quality reclaimed water in a small radius around a well in a brackish aquifer, relying upon organics reduction processes in the aquifer to provide further treatment of whatever small portion of the stored water that may not be recovered seasonally.

The key element of a suggested ASR regulatory strategy is the need for Congressional action directing the Environmental Protection Agency to modify current Underground Injection Control regulations to provide for beneficial ASR applications such as storage of high quality waters in brackish aquifers. Possible approaches include:

1. Establishing a Class VI category for ASR wells, with associated regulations.
2. Establishing a Class VI category for ASR wells and delegating their regulation to the individual states by rule.
3. Delegating the Aquifer Exemption process for Class V ASR wells to the individual states.
4. Developing the concept of a mixing zone surrounding ASR wells and wellfields, to accommodate ASR operations instead of regulating water quality at the wellhead during recharge and recovery.

Other approaches may also be possible. To achieve such Congressional action will require coordinated effort by major water users who recognize the potential benefits of the proposed changes in the ASR regulatory framework.

Some additional elements of a suggested revised ASR strategy for seasonal storage of reclaimed water in brackish aquifers are as follows:

1. Aquifers with TDS concentrations exceeding 1000 mg/L may be appropriate for reclaimed water (oxidized, filtered, chlorinated) storage without prior organics removal for most typical reclaimed water sources. A balancing of risks and benefits would need to be assessed at each site to ensure adequate protection of water quality, public health, and the environment.
2. Wastewater sources would be screened to identify and divert potential industrial and commercial contaminants or toxic wastes, and thereby minimize the likelihood that complex organic compounds may be present in the reclaimed water stream.
3. Reclaimed water quality would be characterized as to primary and secondary drinking water standards. Those parameters with concentrations exceeding potable standards would be identified and their geochemical fate and relative environmental risks during seasonal aquifer storage would be evaluated. For example, coliforms are generally eliminated after a few days of storage in brackish aquifers; sodium, chloride, and TDS concentrations may exceed potable standards without contaminating a brackish aquifer; turbidity, odor, and color in the reclaimed water may not be significant if the water is seasonally recovered and used for irrigation purposes.
4. Sufficient redundance would be built into the treatment processes to enhance treatment reliability so that the risk of failure to meet reclaimed water quality standards is acceptably low.
5. ASR wells would be designed and operated to recover the water stored each season. It will generally be desirable to maintain as little stored water in the aquifer as possible at the end of the recovery season. Carry-over storage

from wet years to dry years would be allowable only to the extent that monitoring shows that migration of the storage bubble away from the well is not significant under the natural hydraulic gradient prevailing at the site.

6. Treatment of reclaimed water prior to injection would be as needed to meet quality standards and minimize plugging of the well due to particulates or other constituents. Treatment would generally need to include oxidation, filtration, and disinfection. At most sites, aquifer hydraulic characteristics will be such that high quality reclaimed water will be necessary in order to minimize plugging and ensure satisfactory long-term well operation.

7. Concentrations of organic compounds in the reclaimed water would be reduced during seasonal aquifer storage. Since native water in the aquifer exceeds potable TDS standards, any recovery of residual stored water blended with native aquifer water in a nearby domestic supply well would probably require membrane treatment to meet potable standards. The time period for movement between the injection well and the nearest domestic supply well, combined with the lateral distance between these wells, would constitute effective barriers to protect public health. In addition, membrane treatment of the brackish water recovered from the domestic supply well would provide a third organics removal barrier to enhance public health protection. The regulatory process would define the extent of these barriers by identifying lateral distances between the ASR well and other domestic wells potentially using brackish water from the same zone for drinking purposes.

8. Nutrients in the reclaimed water may benefit irrigation water users; however, during ASR storage they would probably promote bacterial activity in and around the well, potentially causing well clogging. Whether this can be effectively controlled by maintaining a chlorine residual in the well, as with potable water ASR, remains to be confirmed. Until such time as this question is resolved, a suggested approach is to maintain a chlorine residual in the reclaimed water being recharged, while that water is moving through the well and gravel pack. During storage periods, trickle flow a small chlorinated supply of water into the well at a rate sufficient to maintain this residual in the well. The required flow rate and residual concentration can be estimated by periodically pumping samples from the well following a typical cessation of recharge, to determine the decay rate of the chlorine residual in the well. The intent is to preclude bacterial activity in the immediate vicinity of the well, where it would have the greatest potential for clogging.

9. As discussed in Section 4.6, Disinfection Byproduct Reduction, initial formation and subsequent significant reduction of trihalomethanes and haloacetic acids should occur in the aquifer if the water is stored for a period of several weeks, particularly at sites where anaerobic conditions develop during storage and where a carbon source is available to support bacterial activity.

An appropriate mechanism for further investigation of these issues would be to implement a few such projects at selected sites with brackish water storage zones, monitoring organics concentrations at the ASR well and also at surrounding monitor wells in the reclaimed water storage zone. If results from these sites are positive, the applicable regulations may be amended to specifically provide for this practice at other sites where appropriate criteria are met.

The benefit to be gained from consideration and implementation of this suggested alternative regulatory strategy may be substantial. Many water users experience declining aquifer levels, saltwater intrusion, and unreliable surface water supplies. They would benefit immensely from recharge of brackish aquifers, but are unable to accomplish this with surface recharge facilities due to unsuitable hydrogeology or limited land availability. ASR would be more cost-effective if organics removal treatment requirements were eased. However, reclaimed water ASR is probably not cost-effective for irrigation and other purposes if these requirements remain. Possibly, the rationale presented above can help to achieve these benefits while at the same time protecting the quality of potable water supplies.

Future Directions

"Would you tell me, please, which way I ought to walk from here?"
"That depends a good deal on where you want to get to," said the Cat.
"I don't much care where—" said Alice.
"Then it doesn't matter which way you walk," said the Cat.

Alice's Adventures in Wonderland, Lewis Carroll

It is interesting to look back and see the progress that has been made during the past few years with the development and implementation of ASR. From a concept not many years ago it has evolved into a proven, cost effective water management tool today. The ability to use wells to store and recover a vast amount of water underground at low cost is a significant advancement in water management. A "sea change" has occurred in how we manage water. The change is already apparent to some, and will become apparent to others in future years as the locations and applications of ASR technology become more diverse. It is pertinent, therefore, to consider what lies ahead. What new developments may affect the locations and methods of ASR implementation in the future? What should be the directions for future research in order to expand the benefits associated with ASR? These and other issues are discussed in Chapter 8.

8.1 TECHNICAL DEVELOPMENTS

It seems reasonable to expect that the next few years will bring improvement in understanding of water quality changes occurring underground during ASR storage. It is becoming increasingly clear that bacterial, geochemical, physical, and other processes occurring in both saturated and unsaturated formations are effective at improving water quality. Better understanding of these mechanisms can enable improved design and operation of ASR systems to achieve specific water quality goals. Until better understanding is achieved, however, it is appropriate to take a conservative position with regard to planning and regulation of ASR systems to store water not meeting potable standards. Once contaminated, aquifers may require a long time to restore water quality.

As an example of water quality improvement, organics such as disinfection byproducts (DBPs) have been shown to decline during ASR storage as shown in Section 4.6, Disinfection Byproduct Reduction. The mechanisms governing this have not been confirmed, although bacterial activity is believed to be the driving force. Dilution and adsorption also may be important mechanisms at some sites. Ongoing laboratory investigations are directed at improving the level of understanding of the primary mechanisms, thereby providing a stronger basis for design and operation of ASR systems to maximize DBP reduction where this is a primary goal. Consideration should be given to the following future endeavors to advance understanding of ASR organics reduction mechanisms and applications:

- Replicate DBP testing described in Chapter 4 at additional sites, with minor adjustments to the testing program to focus on key issues that became evident during the initial testing at five sites.
- Obtain wireline cores at one or more new ASR well sites, with emphasis t_r on preserving the integrity of the core material for subsequent bacterial testing. Simultaneously, run an ASR DBP test cycle at the well and parallel laboratory batch and/or column tests on core samples in order to confirm mechanisms responsible for DBP reduction.
- At locations likely to be impacted by the new EPA DBP regulations, consider how to operate an ASR system in order to achieve water quality standards in water going to the distribution system during recovery periods. At some locations, a combination of treatment and ASR alternatives may be worthwhile.
- Where possible, consider recharge with a free chlorine residual instead of a combined residual. This may reduce clogging potential and may also reduce the time required for DBP reduction to any target concentration.

- At new ASR sites storing water from non-potable sources, and in laboratory column and/or batch testing investigations, gather data regarding reduction under saturated conditions of other organics compounds besides DBPs.

Arsenic reduction during ASR storage is also expected to receive attention in the near future, in response to new EPA regulations governing arsenic concentrations in drinking water. At one ASR site work is beginning to investigate the potential for operation of the well in such a way as to enhance arsenic reduction during aquifer storage. If this is successful in achieving the low concentrations proposed by EPA, the benefit to the water industry will be substantial.

pH adjustment to control iron and manganese concentrations in recovered water should receive continuing attention as it is applied at new sites. Manganese reactions tend to be slow in reaching equilibrium, so the long-term success of this approach needs to be confirmed over a period of several years and several operating cycles.

Nutrient reduction during ASR storage has received limited attention at a few sites. With the anticipated extension of ASR technology to the storage of reclaimed and agricultural waters, greater emphasis upon underground reduction of the various phosphorus and nitrogen forms would be valuable. Ammonia and phosphorus reduction and denitrification have been shown to occur; however, the data are limited.

Mechanical design of ASR wellhead facilities has evolved during the past few years. While this is not "high" technology, it is also not "low" technology. Several key features are advisable to achieve testing and operational objectives, as discussed in Chapter 3, Design of ASR Systems. It is anticipated that the downhole control valve will be commonly applied where the depth to water level is substantial or where available recharge flow rates are highly variable.

Better understanding of well clogging rates and backflushing frequencies will be achieved as the number of ASR systems increases. This will improve the opportunity for enhancement of the predictive model developed in Section 4.2, Well Clogging and Redevelopment, which will then provide better estimates of ASR system performance to aid in design and operational planning.

Similarly, better understanding of wellhead filtration techniques, economics and performance will provide an improved basis for ASR pretreatment facilities design in order to minimize the entry of particulates and other suspended solids into the wells. A number of alternatives are available, as discussed in Section 4.3, Wellhead Filtration. By keeping solids out of the wells, ASR performance will improve. This is an important area for future work, in order to identify cost-effective filtration alternatives that can be utilized by agricultural and other operations for ASR storage of available waters that have not been treated to potable standards.

Finally, it is anticipated that well hydraulics during recharge and recovery will be investigated at several new sites to shed light on mechanisms responsible for differences in specific capacity and specific injectivity, as discussed in Section 4.2, Well Plugging and Redevelopment. Clear resolution of this technical issue will affect subsequent wellhead design.

8.2 REGULATORY ISSUES

In the next few years, attention is expected to focus increasingly upon regulatory issues associated with ASR. Storage of treated drinking water is not the main issue. The risks and benefits of storing water from high quality non-potable sources in brackish, and possibly fresh aquifers will undoubtedly be debated in many forums in different states and also at the national level within the U.S. Similar discussions are already occurring in Australia and probably in other countries.

It seems likely that the outcome of these discussions will differ, reflecting the opportunities, needs, and constraints in each area. For water short areas, such as many parts of the Middle East, the rapidly approaching failure of groundwater resources is likely to precipitate creation of strategic water reserves, not only for potable water but also for reclaimed water. These would provide water for drinking and also for irrigation in the event of national emergencies. Under such circumstances, the risks associated with the use and underground storage of reclaimed water are more likely to be outweighed by the benefits. In other areas with more plentiful water supplies, public opinion may support regulatory restrictions upon ASR storage of non-potable water so that the potential benefits will be lost or substantially reduced, reflecting an unwillingness to accept the perceived risks associated with storage of non-potable waters.

It is not easy to project what regulatory path will be followed in the U.S. A suggested direction is as follows:

- Develop a model code for regulation of potable water ASR systems. The code would incorporate various key elements essential to ASR success, while providing the flexibility to fit in with different legal frameworks for water regulation in each state.
- Devise a regulatory framework for ASR storage of non-potable, high quality waters in brackish aquifers that will work in Florida and California. These two states will provide a pattern for other states to consider as they subse-

quently tackle the same regulatory issues. This would include concepts presented in Section 5.3, Legal and Regulatory Issues, and Section 7.3, Reclaimed Water.

• Seek the support of state environmental groups and major water users from agriculture, industry and public water suppliers for the proposed regulatory framework.

• Work with state regulatory agencies and elected representatives to implement rule changes as appropriate to establish the desired regulatory framework at the state level.

• Seek Congressional action directing the Environmental Protection Agency to modify existing Underground Injection Control regulations to provide for ASR storage of high quality, but non-potable water in brackish aquifers, developing a regulatory process that more closely reflects the benefits that may be achieved without adverse environmental and water quality effects.

• Work with the Environmental Protection Agency to develop an appropriate ASR regulatory process. This will require careful negotiations to establish consistency with the intent of the 1974 Safe Drinking Water Act, even though the Underground Injection Control regulations established pursuant to the Act remain in effect. This will probably require development of additional federal regulations pertaining to permitting of ASR facilities.

• Select sites for demonstration ASR projects storing surface water, groundwater, and reclaimed water, using the permitting process for these sites as the vehicle for bringing about the necessary regulatory changes discussed above.

This process may require several years, with uncertain outcome. However, the competition for available water supplies will continue to build, increasing the likelihood that changes will be necessary in the way we manage water. These will include both technical and regulatory changes.

8.3 GLOBAL APPLICATIONS OF ASR

Current ASR activity is primarily concentrated in the U.S. Challenging water situations, however, are found throughout the world. It is only a matter of time before ASR technology is applied worldwide, the principal constraint being lack of understanding that the technology exists and is able to resolve a wide variety of water needs at low cost.

Driving Forces

Several factors favor the rapid implementation of ASR in other countries:

• very low cost relative to other water supply and treatment alternatives
• relatively simple technology to design, construct and operate

- proven success in a wide variety of hydrogeologic settings and water supply applications
- proven performance in the U.S., Holland, Israel, and England
- enhanced reliability of water supplies, at a time when many existing sources are becoming less reliable to sustain growing water needs
- strong support of environmental interests

Constraints

A few factors tend to slow down or discourage ASR implementation, most of which are considered short-term or transitional:

- Lack of awareness or sufficient understanding of ASR technology to realize how it may be applied to meet local needs.
- Political or financial momentum associated with alternative, more visible water resources projects such as major dams, reservoirs, pipelines, and treatment plants. All too frequently this is a major factor for delaying progress on less-expensive alternatives such as ASR. It is difficult to mount a bronze plaque on a well.
- Lack of "local" ASR experience. Few individuals or agencies are willing to be the first to try a new technology in their area, even if similar ASR projects are successfully operating in other areas.
- Unsuitable hydrogeologic conditions for water storage. This is probably a constraint for only a very few areas, based upon experience to date.

Opportunities

ASR implementation opportunities are boundless. There are probably few, if any, countries where ASR cannot be applied beneficially to meet a variety of needs. Chapter 9 discusses several sites where ASR has been applied in the U.S. and other countries, while the various potential types of applications are presented in Section 1.5, ASR Applications to Meet Water Management Needs.

In countries with ample water resources, ASR is more likely to be applied to defer costly expansion of water treatment and conveyance facilities by making more efficient use of existing facilities. Where water resources are unreliable due to seasonal variability in flow or quality, ASR can be used to enhance reliability and supplement the yield. Where water resources are scarce, ASR may be used to store limited available supplies underground, eliminating evaporation and seepage losses associated with surface reservoirs. Water systems subject to emergency loss from natural or man-made causes can store water to meet these needs, thereby reducing system vulnerability. Regardless of the use or the source of the stored water, aquifer storage is usually relatively inexpensive compared to other water supply, surface reservoir storage or treatment alternatives.

Engineers tend to be motivated by dreams or "grand plans" to resolve challenges facing humanity. Occasionally these dreams turn into reality, as evidenced by some of the major engineering feats of the past century such as the Suez and Panama Canals, the Golden Gate Bridge, and the tunnel now connecting England and France.

One such dream is the "Peace Pipeline" connecting the abundant water supplies in Turkey with the water short countries of the Arabian peninsula. Such a pipeline would have to cross some of the most politically volatile countries in the Middle East. The huge cost of this pipeline, combined with the resulting national vulnerability associated with any significant dependence upon an outside water source controlled by several other nations, would suggest that this pipeline will never be much more than a dream.

However, the stakes are high and getting higher. Groundwater resources, which supply a large percentage of the current water withdrawals of the Arabian Gulf countries, will soon be exhausted. Estimates of when this may be expected seem to range from about 15 to 50 years. Oil reserves are substantial; however, oil revenues are insufficient at current world oil prices to support the water needs of a region with the most rapidly growing population in the world. Due to the very high capital and operating costs, desalination of seawater is not really a viable solution for the quantities of water that are required for all uses, whether now or in the future. Interest in the possibility of a pipeline to import water is therefore expected to grow.

Integration of ASR technology with the pipeline can help to ease many of the political and technical shortcomings of the current plan, thereby enhancing the chances of its implementation. At key locations along the pipeline such as major pumping stations, delivery points, and locations with particularly suitable hydrogeologic characteristics, ASR storage reservoirs would be constructed. These would become the new "oases" in the desert, storing vast quantities of water to meet variations in water demand, whether seasonal, long-term, or for national emergencies. These strategic reserves would help to defuse tensions related to potential short-term loss of water from the pipeline, buying time for measures, political and otherwise, to restore water supplies. For example, a reasonable goal would be the development of a one-year reserve of drinking water at each delivery point, following which further supplies would be utilized to meet lower priority needs and seasonal variations in demand.

This is but one "dream." There are many others, limited only by the ability and willingness of water managers to integrate ASR technology into their water supply planning.

CHAPTER 9

Selected Case Studies

9.1 PEACE RIVER, FLORIDA*

The Peace River/Manasota Regional Water Supply Authority serves water from the Peace River to portions of Sarasota, Charlotte and DeSoto counties in southwest Florida. During 1993, maximum day and average day production from the water treatment plant were 36 megaliters/day (9.6 MG/day) and 26 megaliters/day (6.9 MG/day), respectively.

The Peace River is highly variable in both flow and quality such that with the current regulatory diversion schedule, periods of up to 2 months with no allowable diversions are relatively normal, while periods of up to 7 months may occur, as shown in Figure 2.2. Average river flow is about 2450 megaliters/day (1000 cfs). An offstream reservoir with a capacity of 2.4 Mm³ (1920 acre ft) is utilized to meet demands during periods of no diversion from the river, and also to improve raw water quality.

A 45 megaliters/day (12 MG/day) water treatment plant provides water to the service area using alum coagulation, filtration and disinfection with chloramines. Water is treated to meet system demands and also to recharge a system of six ASR wells at the plant site, the yield of which is about 17 megaliters/day (4.5 MG/day). These wells were constructed in two lime-

---

* Peace River/Manasota Regional Water Supply Authority, 1451 Dam Road, Bradenton, Florida 34202

stone artesian aquifers that contain brackish water. Well T-1 was constructed in the Tampa zone, which occurs at a depth of about 122 to 152 m (400 to 500 ft), while the remaining five wells (S1, S1A, S4, S5, S6) were constructed in the Suwannee zone, which occurs at a depth of about 174 to 274 m (570 to 900 ft). Water is stored in the ASR wells during low demand months and is recovered as needed to meet system peak and long-term demands. Figure 4.18 shows the layout of the Peace River facilities.

Operation of the first two ASR wells began in 1985. Three wells were added and an observation well converted to ASR use in 1988. Three additional ASR wells should become operational in 1994, increasing system recovery capacity to between 26 and 30 megaliters/day (7 and 8 MG/day). To meet seasonal and long-term variations in water supply and demand, the target storage volume for each well is 1.33 Mm^3 (350 MG) per MG/day recovery capacity. As of the end of 1993, this target volume has almost been reached since the Authority has 5.7 Bm^3 (1.5 BG) of treated drinking water stored in the ASR wells.

A third zone in the Avon Park aquifer is located at a depth of about 396 to 427 m (1300 to 1400 ft). This zone is being tested to determine its feasibility for storage of untreated or treated water. If feasibility is confirmed, all three zones would be utilized to store water beneath the treatment plant, thereby "stacking" the stored water vertically and concentrating piping and wellfield operations in a small area. This is quite cost-effective. Figure 2.7 shows the hydrogeologic cross-section at this site.

Estimated hydraulic characteristics of the various aquifers and confining layers have been estimated from several pumping tests. For the Tampa zone, transmissivity is about 455 m^2/day (4900 ft^2/day); storativity is 0.0004, and leakance is 0.0001/day. Porosity is estimated at 15%. Static water level is about 22 ft above the measuring point, which is at an elevation of 27 ft above mean sea level. The average regional gradient of the potentiometric surface in this area is about 0.0002 to the WNW; however, production wells in the vicinity may affect the local gradient around the single ASR well in the Tampa zone. This well is open hole in limestone from 380 to 480 ft in depth. The lithology for Well T-1 is generally limestone, poorly to well consolidated, light grey to cream colored, soft to hard, and fossiliferous.

For the Suwannee zone, transmissivity is about 560 m^2/day (6000 ft^2/day); storativity is about 0.0001, and leakance is 0.0085/day. Most of the leakance is believed to occur through the underlying confining layer separating this zone from the Avon Park formation. Static water level is about the same as for the Tampa zone, and the regional gradient is about 0.0003 to the WNW. All but one of the existing operational ASR wells are

TABLE 9.1
RECHARGE AND NATIVE WATER QUALITY: PEACE RIVER, FLORIDA

	Suwannee Zone	Tampa Zone	Recharge Water		
			Maximum	Minimum	Average
Conductivity (μmhos/cm)	1,290	1,150	959	340	471
Chlorides (mg/L)	184	151	162	30	55
Sulfate (mg/L)	224	222	175	32	83
Total dissolved solids (mg/L)	800	700	470	170	247
Alkalinity (as $CaCo_3$, mg/L)	144	142	100	38	50
Calcium (mg/L)	114	80	88	24	33
Magnesium (mg/L)	48	55	—	—	—
Total hardness (as $CaCo_3$, mg/L)	482	424	300	100	133
Non-carbonate hardness (as $CaCo_3$, mg/L)	338	282	100	54	66
Sulfide, total (mg/L)	4.0	3.7	0	0	0
pH	7.4	7.55	8.60	7.90	8.24
pHs	7.61	7.75	8.50	8.25	8.38

Note: Data from test program, Cycles 1–6.

in the Suwannee zone. Lithology is limestone, similar to the description above for the Tampa zone.

The initial two ASR wells and the converted observation well have plain steel casings, while the remaining three wells have epoxy-coated steel casings. All casings are 300 mm (12 inch) diameter except the converted observation well, S1A, which is 200 mm (8 inches) in diameter. All wells are equipped with vertical turbine pumps and recharge is down the annulus. Typical recharge and recovery rates are similar at about 0.5 to 1.0 MG/day. Wells are redeveloped by pumping at the beginning of seasonal recovery but are not otherwise backflushed periodically during recharge.

Table 9.1 shows typical water quality for the recharge water, and also for native water in the two storage zones.

The principal objective of the ASR system is to provide seasonal storage. During early years, additional water is also being stored long-term to meet demands in later years when higher demands will limit the amount that can be stored. These higher demands will also increase the amount that needs to be recovered. Long-term storage, or "water banking," is a secondary objective. By substituting ASR storage for expansion of the offstream surface reservoir, capital costs for system expansion to meet projected future regional demands can be reduced by an estimated 60%. Once the target initial storage volume has been reached, it should be possible to meet projected water demands with high reliability (95%) and with excellent water quality meeting all applicable standards, despite the great variability in flow and quality of the Peace River. Operating experience to date has been satisfactory. **Figure 9.1** shows operating data from well S6 from 1989 to 1991, including monthly recharge and recovery volumes, cumulative volume, and TDS of recharged and recovered water.

A key part of the planning process for this site has been the development and application of a computer simulation model of flow and quality, using a monthly time step and the historic period of record to route flows from the river to the offstream reservoir, treatment plant, ASR wells and distribution system. This model has been used to determine the most cost-effective expansion path for major facility components to meet increasing levels of regional water demand during the planning period. Table 4.8 shows the expansion path, from which it is apparent that incremental expansion of relatively inexpensive ASR wells can defer the next expansion of the 45 megaliters/day (12 MG/day) treatment plant until maximum day demand reaches about 68 megaliters/day (18 MG/day). If ASR wellfield capacity and potential storage volume is sufficient as expected, offstream reservoir expansion can be deferred for many years.

When used to meet seasonal demand variations, an approximate annual mass balance of water stored and recovered can usually be achieved with allowance for some inexpensive factor of safety. When used to meet long-term demand variability such as an increasing trend of water demand, ASR system reliability depends not only upon facilities capacity but also upon the shape of the demand projections curve. A water system with sharp initial growth after new facilities become operational will have less opportunity to store water in early years to meet long-term needs. Conversely, a system with little growth in demand after a new facility becomes operational will have ample opportunity to store water for long-term needs. Consequently, careful attention to demand projections, and periodic updating of these projections, can help to ensure that ASR economies are achieved without jeopardizing system reliability.

The ultimate ASR plan at Peace River includes over 40 ASR wells in the Suwannee zone, meeting system peak demands in excess of 151 megaliters/day (40 MG/day). No further development of the Tampa zone is planned due to relatively low well yields and also due to the use of this zone by others in the surrounding area for water supply and irrigation purposes. If the Avon Park zone is shown to be feasible, it should be possible to reduce the number of Suwannee zone wells and the associated

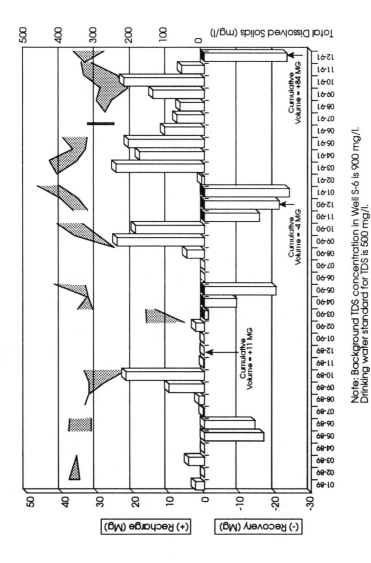

Figure 9.1 ASR Well S-6, monthly recharge and recovery volumes and TDS concentrations, Peace River, Florida.

aereal extent of the wellfield. Each well site would include a pair of wells, one in each zone.

9.2 COCOA, FLORIDA*

The City of Cocoa on Florida's east coast provides water to Cape Canaveral Air Force Station, Patrick Air Force Base, Kennedy Space Center, several communities and other wholesale customers in central Brevard County, in addition to the city's own residents. The primary source of water is the upper Floridan aquifer, from wells located inland in Orange County, since this aquifer is brackish in Brevard County. An overlying secondary artesian aquifer provides a small supplemental supply of water.

Maximum day demand reached 144 megaliters/day (38 MG/day) in 1989 while average demand was about 91 megaliters/day (24 MG/day). Since then, slightly lower demands have occurred due to implementation of water conservation and reuse measures and regulatory restrictions, despite an increase in the number of connections.

Water supply facilities include 24 wells in the limestone artesian aquifer and three wells in the secondary artesian aquifer, with a combined permitted maximum day production capacity of 182 megaliters/day (48 MG/day). Current actual production capacity is about 167 megaliters/day (44 MG/day), since some of the permitted wells have yet to be constructed and hydraulic limitations in the collection system throttle the flows. Wellfield expansion is planned to increase the yield and also to shift the pattern of production, thereby reducing potential saline water intrusion.

Water from the wellfield is pumped to the Wewahootee Pumping Station where it is aerated to remove hydrogen sulfide and also chlorinated. It is then pumped to the Claude H. Dyal Water Treatment Plant which has a rated capacity of 167 megaliters/day (44 MG/day). During months when demand is below wellfield capacity, excess water is treated at the plant and stored in six ASR wells located on the plant site. During peak demand months, the stored water is recovered to meet system demands. From the plant, the water is pumped to the distribution system. The pipeline distance from the wellfield to the end of the distribution system is about 80 km (50 miles). **Figure 9.2** shows the general layout of the Cocoa water supply and distribution facilities, while **Figure 9.3** shows the layout of the ASR wellfield at the treatment plant site.

The aquifer being recharged is the same aquifer from which the water is withdrawn; however, the ASR location is about 24 km (15 miles) to the

---

* Department of Utilities/Public Works, City of Cocoa, 600 School Street, Cocoa, Florida 32922

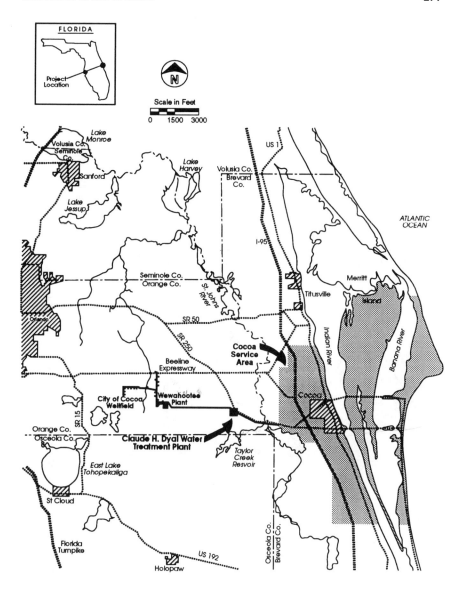

Figure 9.2 Location of water supply facilities, Cocoa, Florida.

east of the main portion of the wellfield. It is a confined limestone artesian aquifer that contains brackish water with typical TDS concentrations of about 900 to 2000 mg/L. ASR operations began in 1987 at the first ASR well. Five additional ASR wells began operation in 1991, increasing recovery capacity to 30 megaliters/day (8 MG/day). At such time as the raw water supply wellfield is expanded to its full permitted capacity, the

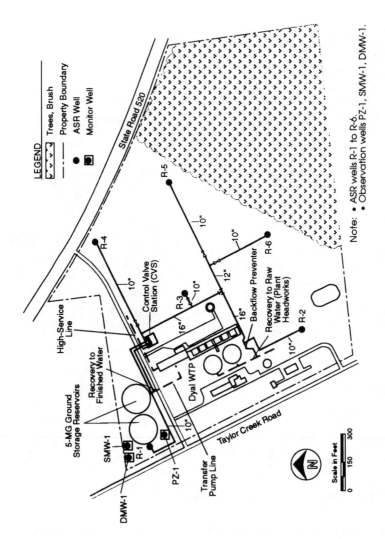

Figure 9.3 Water treatment plant and ASR wells, Cocoa, Florida.

combined wellfield, treatment plant, and ASR system should be able to meet peak demands up to about 212 megaliters/day (56 MG/day) using the existing six ASR wells. Provision for four additional ASR well sites adjacent to the plant is underway.

Each of the ASR wells is completed with open hole construction, including a 400 mm (16 inch) PVC casing to depths ranging from about 85 to 91 m (280 to 300 ft). Holes reach depths of about 113 m (370 ft) and are acidized to improve yield and specific capacity. Each well is equipped with a vertical turbine pump. Recharge occurs down the annulus between the pump column and the well casing at design rates ranging from 3.0 to 4.5 megaliters/day (0.8 to 1.2 MG/day). The maximum system recharge rate is 23 megaliters/day (6.0 MG/day), although lower rates of about half this amount can be effectively utilized. Recovery rates for individual wells range from 3.8 to 6.0 megaliters/day (1.0 to 1.6 MG/day), totaling 30 megaliters/day (8.0 MG/day). Flow rates during recharge and recovery must be balanced between the wells in order to control subsurface lateral movement of the stored water around each well under the influence of hydraulic interference. Over a period of several years, it is anticipated that the storage bubbles around each well will tend to coalesce, reducing the initial sensitivity of the system to variations in recharge and recovery rates and volumes at individual wells.

Figure 2.5 indicates the hydrogeology of the ASR site, including the initial ASR test facilities. Pumping tests in the upper Floridan aquifer at the plant site indicate that aquifer hydraulic characteristics are as follows:

Transmissivity	36,000 to 101,000 G/day/ft
	447 to 1254 m²/day
Storativity	2.3×10^{-4} to 3.0×10^{-4}
Leakance	2.5×10^{-3} to 3.7×10^{-3}/day
Specific capacity	16 to 60 G/min/ft at about 1000 G/min

Static water levels are typically about 11 ft below land surface, which is at a mean sea level elevation of about 44 ft. During recharge, interference between wells is such that wellhead pressures can reach about 28 psi while during recovery, drawdowns of up to about 20 m (65 ft) can occur in individual wells.

Following expansion of the ASR system from one to six wells, a system performance test was performed, recharging all six wells at once and monitoring differences in recharge rates and volumes and recovered water quality from each well. Extensive data were collected, some of which is presented in **Table 9.2.** It is pertinent to note in this table the relative water quality at the end of recovery between Well R-1, which had been in ASR operation for four years, and the remaining wells that were being placed into service. The improvement in quality with successive cycles is apparent.

TABLE 9.2   SYSTEM OPERATIONAL TEST FOR SIX ASR WELLS: COCOA, FLORIDA

Well	Recharge Volume (MG)	Recharge Chloride (mg/L)	Recovery Volume	Recovery Chloride (mg/L)	Background Chloride (mg/L)
R-1	11.2	151	11.2	166	342
R-2	11.9	151	11.7	492	410
R-3	7.2	151	7.7	268	350
R-4	11.9	151	13.7	342	370
R-5	10.3	151	11.4	532	570
R-6	7.6	151	7.9	642	850
Blend	60.1	151	63.6	414	—

Note: Recharge chloride concentration is average of 10 samples ranging from 128 to 194 mg/L. Recovery chloride concentration is at end of recovery on the first operational test cycle. Well R-1 in prior operation for 4 years.

Since ASR operations at all six wells began in 1991, data have been gathered to monitor system performance. **Figure 9.4** shows monthly and cumulative storage volume for all wells for the period September 1992, to August 1993, and variations in recharge and recovery water quality, as indicated by chloride concentrations. Performance is satisfactory except for Well R-2, which consistently shows higher turbidity compared to the other wells. This continues a pattern of abnormal response in Well T-2 that began during well construction and testing. Such differences in well performance in karst limestone aquifers are not unusual. Plugging has not been observed during ASR operations to date.

The target storage volume for each ASR well is 100 MG (378 megaliters) per MG/day recovery capacity, thereby providing about 100 days of recovery. Full recovery efficiency has been demonstrated in this brackish aquifer. Considering that the seasonal variability in supply from the wellfield is small, the only real variability is in demand. Such a target storage volume should enable the city to meet seasonal demand variations up to a peak demand of 212 megaliters/day (56 MG/day).

Further increases in demand will be met from a new surface water source, Taylor Creek reservoir, which is located about 3 miles from the water treatment plant. The supply of water from this source is highly variable, reflecting regulatory and environmental restrictions upon streamflow diversions. In many years, periods of no diversion will extend for about 1 month, while a period of up to about 18 months is possible. As a result, the ASR target storage volume associated with this source will be substantially greater than 378 megaliters (100 MG) in order to meet seasonal and long-term variability in both supply and demand.

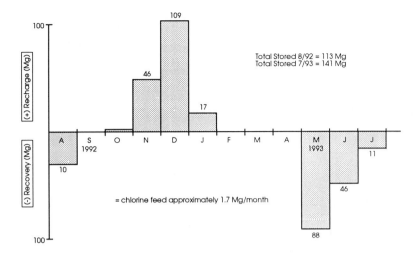

Figure 9.4 Recharge and recovery volumes (8/92-7/93), Cocoa, Florida.

The combination of surface supply, wellfield supply, and ASR storage provides a reliable water system meeting regional water demands at relatively low cost. During wet years it is anticipated that more surface water will be diverted, treated, consumed and stored. During drought years, more groundwater will be produced, treated, and consumed, recovering water from ASR storage to help meet peak demands.

9.3 MARATHON, FLORIDA*

The Florida Keys Aqueduct Authority initiated ASR investigations in 1987 with the objective of determining whether it is feasible to store treated drinking water in a highly saline aquifer for emergency water supply purposes. The Marathon Pumping Station site was selected so that recovered water could be directed either into the local distribution system or into the regional transmission system. The recharge water source is a wellfield and water treatment plant at Florida City on the mainland, from which treated drinking water is pumped about 193 km (120 miles) to Key West, crossing 45 bridges. Concern regarding the vulnerability of this system to disruption during hurricanes prompted the investigation of ASR as a potential alternative to very costly plans for seawater desalination or above-ground storage to meet emergency needs. **Figure 9.5** shows the location of the Marathon site.

Facilities at this site include one ASR test well, two storage zone observation wells, a shallow drainage well for interim disposal of recov-

---

* Florida Keys Aqueduct Authority, 1100 Kennedy Drive, Key West, Florida 33041

ered water, a submersible pump, and wellhead piping. Recharge occurs down the annulus. These facilities were constructed during 1990 following earlier preliminary studies to select the storage zone. Testing began in August 1990. Results of a pumping test conducted at the beginning of the cycle testing indicated that the storage zone is a leaky confined aquifer with the following hydraulic characteristics:

Transmissivity	17,100 G/day/ft (212 m²/d)
Thickness	40 ft (12 m)
Storativity	3.2×10^{-4}

Subsequently, it was determined that the well was not fully developed. Additional development was conducted, improving well specific capacity to about 2.7 G/min/ft at typical pumping rates of 235 to 350 G/min. No subsequent hydraulic testing has been performed to reevaluate hydraulic characteristics or their response to several cycles of ASR operation. No signs of well plugging have been observed.

The ASR well and two observation wells are constructed as shown in Figure 3.1. 400 mm (16 inch) Schedule 80 PVC casing was cemented to a depth of 118 m (387 ft). A 250 mm (10 inch), 0.025 slot stainless steel 316, wire-wrapped screen was then installed to a depth of 130 m (427 ft), plus a 1.5 m (5 ft) sump. The screen was extended with a blank section

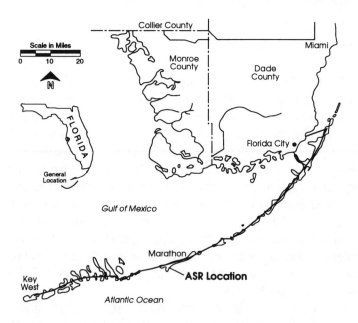

Figure 9.5 Marathon, Florida, ASR location.

inside the casing to a depth of 110 m (362 ft) and the annular space was filled with 6/20 gravel using a 1-inch gravel tube.

During well construction, continuous wireline cores were collected between 107 and 137 m (350 and 450 ft). The storage zone is composed of clean, coarse quartz sands with a porosity of 26 to 35% with minor amounts of carbonate, calcite, and dolomite and traces of smectite, illite/mica, and kaolinite. These cores were analyzed to determine their physical, geochemical, and other properties to provide a basis for ASR zone selection and screen design. From this information and associated geophysical logs, observation well OW-1 was equipped with three 0.5-inch sampling tubes, isolated with packers into production intervals at the top (387 to 405 ft), middle (405 to 418 ft), and bottom (418 to 428 ft) of the storage zone. Sample ports are set at depths of 122, 125, and 130 m (400, 410, and 426 ft).

The storage zone contains seawater. An observation well core description of the storage interval and adjacent portions of the overlying and underlying confining layers is shown in **Table 9.3. Figure 9.6** shows the water quality response at observation well OW-1 during Cycle 2. It is of some interest that the freshest water occurs at the bottom sampling interval during the recharge, storage, and most of the recovery period. During Cycle 2, 37 megaliters (9.7 MG) were injected over a 42-day period. The water was stored for 34 days and then 13 megaliters (3.5 MG) were recovered. Recovery efficiency was about 23% during this cycle before water quality exceeded potable standards. Evidently the vertical hydraulic conductivity within the storage zone is sufficiently low that density stratification occurs quite slowly. In subsequent cycles, recovery efficiency steadily improved as the storage zone was developed, reaching about 72% during the test program.

Static water level in the ASR well typically ranges from 0.9 to 1.2 m (3 to 4 ft) below the measuring point, which is close to land surface elevation of about 1.5 m (5 ft). During recharge and storage periods, freshwater displaces seawater in the storage zone, increasing static head above land surface. Daily tidal variation is about 0.1 m (0.4 ft).

A series of eleven ASR cycles has been conducted to determine the relationships between recharge and recovery rates, volumes, storage periods, and recovery efficiencies. **Table 9.4** shows water quality of the recharge water and the native groundwater at this site. The storage zone TDS is 37,200 mg/L. Simulation modeling was conducted prior to the test program to provide a basis for design of the test cycles.

From the results of the cycles, it is apparent that treated drinking water can be stored in a thin, confined, low permeability aquifer containing

TABLE 9.3 CORE DESCRIPTION: MARATHON, FLORIDA

Sample Run	Depth	Length Attempt	Length Recovery	% Recovery	Color	Lithologic Description	Comments
CR6	378.6–383.6	5.0	5.4	108%	Light olive (10y 5/4)	Sandstone vf grained in calcareous clayey matrix, well consolidated little-same polecypod shell subrounded-Moldic porosity throughout sample. Trace medium coarse silica.	5ft in 1¹/₂ min.
CR7	383.6–388.6	5.0	3.9	77%	Light olive (10y 5/4) and yellowish Gray (5y 7/2)	As above; slightly more phosphorite grains in lime-mud matrix. Coarse-grained silica angular-rounded. Well rounded calcareous sand grains (fine-grained).	1 ft ss; 1¹/₂ soft, 1/2 ss, 2 soft. Middle 1¹/₂ soft has some hard lenses less than .05 ft thick. Sand sample from core-catcher obtained 5 ft in. 1¹/₂ min.
CR8	400–405.5	5.5	2.4	44%	Light olive (10y 5/4)	Sand, vf-med, poorly sorted silica slightly effervesent in HCL angular-subangular. Little vf. black phosphorite in matrix. Little white (N9) limestone fragments in upper 0.6 ft of core. Some lime-mud matrix.	5¹/₂ ft in 2¹/₂ min. 1-ft sample to Mineralogy Inc. for analysis.
CR9	406–410.5	4.5	3.2	71%	As above	As above; limestone frags occur throughout core. Iron staining but may be from core barrel.	No core return from 405.5–406, 4.5 ft in 3¹/₂ min.
CR10	411–413.5	2.5	2.4	96%	As above	As above — with gravel (<.4mm) Subangular to subrounded.	2.5 ft in 5 mins. 2-ft sample to David Pyne.

Sample	Depth		Recovery	Color	Description	Notes
CR11	413.5–416.5	3.0	3.3	110%	As above	As above	More coarse with depth 3.0 ft in 5 min 40 sec. Mud viscosity = 45 sec. Mud weight = 8 3/4 lb/gal. 1 ft sample to Mineralogy Inc.
CR12	416.5–417.4	0.9	0.9	100%	As above	As above	Lost circulation. Retrieved core barrel which has this sample in it. Sample 420–420.4.
CR13	417.4–420.4	3.0	2.0	67%	As above	As above. Appears finer grained. Less phosphorite, calcareous matrix, mottled streaks.	
CR14	422–427	2.0	1.8	90%	As above and pale olive (10y 6/2)	As above. Less calcareous material subrounded-rounded. Predominantly translucent silica grain. Poorly sorted.	Mud viscosity = 60 sec. Mud weight = $8^3/_4$. Casing set but it just was set down into sand. Core obtained most likely from 424'–427'. 5 ft in 4 min.
CR15	427–428.5	1.0	0.0	0%	As above	As above	Switch to conventional shoe for next run because we're in hard material at 428.5 (may be clay lens).
CR16	428.5–433.5	5.0	1.8	36%	As above	As above	Core indicates a cleaner sand below the olive sand we've encountered thus far. Bottom of core is very hard sandy clay.
CR17	433.5–438.5	5.0	4.7	94%	Light olive (10y 5/4) olive and gray (5y 3/2)	As above; thin (<2") clay lenses @434' and 436'–438'. Clay is highly plastic (CH).	Predominantly sand with thin clay lenses.

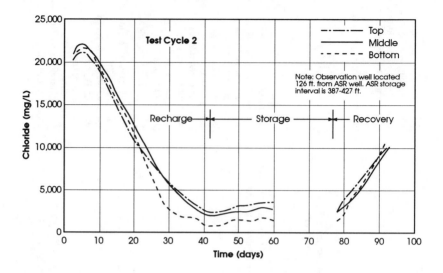

Figure 9.6 Observation well chloride concentration vs. time, Marathon, Florida.

TABLE 9.4
RECHARGE AND NATIVE WATER QUALITY: MARATHON, FLORIDA

Constituent	Recharge Water (mg/L)	Native Water (mg/L)
pH	10.3	7.60
Total alkalinity	23.1	120
Conductivity (μmhos/cm)	397	49,000
Carbonate hardness	110	1,390
Non-carbonate hardness	95.0	6,480
Turbidity (NTU)	<0.2	<0.2
Total dissolved solids	212	37,200
Total suspended solids	<1.0	4.2
Calcium	33.8	398
Magnesium	3.75	1,250
Sodium	20	11,000
Potassium	11.4	385
Silica	4.7	9.43
Aluminum	<0.5	<0.5
Iron	0.05	<1.0
Chloride	41.8	20,800
Fluoride	0.8	0.84
Sulfate	91.1	2,910
Nitrate and nitrite	<0.02	<0.02
Carbonate	16.8	0
Bicarbonate	23.1	146

Note: Recharged water sampled April 3, 1990. Native water sampled May 4, 1990.

seawater. The recovery efficiency declines with increasing duration of storage, as might be expected, reflecting density stratification of the stored water in the aquifer. Figure 4.3 shows recovery efficiency as related to storage time between the midpoint of recharge and the midpoint of recovery, based upon data from the series of test cycles. Storage volumes during this series were typically either about 19 megaliters or 57 megaliters (5 MG or 15 MG). Testing at the larger storage volume showed that if a small trickle flow of about 3 L/sec (50 G/min) is added to the well between the end of recharge and the beginning of recovery to counteract mixing losses due to density stratification, it is then possible to ensure that a given target recovery volume will be available when needed. The longer the period of trickle flow to maintain the target recovery volume, the lower the overall recovery efficiency, although the ability to recover the target volume remains unchanged.

Under optimum operating conditions, it is expected that overall annual recovery efficiency will range from about 45 to 65%, including the trickle flows required during storage periods. With these trickle flows, it should be possible to store 57 megaliters (15 MG) per year and recover about 38 megaliters (10 MG) per year from each ASR well to help meet peak seasonal demands or to meet emergency needs if they arise.

In an operating mode, ASR storage occurs during May and June, immediately prior to the hurricane season that occurs between late June and early November. The recharge rate is typically about 1.1 to 1.6 megaliters/day (200 to 300 G/min). Once the desired storage volume of about 57 to 76 megaliters (15 MG to 20 MG) is attained in each well, recharge continues at a trickle feed rate of about 3 L/sec (50 G/min) to offset density stratification losses in the storage zone. In the event of a hurricane emergency that interrupts water supplies, the ASR well is pumped at a rate of about 1.35 to 1.89 megaliters/day (250 to 350 G/min) until water quality exceeds potable standards. If no emergency occurs, the stored water is recovered to help meet water demands during January and February, which are usually the peak demand months. An alternative operating scenario under consideration is to store additional water in each well prior to the hurricane season, recovering a portion to help meet a secondary peak demand period during August while leaving the hurricane emergency storage volume in the wells.

The water supply plan for the Authority includes expansion of ASR facilities at the Marathon site to a recovery capacity of 3 MG/day. At this operating scale, the annual value of the water stored but not recovered is estimated at about $81,000 (1993 dollars). Other operation and maintenance costs are expected to increase the annual total cost to about $118,000.

The capital cost is estimated at about $3 million, or $1.00 per gallon of installed emergency capacity. Both the capital and operating costs are significantly less than the cost of alternative emergency water supply sources available in the Florida Keys. Furthermore, the ASR system will help meet peak demands, potentially deferring the need for expansion of some treatment and transmission facilities. An added advantage is the opportunity to phase in construction over a period of a few years.

This is believed to be the first ASR system in the world to store drinking water in a seawater aquifer. The first ASR well is fully permitted and currently awaits installation of a chlorinator so that recovered flows can be directed into the distribution system. Seven additional wells are planned at this site to increase recovery capacity to 11 megaliters/day (3 MG/day).

9.4 PORT MALABAR, FLORIDA*

Port Malabar is a residential community located on Florida's east coast. At the time ASR investigations began in 1987, the community water supply was furnished by a private utility, General Development Utilities, Inc. Recently the Town of Palm Bay acquired the utility and now serves the people of Port Malabar with water. Facilities include a shallow secondary artesian aquifer wellfield comprising over 31 wells located throughout the service area, a 38 megaliters/day (10 MG/day) water treatment plant, raw water collection and treated water distribution system, an ASR well and three observation wells. Inadequate raw water supply is the greatest constraint upon system operations. Existing wells can barely meet a maximum day demand of about 21 megaliters/day (5.5 MG/day).

The ASR storage zone is a semi-confined, brackish artesian aquifer located at a depth of 91 to 113 m (298 to 370 ft). The system became operational in 1989 and was intended to stretch the capacity of a smaller water treatment plant to meet increasing system demands until such time as a treatment plant and wellfield expansion could be completed. Although the plant has been expanded, wellfield capacity is still insufficient and seven new wells are planned.

Testing at the ASR well indicated the following aquifer characteristics:

Transmissivity	17,200 G/day/ft
	213 m²/day
Storativity	0.00015
Leakance	0.00025 day⁻¹
Specific capacity	13.0 G/min/ft at 610 G/min (recovery)
	10 to 15 G/min/ft during recharge (Cycles 4,5)

---

* Town of Palm Bay, 1101 Troutman Boulevard, NE, Palm Bay, Florida 32905

The static water level during the test program was a few feet below land surface, which is about 15 ft above sea level. The storage zone is a production interval at the top of the Ocala formation, confined above by clays of the Hawthorn formation and below by low permeability limestones. Construction details of the ASR well, a production well in the overlying secondary artesian aquifer, and ASR observation wells are shown in **Figure 9.7.** The ASR well is constructed of 300 mm (12-inch) PVC casing to 91 m (298 ft), and is open hole to 113 m (370 ft). The two production zone observation wells are located at distances of 15 and 30 m (50 and 100 ft) from the ASR well while the deep monitor well in an underlying production zone is located 12 m (39 ft) from the ASR well. The well is equipped with a vertical turbine pump and a chlorinator, and recharge occurs down the annulus.

Storage zone native water quality constituents of interest included TDS = 1320 mg/L, chloride = 588 mg/L, and hydrogen sulfide = 3.4 mg/L. Recharge water chloride concentration averaged about 180 mg/L. Since the drinking water standard for chloride is 250 mg/L and the ASR well is located within the distribution system, the opportunity for blending the recovered water with treated water from the wellfield is limited.

Recharge occurred at variable rates between 1.1 and 3.8 megaliters/day (200 and 700 G/min), depending upon flow available from the water treatment plant. Recovery also occurred at variable rates, ranging between 1.4 and 4.1 megaliters/day (260 and 750 G/min), depending upon system operational requirements. **Figure 9.8** shows the improvement in chloride concentrations occurring during the first three ASR test cycles, which included small storage volumes of about 13 megaliters (3.5 MG). The first cycle recovered 170% of the injected volume, in order to trace the initial mixing characteristics for this well. The second and third cycles recovered water until chloride concentrations reached about 275 mg/L, leaving a small buffer zone underground to enhance recovery efficiency in subsequent cycles.

Based upon the successful results of these cycles, two larger cycles were then conducted. In cycle 4, 159 megaliters (42 MG) were injected; however, only 61 megaliters (16 MG) were recovered due to low demands in the water system at this time. The balance was left in the aquifer to enhance recovery efficiency in subsequent cycles. Chloride concentration at the end of recovery was 196 mg/L. In cycle 5, 227 megaliters (60 MG) were injected and 163 megaliters (43 MG) were recovered before chloride concentration reached about 250 mg/L. At 100% recovery, chloride concentration reached 310 mg/L, compared to 588 mg/L for the native water in the aquifer.

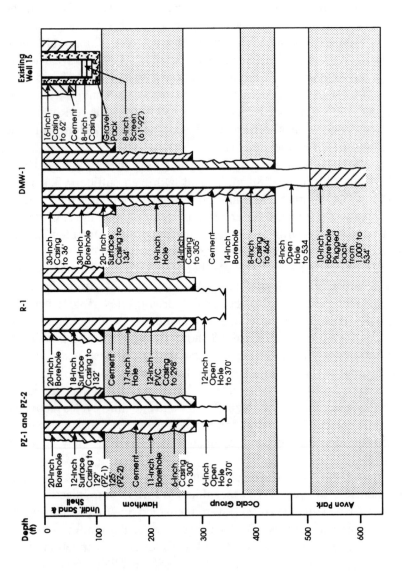

Figure 9.7 Construction details of the ASR facility, Port Malabar, Florida.

Extensive monitoring in the distribution system was conducted during Cycle 5. The testing focused upon trihalomethane concentrations and stability, verifying that no problems occurred due to mixing between the recovered water and the product water from the water treatment plant. Chlorination of the recovered flows was unable to form an adequate combined residual since the reduced ammonia concentration in the recovered water was destroyed by the free chlorine. However, a combined residual was formed in the distribution system at the point of blending, combining with ammonia present in the flow from the treatment plant.

Upon conclusion of the test program, it was recommended that the well should be backflushed to the raw water collection system during recharge periods, with a frequency of about once every month, a flow rate of about 3.8 megaliters/day (700 G/min), and a duration of about 20 min each time. A target storage volume of about 341 megaliters (90 MG) appeared appropriate, to provide seasonal storage and recovery and thereby augment the water treatment plant capacity during peak demand periods.

Of some interest at this site is that, upon completion of testing in 1989, the target storage volume was rapidly achieved, following which the ASR system shifted into a maintenance mode with a trickle flow of about 0.3 to 0.6 L/sec (5 to 10 G/min) to maintain a chlorine residual in the well and thereby control bacterial activity. Shortly thereafter, the demand upon the

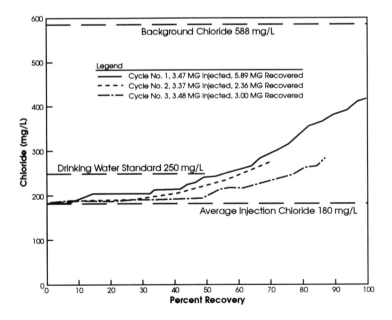

Figure 9.8 Chloride concentrations observed during ASR cycle testing, Port Malabar, Florida.

water system was reduced by about 30% due to loss of a major industrial water user that changed its water source to onsite wells. With the reduction in system demand, no recharge or recovery occurred for about three years other than the trickle flow. Rapid growth in water demand then commenced, precipitating the need for recovery of the stored water. However, recovery efficiency was less than expected. Analysis of available data suggests that the redistribution of pumping in the area probably pulled some of the stored water away from the ASR well towards the industrial supply wells, which are located about 1859 m (6100 ft) from the ASR site. Neglecting dispersion, the theoretical radius of the stored water bubble around the ASR well is about 488 m (1600 ft). With seasonal operation as originally planned, it is unlikely that any reduction in recovery efficiency would have been noted. However, long term storage in this zone for periods of several years may be at the expense of some loss in recovery efficiency. An alternative would be to use a deeper producing interval at a depth of 141 to 163 m (464 to 534 ft) for ASR storage at such time as the system may be expanded.

9.5 BOYNTON BEACH, FLORIDA*

The City of Boynton Beach is located in Palm Beach County along Florida's southeast coast. Water supplies are obtained primarily from shallow wells completed in the Biscayne aquifer, water from which is treated at the city's East Water Treatment Plant which has a capacity of about 83 megaliters/day (22 MG/day). A new reverse osmosis water treatment plant with a capacity of 15 megaliters/day (4 MG/day) has recently been completed and is located several miles from the East Water Treatment Plant. Water for the reverse osmosis plant is obtained from the brackish Floridan aquifer underlying the Biscayne aquifer.

Treated drinking water from the East Water Treatment Plant is utilized during offpeak months to supply water to an ASR well located on the plant site. ASR facilities include the ASR well, vertical turbine pump, motor and wellhead piping, and an observation well in a saline aquifer overlying the storage zone. Water is stored in the lower Hawthorn formation, a sandy limestone production interval overlying the Floridan aquifer between 244 and 274 m (800 and 900 ft) below land surface. Facilities construction occurred during 1992 and cycle testing was conducted during 1992 to 1993. The system is now operational.

Pumping tests were conducted upon completion of the well, with the following aquifer characteristics:

---

* City of Boynton Beach, 124 SE 15th Avenue, Boynton Beach, Florida 33425

Transmissivity	70,000 G/day/ft
	870 m²/day
Specific capacity	29 G/min/ft @ 800 G/min (initial pump test)

During the first three ASR test cycles, recharge and recovery specific capacities were similar, ranging from 13 to 19 G/min/ft.

Static water level prior to the beginning of ASR cycles was about 14 m (45 ft) above the measuring point elevation, which was at 6 m (19 ft) above mean sea level. As the storage zone around the well was alternately displaced with freshwater and brackish water during ASR cycle testing, the density of the water column in the well varied. Consequently, the static water level rose in the well during recharge and declined during recovery, to a greater extent than would have occurred without density changes. During recharge, the injection pressure reaches about 28 m (93 ft) above mean sea level while during recovery, water levels reach about 8 m (27 ft) below mean sea level.

The storage zone is characterized as yellowish and pinkish grey, fossil-iferous biomicritic limestone of Oligocene age, overlain by calcareous clay, fossiliferous limestone, chert, and shell. The overlying sediments and the underlying limestone are not productive.

The ASR well is designed with 400 mm (16-inch) diameter steel casing to a depth of 245 m (804 ft), below which the well is open hole to a depth of 277 m (909 ft). Originally it was drilled to a depth of 1260 ft to determine the location and quality of water from potential deeper storage intervals. Once the ASR interval was selected, the hole was plugged back to its final depth.

Large quantities of fine sand were pumped from the well during an extended period of airlift development; however, concentrations showed a reducing trend. Sand production ceased when the well was subsequently equipped with a vertical turbine pump.

Recharge occurs down the annulus at flow rates of up to about 5.4 megaliters/day (1000 G/min). Recovery occurs at about 3.2 to 5.4 megaliters/day (600 to 1000 G/min). **Table 9.5** shows typical recharge and recovery water quality. The storage zone is a semi-confined limestone artesian aquifer with a native water TDS concentration of 3910 mg/L.

Cycle volumes have not exceeded 227 megaliters (60 MG). Backflushing occurring at the beginning of recovery has been sufficient to maintain recharge specific capacity at original values. During the first seven cycles, recovery efficiencies at this volume have reached 80% and are showing steady improvement. **Table 9.6** shows the volumes, flow rates, recovery efficiency and specific capacity results for the first three cycles. Despite considerable effort, the reason for the unusual pattern of variation in

TABLE 9.5 RECHARGE AND RECOVERY WATER QUALITY: BOYNTON BEACH, FLORIDA

	pH	Alkalinity "T"	Hardness "Total"	Hardness "CA"	Turbidity (NTU)	CL (mg/L)	Chlorine Residual Free (mg/L)	Chlorine Residual Total (mg/L)	NH3 (mg/L)	Color	Sulfate (mg/L)	Iron (mg/L)	Sodium (mg/L)	TSS (mg/L)	TDS (mg/L)	DO (mg/L)	Organ. THM (µg/L)
Background	7.60	130.00	930.00	94.50	0.00	1920	NA	NA	0.73	0.00	436.00	0.02			3910.00		NA
Cycle 1 Recharge																	
Minimum	8.80	29.00	80.00	73.00	0.03	49	2.50	3.00	0.00	1.00	30.50	0.00	17.00	1.00	156.00	9.50	152.00
Maximum	9.60	47.00	122.00	100.00	0.10	150	4.00	4.20	0.00	1.00	30.50	0.00	17.00	23.00	178.00	10.70	170.40
Average	9.06	35.64	92.07	85.14	0.08	58	3.25	3.60	0.00	1.00	30.50	0.00	17.00	12.00	167.00	10.10	160.37
Cycle 1 Recovery																	
Minimum	8.10	33.00	102.00	93.00	0.07	57	0.00	0.00	0.00	2.00	42.90	0.05	42.00	0.00	288.00	2.00	113.00
Maximum	8.90	80.00	462.00	369.00	0.19	756	3.50	0.00	0.20	5.00	206.00	0.45	382.00	23.00	1422.00	6.20	202.00
Average	8.39	53.63	237.00	198.63	0.12	168	1.75	0.00	0.06	3.60	103.60	0.19	212.00	11.50	855.00	3.53	162.33
Cycle 2 Recharge																	
Minimum	8.50	28.00	79.00	72.00	0.03	44	1.50	4.10	0.10	2.00	22.60	0.00	16.50	0.00	190.00	4.90	0.00
Maximum	9.50	77.00	132.00	130.00	0.70	72	5.00	5.90	0.35	11.00	28.70	0.00	21.70	5.00	329.00	5.50	0.00
Average	9.08	41.17	93.86	87.50	0.11	51	3.50	4.83	0.14	7.29	25.77	0.00	19.58	1.00	229.86	5.20	0.00
Cycle 2 Recovery																	
Minimum	8.00	39.00	86.00	80.00	0.00	50	0.00	0.00	0.52	3.00	40.00	0.05	10.00	0.00	349.00	1.10	69.60
Maximum	9.20	49.00	246.00	207.00	6.64	424	0.00	0.20	1.60	4.00	58.60	0.06	118.50	6.00	598.00	7.50	72.10
Average	8.62	44.50	161.63	141.75	0.56	255	0.00	0.03	1.34	3.67	47.53	0.06	56.50	3.00	450.67	4.30	70.85
Cycle 3 Recharge																	
Minimum	8.50	32.00	74.00	68.00	0.04	40	0.00	3.70	0.10	6.00	22.15	0.00	20.34	0.00	161.00	5.90	0.00
Maximum	9.40	77.00	112.00	104.00	1.65	52	0.00	5.10	0.20	10.00	27.55	0.00	22.00	6.00	263.00	5.90	0.00
Average	9.11	45.81	91.10	85.21	0.14	46	0.00	4.85	0.13	8.00	24.71	0.00	20.98	3.00	205.17	5.90	0.00
Cycle 3 Recovery																	
Minimum	8.30	46.00	115.00	103.00	0.11	50	0.00	0.00	0.22	3.00	24.77	0.00	21.00	0.00	233.00	0.00	87.00
Maximum	8.90	66.00	228.00	199.00	1.42	300	0.00	0.00	1.02	6.00	38.60	0.00	96.40	4.00	667.00	0.00	87.00
Average	8.43	53.17	162.00	141.43	0.21	162	0.00	0.00	0.56	4.80	30.69	0.00	54.33	2.00	440.00	0.00	87.00
Max. conc. for drinking water	6.50 to 8.50	NA	NA	NA	1.00	250	3.00	5.00	NA	15.00	250.00	0.30	160.00	NA	500.00	NA	0.10

Note: 1. 0.00 means the parameter was EITHER not analyzed OR not detected above method detection limits. 2. NA: not applicable. 3. Temperature and conductivity measured but not shown in this table. Cycle 1 recharge total organic carbon averaged 15.3 mg/L. Temperature and conductivity measured but not shown in this table. Cycle 1 Recharge total carbon averaged 15.3 mg/L against a background of 1.70 mg/L. Recharge total organic carbon averaged 1.7 mg/L. Cycle 1 Recharge total carbon averaged 15.3 mg/L against a background of 1.70 mg/L.

TABLE 9.6
ASR CYCLE TESTING SUMMARY: BOYNTON BEACH, FLORIDA

Cycle	Total Duration (days)	Recharge Recovery Volume (G)	Recharge/ Recovery Rate at End of Cycle (G/min)	Specific Capacity At End (G/min/ft)	Recovery Efficiency (%)
Cycle 1					
Recharge	14	12,525,510	720	19	
Recovery	8	9,578,934	934	13	55
Cycle 2					
Recharge	44	57,323,180	940	15	
Recovery	31	26,099,896	968(a)	15	42
Cycle 3					
Recharge	43	58,342,670	930	17	
Recovery	25	32,240,565	1,032	17	50+(b)

(a) Recovery during Cycle 2 was performed over a range from 330 to 1,000 G/min to estimate water quality sensitivity to recovery rate.
(b) Recovery terminated at chloride concentration of 270 mg/L compared to 350 mg/L on previous cycles.

recovery efficiency during the first two cycles (55 and 42%, respectively) was not determined. Recovery efficiency is expected to continue increasing in subsequent cycles. The target water quality during recovery was a TDS of 350 mg/L.

The objective of the ASR system at this site is to augment seasonal water supplies. The city currently has sufficient treatment capacity; however, raw water supplies are limiting, reflecting declining capacity of existing wells and difficulties in locating suitable new well sites in this urban area. Recovery of stored water from the ASR well during peak demand months provides a significant benefit to the city. Consideration is being given to future construction of a second ASR well in the distribution system, to augment peak supplies from the wellfield and also to resolve distribution system hydraulic constraints.

9.6 OKEECHOBEE, FLORIDA*

Located in southeast Florida, Lake Okeechobee is the largest lake in Florida and provides a wide variety of benefits for water supply, flood control, fishing, recreation, environmental, and other water uses. In recent years biologists have observed that the lake is highly eutrophic, reflecting large annual loadings of phosphorus in tributary waters. Studies indicated that 29% of the phosphorus was entering the lake with 4% of the flow. This

* South Florida Water Management District, PO Box 24680, West Palm Beach, Florida 33416

was from Taylor Creek-Nubbin Slough, which drains a dairy farming area. Efforts were therefore initiated by the South Florida Water Management District to improve the condition and operation of Lake Okeechobee. Among these efforts, the District investigated ASR as a method for storing seasonal peak flows from Taylor Creek-Nubbin Slough in a brackish aquifer, recovering the water to meet agricultural irrigation needs and keeping it out of the lake.

The ASR test well and a dual-zone observation well are adjacent to the L-63N Canal near the town of Okeechobee. Canal water quality during the testing period varied from TDS values of 268 mg/L to a high of 996 mg/L, depending upon local rainfall. Water is obtained from the canal and is pumped into a shallow basin for preliminary settling and to provide some chlorine detention time, if chlorination is needed. It is then repumped down the ASR well at a design rate of 19 megaliters/day (5 MG/day). The shallow basin can be bypassed, if desired. Well hydraulics are such that recharge flows at double this rate could easily be achieved with the appropriate pumping equipment. Recharge occurs in the upper Floridan Aquifer System at depths of 366 to 518 m (1200 to 1700 ft). The water is recovered by gravity flow at initial rates of about 17 megaliters/day (4.5 MG/day), declining to about 13 megaliters/day (3.5 MG/day) as the density of the water column in the well increases and as recovery hydraulic effects approach equilibrium. ASR well testing began in May 1989 and continued intermittently until September 1991.

An ASR well pumping test indicated the following characteristics for the Upper Floridan Aquifer:

Transmissivity	4.38 MG/day/ft
	54,400 m^2/day
Storativity	1.25×10^{-3}
Specific capacity	838 G/min/ft at 5450 G/min

This is a highly transmissive, semi-confined limestone artesian aquifer. Transmissivity is the highest of any ASR site tested to date. Static water level is about 2 m (7 ft) above land surface, or about 6 m (20 ft) above mean sea level. Production intervals were identified during construction at several depth intervals, and showed a trend of deteriorating water quality with depth in the storage zone. Subsequent straddle packer tests indicated that at a depth interval of 1288 to 1354 ft, the TDS concentration was 4000 mg/L while at 1540 to 1662 ft, the TDS concentration was 6710 mg/L. Background TDS concentration of water pumped from the well during a pumping test was 6870 mg/L, suggesting significant inflow from the bottom of the well. **Figure 9.9** shows formation lithology at this well site.

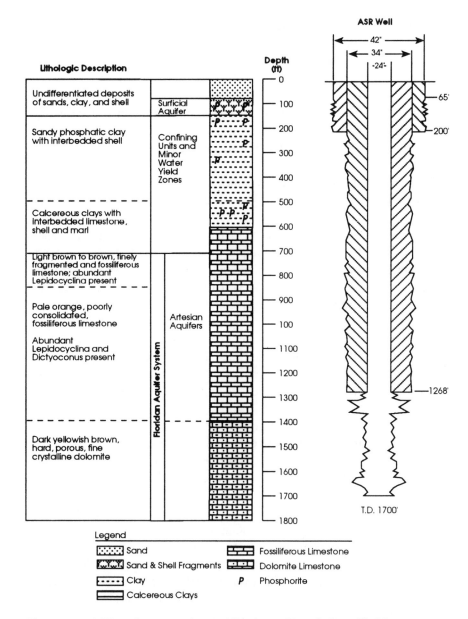

Figure 9.9 ASR well construction and lithology, Okeechobee, Florida.

The ASR well is constructed with 600 mm (24-inch) black steel casing to a depth of 386 m (1268 ft), below which the well is open hole to 521 m (1710 ft). Recharge occurs down the well casing, which has no pump, injection tubing, or other internal piping. Reflecting the high transmissivity of the storage zone, no well plugging has been experienced during testing to date.

During initial ASR cycle testing, small volumes were stored for short periods to ascertain the effect of the brackish water storage zone on the reduction of coliform concentrations in the recharge water. In general, it was shown that coliform concentrations reduced to acceptable levels within 1 to 3 days of storage in the brackish aquifer. Chlorination of recharge flows was not found to be necessary to protect water quality, public health or welfare, and was also not necessary to prevent well plugging due to bacterial activity, reflecting the high transmissivity of this well.

The ASR well was designed to achieve maximum recharge rates at the expense of recovery efficiency. The combination of substantial storage zone thickness, high transmissivity and poor native water quality is a difficult combination if the objective is to achieve high recovery efficiency in an ASR well. Test cycles completed to date have shown recovery efficiencies increasing to 35%, associated with steadily increasing storage volumes of up to 1.9 Mm^3 (500 MG). Under normal seasonal operation, larger storage volumes would be expected. Improvement in quality with successive small volume ASR test cycles at this site suggests that larger volumes would probably achieve recovery efficiencies in the range of 40 to 60%, or possibly higher. Plugging back the well to eliminate productive, very brackish intervals at the bottom of the well would improve recovery efficiency but would probably reduce potential recharge and recovery rates.

While this recovery efficiency may appear to be quite low, further analysis has shown that any recovery efficiency greater than about 40% is a net gain to the regional water management system. Evapotranspiration and seepage losses for water stored in Lake Okeechobee and the adjacent Water Conservation Areas, or conveyed from the lake to urban areas through the canal system, range from about 60% near the lake, to at least 90% during conveyance to more distant parts of the region. Furthermore, water stored but not recovered may serve as a source of recharge to the brackish upper Floridan Aquifer System, which is being developed as a regional water source for brackish water to supply reverse osmosis water treatment plants along the coast. Over a period of several decades it is expected that an expanded ASR system recharging canal water, either directly or from adjacent shallow wells, would recharge the Upper Floridan Aquifer and also tend to regionally reduce TDS concentrations.

The original objective of the ASR facilities at this site was to divert as much phosphorus-laden water as possible out of Lake Okeechobee. A secondary objective was to recover water for agricultural irrigation purposes during the dry season. Economic analyses conducted during 1989 showed the favorable low cost of an expanded ASR system to meet regional objectives at this site, including an assumed 30-well system to divert canal water at rates up to about 1140 megaliters/day (300 MG/day). The ASR well system has not been expanded at this site, reflecting changing water management priorities, needs and opportunities. Instead, similar ASR facilities are being developed at five other sites in southeast Florida, recharging from canals and shallow wells.

By demonstrating the feasibility of ASR in the Upper Floridan Aquifer System using canal water as a recharge source, this ASR project has played an important role in focusing attention in Florida upon ASR as a tool for regional water management. Further testing at the well has been terminated due to expiration of the temporary operating permit that was issued for the test program. Efforts continue at other sites, particularly with regard to resolution of important permitting and regulatory issues associated with recharge of high quality waters that may not quite meet all potable standards. This is discussed in greater detail in Chapter 7, Alternative ASR Applications.

9.7 CHESAPEAKE, VIRGINIA*

Between 1988 and 1990, the City of Chesapeake, VA, completed construction and initial testing of an ASR well, two observation wells, a pipeline, an access road, wellhead facilities, a disinfection system, and a wellhouse at the city's Western Branch wellfield in the northwest corner of the city's water service area. Recharge water is obtained from the Northwest River water treatment plant which has a capacity of 38 megaliters/ day (10 MG/day). Due to movement of saltwater up the estuary, the Northwest River is not available as a reliable water source during dry summer months. At such times, water supply has historically been a challenge for the Chesapeake area, requiring greater reliance upon wholesale water purchases from surrounding utility systems. Seasonal storage of available water supplies during wet months was seen as a potentially viable and cost-effective water supply alternative.

Facilities include an ASR well equipped with a 19 megaliters/day (5 MG/day) vertical turbine pump. The casing is 600 mm (24 inch) in diameter and constructed of epoxy-coated steel to a depth of 119 m (390 ft). The storage zone includes the Upper and Middle Potomac aquifers of Cretaceous Age. These are clayey sand aquifers in the Atlantic Coastal Plain. The upper screen section is from 171 to 189 m (560 to 620 ft), while the lower section is from 203 to 221 m (665 to 725 ft). The total well depth

* Department of Public Utilities, City of Chesapeake, PO Box 15225, Chesapeake, Virginia 23320

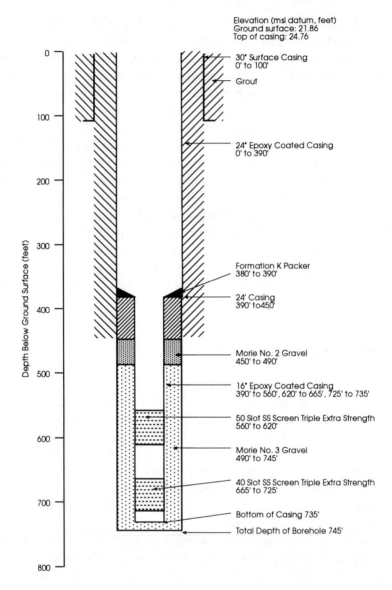

Elevation (msl datum, feet)
Ground surface: 21.86
Top of casing: 24.76

30" Surface Casing
0' to 100'

Grout

24" Epoxy Coated Casing
0' to 390'

Formation K Packer
380' to 390'

24' Casing
390' to450'

Morie No. 2 Gravel
450' to 490'

16" Epoxy Coated Casing
390' to 560', 620' to 665', 725' to 735'

50 Slot SS Screen Triple Extra Strength
560' to 620'

Morie No. 3 Gravel
490' to 745'

40 Slot SS Screen Triple Extra Strength
665' to 725'

Bottom of Casing 735'

Total Depth of Borehole 745'

Depth Below Ground Surface (feet)

Figure 9.10 ASR well construction, Chesapeake, Virginia.

is 227 m (745 ft). **Figure 9.10** shows the well construction. Operation began in 1990.

Aquifer hydraulic characteristics were determined from pumping tests, with the following representative results:

Transmissivity	9,810 to 12,370 ft²/day
	911 to 1149 m²/day
Storativity	0.00004 to 0.00005

Leakance   negligible
Specific capacity   31 G/min/ft at 1250 G/min (recharge, Cycle 1)
36 G/min/ft at 2200 G/min (recovery, Cycle 1)

During observation well construction, 195 m (640 ft) of continuous wireline cores were collected between 107 and 302 m (351 and 992 ft). The cores and associated geophysical logs helped to identify loose, medium- to coarse-grained sands confined by layers of dense, dry clay to silty clay. The mineralogic analysis of selected cores indicated less than 2% clay, including sodium montmorillonite adhering to the sand grains. Geochemical analysis of the recharge and native waters and also the aquifer minerals indicated low potential for geochemical problems, suggesting that a residual of stored water should be left in the aquifer after each cycle so that any precipitation would not occur near the well. **Table 9.7** shows typical recharge and native water quality in the storage zone.

The static water level in the storage zone is about 27 m (90 ft) below ground level, which is at an elevation of 6.7 m (21.9 ft) above mean sea level. The vertical turbine pump is set at 91 m (300 ft) and has a design production rate of 26 megaliters/day (3500 G/min) at 40 m (130 ft) of drawdown. During recharge, wellhead pressures are limited to 21 m (30 psi, 70 ft) to minimize the potential for physical well plugging that may be difficult to reverse by periodic backflushing. Recharge can occur at variable rates down the pump column or down the annulus. Recovery can also occur at variable rates. Typical recharge rates during cycle testing ranged widely up to 9.7 megaliters/day (1800 G/min) while recovery rates ranged from 10.8 to 17.3 megaliters/day (2000 to 3200 G/min).

During the first three test cycles, volumes recharged were 19, 91, and 33 megaliters (5, 24, and 88 MG), respectively, while recovery volumes were 15, 79, and 38 megaliters (4, 21, and 10 MG). A significant volume was left in the aquifer following Cycle 3 to provide water in storage for the following dry season. Mixing between stored and native water in the aquifer was negligible, as shown on **Figure 9.11.**

Experience during the cycle testing showed a decline in specific capacity which occurred during recharge periods and which could be reversed by periodic backflushing. Experimentation with different backflushing frequencies has lead to a two-week cycle for surging the well by pumping to waste for about an hour during recharge and storage periods.

The primary objective of ASR facilities at this site is seasonal storage. Secondary objectives include longer-term storage and also maintaining distribution system pressures during recovery periods, since the ASR well is at considerable distance from the city's water treatment plant.

Operating experience since 1990 has identified several issues that were not apparent during the test program. In particular, pH of the recharge

TABLE 9.7   RECHARGE AND NATIVE WATER QUALITY: CHESAPEAKE, VIRGINIA (in mg/L except where noted)

Parameters	Groundwater 570–585 ft	Groundwater 675–690 ft	Recharge Water
Total alkalinity (as CaCO₃)	510	440	65
Total dissolved solides	707	580	438
Total suspended solids	1	1	6
Turbidity (NTU)	6	1	1
Color (cu)	35	15	15
Specific conductance (umhos/cm at 25°C)	1,300	1,000	740
Chloride	75	59.4	24
Fluoride	2.26	6.58	0.96
Sulfate	13	11	67
Carbonate alkalinity (as CaCO₃)	<10	<2	58
Bicarbonate alkalinity (as CaCO₃)	510	440	<2
Total silica	9.72	15.9	4.65
Calcium	5.1	3.7	51
Magnesium	1.14	0.60	5.6
Sodium	359	347	93
Potassium	16	8.8	23
Iron	0.12	0.39	<0.05
Aluminum	0.1	<0.1	0.1
Copper	<0.01	0.01	0.01
Manganese	0.01	0.01	0.01
Zinc	0.03	0.03	<0.01
Total hardness (as CaCO₃)	16.2	8.6	148
Calcium hardness (as CaCO₃)	12.6	9.2	128
Nitrate	<0.01	0.03	0.47
Phosphate	0.33	0.51	0.07
Ammonia	0.44	0.36	0.29
Total organic carbon	23	26	8
Total halogenated hydrocarbons	<0.012	<0.012	0.143
Specific gravity of fluid density (at 60°F)	0.998	0.998	0.999
Total coliform (Ml)	16/100	<2.2/100 (a)	<2.2/100
Chloroform	<0.001	0.010	0.068
Bromodichloromethane	<0.001	<0.001	0.021
Dibromochloromethane	<0.001	<0.001	0.012
Bromoform	<0.001	<0.001	0.008
Total trihalomethane	<0.001	0.010	0.109

(a) Below detection limit.

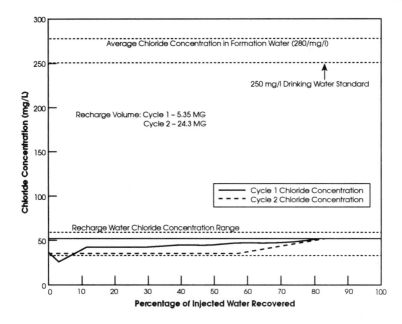

Figure 9.11   Chloride concentrations during cycles 1 and 2, Chesapeake, Virginia.

water subsequent to the test program was considerably lower than values recorded during the test program. The lower pH mobilized manganese present in the aquifer, causing discoloration of recovered water. A pH adjustment system was installed to ensure that recharge water pH remains at least above 8.0. Some of the stored water was pumped out to remove more acidic waters with associated concentrations of manganese. A trailer-mounted manganese removal treatment system was then tested to confirm that any ASR recovered water containing manganese could be treated prior to pumping into the distribution system. The city has continued to recharge the ASR well with water at acceptable pH. Recovered water quality has shown steady improvement with successive cycles, suggesting that pH adjustment is probably sufficient for controlling manganese concentrations once the residual acidity in the aquifer from earlier low pH operations has been overcome.

This experience suggests that pH control is particularly important in clayey sand aquifers, even when core analyses and geochemical analyses suggest that no problems are expected. The difference is due to changing pH values, outside the range of the original analyses. The ten core intervals selected for detailed analysis apparently did not include significant concentrations of manganese-bearing minerals that were present in the aquifer. Furthermore, the geochemical analysis did not reflect subsequent pH variability in water from the treatment plant.

Another important operating issue that arose following testing was residual alum floc in the distribution system. At times the color of the raw water from the Northwest River is such that higher concentrations of alum are used in the treatment process, some of which carries over into the distribution system and flows eventually down the ASR well. Well plugging became apparent, requiring redevelopment backflushing procedures to restore specific capacity.

The ASR test well was located several miles from the water treatment plant. As a result, the logistical issues that had to be resolved during the test program were more complicated than would have been the case if the well was located at the plant. Routine sampling and measurements of flows and water levels during the test program at the ASR site were quite intensive and therefore required a greater level of time and effort. The site selected for testing was superior hydrogeologically but required greater effort to test. This is an important consideration for ASR site selection in other areas.

9.8 SWIMMING RIVER, NEW JERSEY*

New Jersey American Water Company Eastern Division operates the Swimming River reservoir and water treatment plant, which is located in Monmouth County and has a design peak capacity of 159 megaliters/day (42 MG/day). During 1992, average day demand was about 102 megaliters/day (27 MG/day). Peak demands have reached the plant capacity during summer months for several years.

To augment peak capacity, three wells were constructed during 1988 in the Upper Potomac Raritan Magothy aquifer near the plant site; however, the use of these wells was subsequently restricted by the New Jersey Department of Environmental Protection and Energy so that they could only be used for production during a declared state of drought emergency. The site is located within a "critical area" established by the state, within which groundwater production has been excessive, causing groundwater levels to drop significantly and risking saltwater intrusion. In order to gain beneficial use of the investment in these wells, the utility initiated investigations to determine the feasibility of converting the wells to an ASR mode of operation. In this mode, the wells would be recharged during off-peak months and the stored water would be recovered to help meet peak summer demands. Net withdrawal from the aquifer would be zero. Investigations were conducted between 1991 and 1993.

---

* New Jersey American Water Company Eastern Division, 100 James Street, Lakewood, New Jersey 08701

Facilities include the ASR well (Well SR-1); an observation well at a radius of 32 m (105 ft), a pipeline connecting the ASR well to the adjacent water treatment plant, a sulfonator and injection system that was used temporarily at the beginning of testing to pretreat the aquifer, a permanent pH adjustment system to raise recharge water pH to about 9.0, wellhead piping, and a wellhouse. Existing wells SR-2 and SR-3 were also used for observation purposes during aquifer testing.

The storage zone is a clayey sand confined aquifer with the following hydraulic characteristics:

Transmissivity	41,000 to 53,000 G/day/ft
	509 to 658 m²/day
Storativity	9.0×10^{-5} to $6.1\ 10^{-4}$
Leakance	6.1×10^{-5} to 1.1×10^{-3}/day
Thickness	24 m (80 ft)
Specific capacity	21 G/min/ft (baseline during recovery)
Porosity	25% (average from five cores)

Extensive hydraulic testing was conducted during the test program, showing a small increase in specific capacity following initial pretreatment of the well and a subsequent decline to about 13 G/min/ft following six ASR cycles. Specific capacity appeared to stabilize at this level. The static water level is about 27 m (90 ft) below ground surface. The hydraulic gradient locally is about 0.001 ft/ft. **Figure 9.12** shows the core log for the storage zone while **Figure 9.13** shows the well design for Well SR-1, which was converted to ASR.

The ASR well has a 400 m (16-inch) steel casing to 180 m (590 ft). The hole is underreamed to a diameter of 700 m (28 inches) and provided with 21 m (70 ft) of 25 mm (10-inch), 0.065 slot stainless steel wire-wrapped screen and gravel pack. The bottom of the screen is at 204 m (668 ft). The well is equipped with a vertical turbine pump and recharge occurs through the pump column.

Recharge flow rates averaged about 2.2 megaliters/day (400 G/min) during the test program, while recovery rates ranged from 4.9 to 6.5 megaliters/day (900 to 1200 G/min). Recharge volumes ranged from 11 to 179 megaliters (3 to 47 MG) during seven cycles. Percent recoveries ranged from 55 to 90%. For geochemical reasons, a small volume is left underground during each cycle to ensure that any precipitation products such as ferric hydroxide do not plug the area close to the well. Typical recharge water quality is shown in **Table 9.8** while native water quality in the storage zone is shown in **Table 9.9.**

The Swimming River ASR site is one of the most geochemically complex sites completed to date. The native water quality in the storage

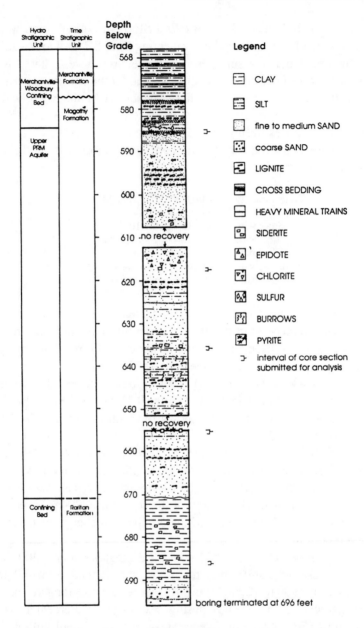

Figure 9.12 Core log of Upper PRM aquifer, Swimming River, New Jersey.

zone indicates an iron concentration of about 13 mg/L, suggesting the presence of siderite, or ferrous carbonate. This was confirmed through detailed analysis of cores obtained during continuous wireline coring of the observation well. Pyrite was also noted in some of the cores selected for analysis, but at concentrations of 1%. Recharge with treated surface

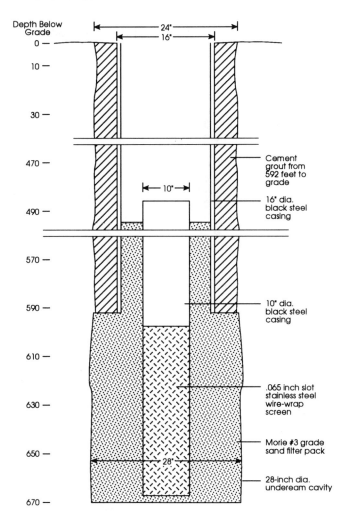

Figure 9.13 Construction of ASR well SR-1, Swimming River,
New Jersey.

water containing oxygen concentrations close to saturation values during winter months would be expected to cause precipitation of ferric hydroxide floc in the aquifer, plugging the well.

The pretreatment and testing plan was developed based upon column testing performed on selected sections of core material, simulating proposed ASR operations. The initial testing approach was to pretreat the aquifer around the well using deoxygenated water at a pH between 2.0 and 4.0. Approximately 121 megaliters (32 MG) of this low pH water was injected into the well, eliminating the siderite around the well and creating a plume of acidization products that was subsequently detected at the observation well. This was followed by a buffer volume of 8 megaliters (2

TABLE 9.8
RECHARGE WATER QUALITY: SWIMMING RIVER, NEW JERSEY

Constituent	November 1988 (mg/L)	January 1990 (mg/L)	NJDEP Drinking Water Standards (mg/L)
pH	8.2	8.7	6.5–8.5
Total dissolved solids	212.0	—	500
Hardness	68.0	—	—
Alkalinity	—	38.0	—
Bicarbonate	30.0	—	—
Calcium	21.21	41.0	—
Chloride	37.43	3.85	250
Fluoride	1.0	—	2.0
Nitrite	<0.01	—	—
Magnesium	3.67	5.00	—
Nitrate as N	<1.0	2.27	10
Sodium	27.65	25.60	—
Sulfate	37.0	38.40	250
Pesticides	<0.0001	—	—
Aluminum	0.059	1.00	—
Antimony	<1	—	—
Arsenic	<0.005	—	0.05
Barium	<0.05	—	1.0
Beryllium	<0.1	—	—
Boron	<0.1	—	—
Cadmium	<0.0002	—	0.010
Chromium	<0.001	—	0.05
Cobalt	<0.1	—	—
Copper	<0.02	—	1.0
Iron	<0.05	0.12	0.3
Lead	<0.005	—	0.05
Manganese	<0.02	<0.01	0.05
Mercury	<0.005	—	0.002
Molybdenum	<0.5	—	—
Nickel	<0.05	—	—
Potassium	3.28	3.50	—
Selenium	<0.003	—	0.01
Silver	<0.0002	—	0.05
Strontium	<0.2	—	—
Thallium	<0.5	—	—
Vanadium	<5	—	—
Zinc	<0.055	—	5.0
Total THMs	0.041	0.095	0.10
Silica	—	5.96	—
Phosphate	—	0.06	—

MG) of deoxygenated water at normal pH, followed by testing volumes using product water from the water treatment plant.

As shown in Figure 4.16, iron concentrations were initially higher than expected during the ASR test cycles that were conducted following the aquifer pretreatment phase. It was concluded that, although the siderite

**TABLE 9.9   NATIVE GROUND WATER QUALITY:
SWIMMING RIVER, NEW JERSEY, ASR WELL SR-1**

Constituent	June 1989 (mg/L)	January 1990 (mg/L)	NJDEP Drinking Water Standards (mg/L)
pH	6.0–6.4	6.15	6.5–8.5
Total dissolved solids	80	—	500
Hardness	24	—	—
Alkalinity	28	19.30	—
Calcium	6.61	8.50	—
Chloride	2.49	77.00	250
Fluoride	<0.2	—	2.0
Magnesium	1.86	3.25	—
Nitrate as N	<0.5	1.5	10
Nitrite	<0.01	—	—
Sodium	1.48	2.05	—
Sulfate	7.0	10.60	250
Phosphate	—	0.20	—
Aluminum	0.008	0.50	—
Antimony	<1	—	—
Arsenic	<0.005	—	0.05
Barium	0.06	—	1.0
Beryllium	<0.1	—	—
Boron	<0.1	—	—
Cadmium	<0.0002	—	0.010
Chromium	<0.001	—	0.05
Cobalt	<0.1	—	—
Copper	<0.02	—	1.0
Iron	13	11.20	0.3
Lead	<0.005	—	0.05
Manganese	0.07	0.28	0.05
Mercury	<0.0005	—	0.002
Molybdenum	<0.5	—	—
Nickel	<0.05	—	—
Potassium	1.74	2.70	—
Selenium	<0.003	—	0.01
Silver	<0.0002	—	0.05
Strontium	<0.2	—	—
Thallium	<0.5	—	—
Vanadium	<5	—	—
Zinc	<0.05	—	5.0
Silica	—	4.16	—
Total THMs	<0.0005	—	0.10

was no longer present around the well, the pyrite in the aquifer had probably been "polished" by the acidization pretreatment process and was therefore highly reactive to oxygen-bearing water, forming a ferric hydroxide precipitate. The iron content of the recovered water reduced with successive cycles, reflecting a probable buildup of oxidation products on the pyrite grains, thereby reducing pyrite reactivity. However, the iron concentrations in the recovered water were still too high. As a result, pH adjustment facilities using sodium hydroxide were installed to raise the pH of the recharge water from about 7.6 to about 9.0, thereby reducing the potential for ferric hydroxide formation. This was implemented following Cycle 5. In subsequent Cycles 6 and 7, iron concentrations in the recovered water declined to below 1 mg/L. At this concentration, the water can be easily blended with water treated at the plant so that the blend meets drinking standards going to the consumers during summer months.

The same reactions also mobilized manganese associated with iron-bearing minerals in the aquifer. As shown in Figure 4.17, manganese concentrations were initially very high but decreased to within drinking standards once the pH adjustment system was placed in service.

Some well plugging was noted during early cycle testing. Detailed analysis of recharge and recovery water quality indicated that the plugging was due to alum floc associated with filter backwash operations at the treatment plant, rather than ferric hydroxide precipitation in the aquifer. The recharge water source was then relocated to a different point in the plant piping to eliminate the solids in the water. This resolved the plugging problem. Nevertheless, operating experience at this site suggests that backflushing to waste at least weekly is required for a period of about 15 min. When backflushing frequency was increased to two or three times per week, overall recharge performance improved. Currently the well is backflushed daily for 15 min during recharge periods.

The operating mode for this site is to recharge at 25 L/sec (400 G/min) for about 240 days each year. This water can be recovered at a rate of about 66 L/sec (1050 G/min) for about 90 days during the period of peak summer demand. Addition of the two remaining wells would increase peak recovery capacity at this site to about 19 megaliters/day (5 MG/day).

9.9 WILDWOOD, NEW JERSEY*

Wildwood is a coastal resort community on the Cape May peninsula of New Jersey. Tourist population is substantially greater during summer months than that of the remainder of the year. Water is obtained from the Rio Grande wellfield located 5 miles inland, comprising 13 wells with a combined capacity of about 51 megaliters/day (13.5 MG/day). During 1968 Wildwood began recharging two coastal wells during offpeak months, recovering the stored water at much higher rates to help meet summer peak demands. This was the first ASR system in the U.S. and has been in operation since that time, adding a third well in 1983 and a fourth in 1985. Each year about 380 megaliters (100 MG) is stored in each of the four ASR wells during offpeak months. During summer months about 303 megaliters

---

* Wildwood Water Utility, City of Wildwood, Pine and New Jersey Avenue, Wildwood, New Jersey 08260

(80 MG) is recovered from each well, while the remainder is left underground to help control saltwater migration into the aquifer.

Recharge rates in individual wells typically range from 2.2 megaliters/
day (400 G/min) at the beginning of the recharge season to about 1.1
megaliters/day (200 G/min) at the end, while recovery rates are about 3.5
to 5.4 megaliters/day (650 to 1000 G/min). Summer water demands average about 38 megaliters/day (10 MG/day) while peaks reach 49 megaliters/
day (13 MG/day). About 13 megaliters/day (3.5 MG/day) of the peak
summer demands are met from the ASR wells. Winter demands average
about 9 megaliters/day (2.5 MG/day). The ASR wells provide system
reliability in case the pipeline from the interior wellfield is broken acciden-
tally. The wells also provide distribution system storage, which has been
partially substituted for elevated storage in the coastal service area.

All four ASR wells utilize the Cohansey aquifer. Static water levels in
this aquifer are about 4 m (12 ft) below land surface, while well depths are
about 76 m (250 ft).

Aquifer hydraulics are not well known. Transmissivity of one well in
the Cohansey aquifer has been estimated at about 1,080 m^2/day (87,000 G/
day/ft). This well has a total depth of 100 m (328 ft) and has a 300 mm (12-
inch) screen (0.045 slot) from 65 m (212 ft) to the bottom. During a
pumping test at a rate of 5.6 megaliters/day (1034 G/min) the specific
capacity was 24 G/min/ft. Recovery specific capacities typically range
from 9 to 13 G/min/ft, associated with drawdowns of about 30 m (100 ft).

The wells are backflushed daily for about 10 min to remove iron floc
formed from reaction of oxygen and chlorine in the water with ferrous iron
present in the aquifer, and also to remove rust from the distribution system,
which enters the wells during recharge. The wells are fully redeveloped
about every 5 years and are chemically treated with surging and acid
treatment about twice per year to maintain the screen capacity.

The ASR system has operated successfully for 26 years and two more
ASR wells are planned.

9.10 KERRVILLE, TEXAS*

The Upper Guadalupe River Authority supplies treated drinking water
to the City of Kerrville, TX, from a small instream reservoir and 19
megaliters/day (5 MG/day) water treatment plant on the Guadalupe River.
These facilities began operation during 1980, supplementing groundwater
withdrawals by the city, which had increased steadily in prior years,
causing groundwater levels to decline by at least 100 m (330 ft). Water
demands for Kerrville vary seasonally, with peak demand occurring dur-

---

* Upper Guadalupe River Authority, 215 W. Water Street, Kerrville, Texas 78029

ing the summer at typically 1.53 times average demand. The Authority provides water at rates up to the plant capacity, and the city meets higher summer peak demands from its wells.

Faced with the need to construct a new offstream reservoir at great expense to meet projected increases in water demand while also meeting downstream minimum flow requirements, the Authority conducted an investigation of ASR feasibility. This investigation was completed successfully in three phases between 1988 and 1992, including a preliminary feasibility investigation; construction and testing of an ASR zone observation well; and construction and testing of an ASR well, wellhead facilities, and two additional observation wells in overlying production zones. The associated facilities are operational and were fully permitted during 1993.

The ASR storage zone is the Hosston-Sligo formation in the Lower Trinity aquifer. The storage zone is a confined sandstone and conglomerate artesian aquifer. Hydraulic testing at the ASR well indicated the following aquifer characteristics:

Transmissivity	7000 G/day/ft (87 m²/day)
Storativity	0.0007
Leakance	5×10^{-7}/day from core permeability
Specific capacity	7.3 G/min/ft recharge at 600 G/min
	6.8 G/min/ft recovery at 900 G/min
Porosity	0.23 from core analyses

During the test period, static water elevation was about 460 m (1510 ft) above sea level, or about 39 m (128 ft) below land surface elevation of about 499 m (1638 ft). Local gradient of the potentiometric surface was about 0.0011. **Table 9.10** shows the general lithology of the storage zone and overlying formations.

TABLE 9.10   GENERAL LITHOLOGY: KERRVILLE, TEXAS

Formation Name	Average Thickness in Kerrville Area (ft)	Primary Materials
Upper Member of Glen Rose Limestone	130	Fossiliferous limestone, shale, and marl
Lower Member of Glen Rose Limestone	210	Fossiliferous limestone, dolomite, marl and shale
Hensell Sand	55	Conglomerate, shale, sand, dolomite and marl
Cow Creek Limestone	35	Sandy limestone
Pine Island Shale	18	Shale with some sand and limestone
Hosston-Sligo Formation	75	Conglomerate, sand, and shale

The ASR well, referred to as Well R-1, is constructed of 16-inch, epoxy-coated steel casing to a depth of 151 m (495 ft). The well is open hole with a nominal diameter of 15 inches to a depth of 187 m (613 ft). The well is equipped with a submersible pump. Injection is through one or both of two stainless steel injection tubes, or through the casing annulus. Recharge can occur at variable rates between 13 and 63 L/sec (200 and 1000 G/min). Design recovery capacity is 50 L/sec (800 G/min).

Analysis of the recharge and native groundwater qualities showed that they are similar. Analysis of cores obtained in the ASR zone indicated that geochemical problems were not expected from ASR operations at this site. Consequently, ASR testing focused on confirmation of hydraulic response of the well and aquifer to ASR operations.

Two ASR test cycles were conducted. During the first cycle, 11 megaliters (3 MG) of treated drinking water were recharged at an average rate of about 43 L/sec (675 G/min). The water was recovered at an average rate of 56 L/sec (890 G/min). During the second cycle, 94 megaliters (25 MG) were stored at an average rate of about 38 L/sec (600 G/min). After a 20 day storage period, all of the water was recovered at a rate of about 57 L/ sec (900 G/min). As shown in **Table 9.11,** specific capacity declined during extended recharge, from 7.2 G/min/ft initially to 4.9 G/min/ft after 29 days. However, no indication of residual well plugging was noted. The decline in specific capacity was attributed primarily to mounding in the storage zone.

The objective of ASR testing at this site was to confirm the feasibility of seasonal and long-term aquifer storage of a sufficient volume of treated drinking water so that the Authority could meet projected water demands without construction of an expensive offstream reservoir. This objective was achieved with issuance of the final permit in 1993. However, some opposition to permit issuance was raised by a whitewater boating organization, on the grounds that diversion of a small percentage of flows for ASR storage during high flow months would adversely affect their recreational activities. This is significant in that it is the only opposition to a treated drinking water ASR project known to date in the U.S. The opposition was on recreational, not environmental grounds.

A significant aspect of the Kerrville ASR project is the operating plan that was developed following completion of the test program. Based upon extensive regional aquifer modeling, the plan entails recharging whenever water is available from the water treatment plant and water level elevations are below a range of 457 to 469 m (1500 to 1540 ft). Model simulations indicated that this would maintain sufficient volume of water in aquifer storage so that, during a repeat of the worst drought on record, the well

TABLE 9.11   SPECIFIC CAPACITY DURING CYCLE TESTING:
KERRVILLE, TEXAS

Day No.	Recharge (G/min/ft)		Recovery (G/min/ft)	
	Cycle 1	Cycle 2	Cycle 1	Cycle 2
1	6.8	7.2	7.2	6.8
2	6.2	6.7	6.8	6.6
3	6.2	6.0		6.3
4		6.0		6.0
5		6.0		6.1
6		5.8		5.8
7		5.7		5.7
8		5.5		5.4
9		5.4		5.6
10		5.3		5.8
11		5.3		5.2
12		5.2		5.6
13		5.2		5.2
14		5.1		5.0
15		5.1		5.4
16		5.1		5.1
17		5.2		4.8
18		5.2		4.9
19		5.2		
20		5.2		
21		5.2		
22		5.1		
23		5.1		
24		5.1		
25		5.1		
26		5.0		
27		5.0		
28		4.9		
29		4.9		

would continue to supply water at an acceptable rate for the 4-year duration of the drought. This drought provides an important reference point for
water supply regulation in Texas. The storage volume required for Kerrville
is about 4.3 Mm^3 (3500 acre ft) to meet projected demands in the year
2040.

With the success of the ASR program, the city was able to save the
approximately $30 million expected cost of an offstream reservoir. In-
stead, ASR wells can be constructed in low-cost increments to meet
increasing demands, at a total expected cost of under $3 million. To meet
projected needs up to the year 2015, an investment of about $1 million is
anticipated.

This project won an Honors Award in the American Consulting
Engineers Council National Engineering Excellence Competition for

1993. Since then, the City of Kerrville has converted an existing water supply well, R-5, to ASR operation, supplementing the yield of Well R-1.

9.11 HIGHLANDS RANCH, COLORADO*

The Centennial Water and Sanitation District supplies water to the Highlands Ranch development in the southern Denver metropolitan area. It currently serves about 22,000 residents. At expected buildout by the year 2022, it is anticipated that the district will serve 36,000 residential dwelling units on 3700 hectares (8889 acres). To meet water demands, the district utilizes surface water from McClellan Reservoir, supplemented with groundwater production from 29 wells penetrating 3 local aquifers. These are the Denver, Arapahoe, and Laramie-Fox Hills aquifers. Typical monthly water demand ranges from a low of 36% of annual average demand in February to a high of 217% in July. To meet increasing peak demands, the Highlands Ranch Water Treatment Plant was originally planned for expansion in 1996. ASR investigations conducted during 1991 to 1993 demonstrated that aquifer storage of treated drinking water would enable the district to defer water treatment plant expansion and reduce capital costs.

ASR facilities include ASR Well A-6, which was converted from an existing supply well in the Arapahoe formation, associated piping and pumping facilities, and an observation well in the same aquifer. Three cycles of testing were conducted during 1992 and full operation began during 1993 when the well was fully permitted.

The storage zone is a confined artesian aquifer, the producing intervals of which are principally made up of loosely cemented sands mixed with potassium feldspar. Aquifer hydraulic characteristics were obtained from two pumping tests, as follows:

Transmissivity	8,500 G/day/ft (106 m²/day)
Storativity	0.000336
Leakance	negligible
Porosity	0.25 estimate
Gradient	0.0008 to 0.0085

This site has probably the lowest transmissivity of the operational ASR sites in the U.S. as of 1993.

The depth to static water level during the test program was about 250 to 268 m (820 to 880 ft). Less than 1% of the mean annual precipitation of 279 to 457 mm (11 to 18 inches) reaches the Arapahoe formation as

---

* Centennial Water and Sanitation District, 62 West Plaza Drive, Highlands Ranch, Colorado 80126

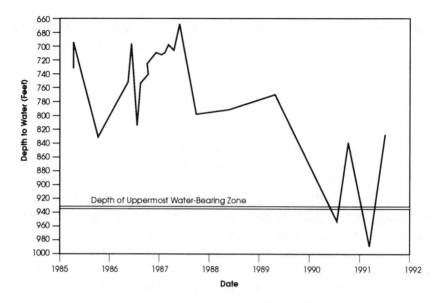

Figure 9.14 Hydrograph for Arapahoe Well A-6, Centennial Water and Sanitation District, Highlands Ranch, Colorado.

recharge. As shown in **Figure 9.14,** water levels in the Arapahoe formation have been declining rapidly in recent years due to increasing groundwater production and are approaching the top of the first water-bearing zone in the aquifer at a depth of about 281 m (922 ft). The aquifer is 142 m (467 ft) thick, of which 61 m (200 ft) comprise 10 water bearing zones in intervals with thicknesses of 3 to 9 m (10 to 30 ft). Ground elevation is about 1600 m (5250 ft). **Table 9.12** shows the results of a vertical velocity profile obtained prior to testing of the ASR well, in order to establish a baseline against which future results might be compared and thereby evaluate any plugging effects. Half of the flow to the well occurs within the top 38 m (100 ft) of the aquifer.

Sidewall cores were obtained during construction of an observation well at the ASR site, indicating that the storage zone is a sandstone aquifer, predominantly quartz at the top and with potassium feldspar increasing to about 5 to 10% with depth. Calcite is present throughout, while a trace of smectite was found in the lowest three of the four cores. Illite/mica was present throughout at concentrations of less than 2%. Pyrite is present as a trace constituent throughout.

Well A-6 was constructed during 1985. It has a 10.75 inch (outer diameter) steel casing with a wall thickness of 0.365 inches. Ten 10-inch diameter nominal pipe size screen intervals are located between depths of 281 m to 413 m (922 to 1354 ft), each of which is provided with 0.020 inch wire-wrapped screen. A 6 m (20 ft) tailpipe is provided below the bottom producing zone to a depth of 419 m (1374 ft). The screen has no gravel pack and the tailpipe has no end cap on the bottom. The well is equipped with a variable frequency drive submersible pump set at a depth of 387 m (1270 ft). The pump is rated at 21 L/sec (330 G/min); however, it can produce at rates exceeding 32 L/sec (500 G/min). The pump column outer diameter is 140 mm (5.5 inches). Considering the column couplings, the annular space in this well is very small and the surface area open to corrosion would be substantial if recharge occurred down the annulus. Therefore, recharge occurs down the pump column.

Table 9.13 shows typical recharge water quality and also average native water quality in the storage zone. Initial geochemical investigations indicated some concern regarding potential plugging due to calcium carbonate and ferric hydroxide precipitation, particularly in the upper intervals of the aquifer.

Three test ASR cycles were conducted during a six month period. During Cycle 1, 5.7 megaliters (1.5 MG) was injected at an average rate of 12.5 L/sec (198 G/min). This was recovered at an average rate of 19.5 L/sec (309 G/min). During Cycle 2, 22 megaliters (5.8 MG) were injected at an average rate of 16.6 L/sec (263 G/min) and 27.2 megaliters (7.2 MG) were recovered at an average rate of 25.9 L/sec (410 G/min). During Cycle 3, 67.4 megaliters (17.8 MG) were injected and 82.3 megaliters (21.7 MG) were recovered at the same rates as in Cycle 2. During recovery, pumping water levels fell below the uppermost screen interval in the well. Head buildup during recharge was substantial, with heads rising to less than 122

TABLE 9.12   VERTICAL VELOCITY PROFILE: HIGHLANDS RANCH, COLORADO

Screened Interval (ft)	Screen Length (ft)	Net Flow per Interval (G/min)	Apparent Hydraulic Conductivity (ft/day)
922 to 942	20	0.51	0.034
954 to 964	10	11.1	1.461
985 to 1,015	30	21.79	0.956
1,061 to 1,081	20	11.54	0.760
1,100 to 1,120	20	8.32	0.548
1,143 to 1,163	20	7.17	0.472
1,209 to 1,219	10	0.3	0.039
1,231 to 1,251	20	2.61	0.172
1,260 to 1,280	20	2.9	0.191
1,314 to 1,344	30	2.58	0.113
Totals	200	68.82	4.75

m (400 ft) below land surface. The substantial water level fluctuations in this well during cycle testing are shown in **Table 9.14.**

Sand was found in the recharge water as determined by a cartridge filter and a Rossum sand tester. However, this was purged from the well during recovery and no evidence of residual well plugging was found between successive cycles. The sand was probably settled in the water distribution system and was carried into the well during reverse flows occurring during recharge. Sand content in the recharge water declined after the first cycle.

Upon completion of testing, it was recommended that the well should be backflushed to waste at least every 4 weeks for a period of 4 to 8 hours to maintain its capacity. It was also recommended that pH of the recharge water should be maintained within a range of 7.5 to 8.3 to minimize potential for precipitation of ferric hydroxide and calcium carbonate.

The ASR well is now in full operation and two additional wells are being converted to ASR use. The objective of the ASR system is to meet seasonal peak water demands through more efficient use of existing supply and treatment facilities. All of the stored water will be recovered. Long-term drought storage is also under consideration and is the subject of an

TABLE 9.13   RECHARGE AND NATIVE WATER QUALITY: HIGHLANDS RANCH, COLORADO

Parameter	Average Recharge Water Quality (a)	Average Groundwater Quality (a)
Alkalinity	92	123
Aluminum	0.05	0.06
Ammonia	0.43	0.17
Boron	0.05	0.10
Calcium	51	32
Chloride	28	3.25
Eh, millivolts (estimated)	400	200
Fluoride	0.95	0.75
Iron	0.015	0.13
Magnesium	10	1.85
Manganese	0.005	0.02
Nitrate	0.05	0.05
pH, laboratory	7.6	8.1
Phosphorus	0.02	0.005
Potassium	2.85	3.75
Silica	4.95	12.85
Sodium	29.5	27.5
Sulfate	91	16.5
Total dissolved solids	288	180

Note: All values expressed in milligrams per liter (mg/L) unless otherwise specified.

(a) To compute the average values, half the detection limit value was used when the value for the parameter was below the analytical detection limits.

TABLE 9.14 WATER LEVEL RESPONSE DURING ASR TESTING: HIGHLANDS RANCH, COLORADO

Test	Rate (G/min)	Observation Well Drawdown/Drawup			Pumping Well Drawdown/Drawup		
		24 Hour	72 Hour	End of Test	24 Hour	72 Hour	End of Test
Pump Test	300	24.2	—	24.5 (26.8 hr)	186.5	—	110.5 (26.8 hr)
Cycle 1 Inj	198	19.6	22.17	23.4 (121.6 hr)	186.5	241.5	272.9 (121.6 hr)
Cycle 1 Rec	309	25.5	32.9	34.0 (86.7 hr)	89.0	96.9	97.8 (86.7 hr)
Cycle 2 Inj	263	25.0	30.5	35.9 (364.8 hr)	195.0	263.7	381.8 (364.8 hr)
Cycle 2 Rec	410	39.0	45.5	55.9 (291.7 hr)	139.1	146.3	156.0 (291.7 hr)
Cycle 3 Inj	263	28.7	32.6	55.6 (1,125 hr)	174.9	225.7	467.5 (1,125 hr)
Cycle 3 Rec	410	36.6	40.6	40.4 (976.7 hr)	151.1	146.0	130.5 (976.7 hr)

ongoing aquifer modeling investigation. Under the terms of the current state permit, stored water must be recovered within 5 years; however, this period may be extended.

An important issue at the Centennial ASR site is the great depth to water level and the concern regarding effective hydraulic control over recharge flow rates. Any cascading of water would rapidly plug the low transmissivity aquifer due to air binding. As discussed in Chapter 4.4, Flow Control and Measurement, a new downhole flow control valve was developed and applied successfully at this site. This valve enabled recharge down the pump column at variable flow rates without cascading. The valve is located at the base of the pump column immediately above the pump. It operates as a packer between the pump column and the casing and is controlled hydraulically from the surface. It does not interfere with pump operation during recovery. The successful development of this valve is an important development that will benefit future well recharge and ASR operations in other areas with sufficient depth to static water level that cascading control is required. The valve is manufactured by Baski Valve Company, Denver, CO. A schematic diagram is included in **Figure 9.15.**

The Centennial ASR site is also significant in that it is the only site to date that has utilized sidewall cores instead of continuous wireline cores to provide a basis for geochemical assessment. Four sidewall cores were obtained in the Arapahoe formation from the observation well. The small volume and the disturbed nature of these cores limited their usefulness; however, X-ray diffraction, cation exchange capacity and acid insoluble residue tests were conducted. An analysis of the data provided a limited basis for inferring the oxidation state of the formation materials and their potential susceptibility to plugging during ASR operations.

Analysis of water quality data collected during the three test cycles indicated that most changes in constituent concentrations could be accounted for through mixing between recharge and native water. Initial concerns regarding iron and calcium precipitation were not borne out in the data. However, pH showed a tendency to reach native water levels more rapidly than would be indicated by mixing, and trihalomethane concentrations also declined very rapidly. Total organic carbon concentrations also declined underground at a rate exceeding that due to dilution effects alone.

During Cycle 3, the ASR well experienced hydraulic fracturing of the aquifer. This occurred when the increase in water level during recharge reached 125 m (410 ft), of which approximately 23 m (75 ft) was attributable to aquifer response and the remainder to wellbore plugging. Subsequently, both the ASR well and the observation well experienced reduced hydraulic response to recharge and recovery operations.

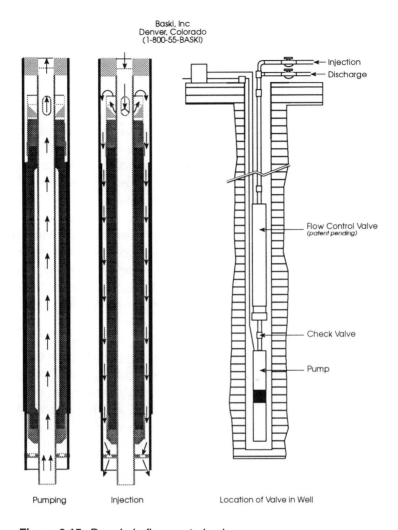

Figure 9.15 Downhole flow control valve.

9.12 LAS VEGAS VALLEY WATER DISTRICT, NEVADA*

Las Vegas is located in the Mohave desert and receives about 100 mm (4 inches) of rainfall per year. Surrounding mountains may receive greater than 500 mm (20 inches) of precipitation, mostly in the form of snow. Water supply to this area is supplied by the Las Vegas Valley Water District, which operates what is currently (1993) the largest ASR system in the world with a recharge capacity of about 300 megaliters/day (80 MG/day) and a recovery capacity probably exceeding 380 megaliters/day (100 MG/day).

Recharge water is obtained from the treated water distribution system, which is supplied from the Colorado River at Lake Mead. Water is low in

* Las Vegas Valley Water District, 3700 West Charleston Boulevard, Las Vegas, Nevada 89153

color and turbidity, but is hard. It is treated in a 1500 megaliters/day (400 MG/day) direct filtration water treatment plant that uses pre- and post-chlorination. As of 1990, about 80% of the local water supply was obtained from the Colorado River, while the remaining 20% was from groundwater. The groundwater system has been overdrafted for several decades due to population growth and development.

The ASR site is about 20 miles from the water treatment plant. ASR operations began at two wells during 1987 and have since been expanded to over 35 wells. In addition, the district operates five single-purpose injection wells and four wells that are used only for production. Nine additional single-purpose injection wells are planned.

The aquifer being recharged is comprised of basin-fill clastic deposits from surrounding mountain ranges consisting of sandy gravels with a trace of sandy clay, poor- to well-cemented conglomerate, cemented sandstone, limestone, dolomite, and traces of calcite to depths of over 300 m (1000 ft). Production zones within these clastic deposits are generally semi-confined although unconfined conditions frequently develop, associated with delayed drainage.

Figure 9.16 shows lithology and construction details for Wells 16 and 17, which are considered to be reasonably representative of the more productive wells in the wellfield, with typical recharge and production rates of about 11.4 megaliters/day (3 MG/day) each. Static water levels are currently about 160 to 240 ft below land surface, which is typically at an elevation of about 2000 ft above sea level in the wellfield area.

Aquifer hydraulic characteristics vary within the following typical ranges:

Transmissivity	1000 to 300,000 G/day/ft
	12 to 3726 m^2/day
Storativity	0.001 in confined areas
	(increasing with time)
Specific yield	5 to 30%, average about 15%

Maximum water level decline from predevelopment conditions near the pumping center has been about 240 ft. As a result, some subsidence has occurred.

Recharge occurs down the pump column of the vertical turbine pumps. The pump impellers are lowered onto the bowl housing, providing sufficient back-pressure so that a solid column of water occurs in the pump column. In some wells the check valve is adjusted in each well to provide for seasonal recharge. In wells that have been converted more recently, bypass piping has been provided around the check valve, equipped with a remote-actuated pressure reducing valve to facilitate recharge. Magnetic

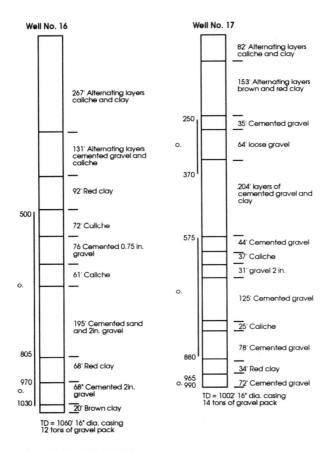

Source: Brother and Katzer, 1990.

Figure 9.16 Construction of ASR Wells 16 and 17, Las
Vegas Valley Water District, Nevada.

flowmeters are used to accommodate bi-directional flow during recharge
and recovery periods.

Recharge rates vary among the wells, within a typical range of about 3.8
to 11.4 megaliters/day (1 to 3 MG/day). Recharge occurs during fall, winter,
and spring months when demand is reduced and the treatment and distribution system has available capacity. Recovery occurs during summer months
to help meet peak system demands. Recovery rates are typically similar to
recharge rates. Well clogging rates have been sufficiently low that no
redevelopment backflushing has been required to maintain recharge rates,
other than that which occurs normally during peak season recovery.

Table 9.15 shows a chemical analysis of groundwater from selected
district wells, while **Table 9.16** shows similar analyses for water from

TABLE 9.15 WATER QUALITY ANALYSES
FROM SELECTED WELLS: VALLEY WATER
DISTRICT, LAS VEGAS, NEVADA

Constituent (mg/L)	Local Well Number and Date of Collection	
	17 8/87	16 8/87
Alkalinity (HCO)	235	228
Calcium	54.1	53.2
Magnesium	27.1	27.3
Chloride	5.0	4.7
Nitrate (N)	0.64	0.68
Potassium	2.26	3.24
Conductivity (μmhos cm^{-1})	461	467
Sodium	7.11	7.04
Sulfate	53.9	63.0
pH (units): lab	8.06	8.10
pH (units): field	7.41	7.54

Source: Brothers and Katzer (1).

Lake Mead, both before and after treatment. Treated water is used for recharge purposes.

The principal objective of this ASR system has been to seasonally store available water to meet peak summer demands and also to meet long-term water requirements. The system has been very successful in this regard, having stored approximately 80 Mm^3 (21 BG, 65,000 acre ft) through 1993. Principal concerns have related to the quality of the recharge water which, while meeting all standards, has a higher conductivity than the native groundwater and a possible trend of increasing conductance in recent years. Other water quality concerns have included the potential for calcite precipitation in the aquifer, which has proven to not be a problem, and the potential formation of trihalomethanes during storage. If THMs form underground and are not reduced, they could possibly exceed expected future standards for drinking water. Investigations conducted by the district have suggested that no THM reduction occurred during ASR storage during the test period at this site.

9.13 CALLEGUAS MUNICIPAL WATER DISTRICT, CALIFORNIA*

The North Las Posas Groundwater Basin is about 29 km (18 miles) long and 7 km (4.5 miles) wide, and is located in southern California near Los

---

* Calleguas Municipal Water District, 2100 Olsen Road, Thousand Oaks, California 91362

TABLE 9.16   RECHARGE WATER QUALITY: LAS VEGAS VALLEY WATER DISTRICT, NEVADA

Constituent (mg/L)	Date of Sample Collection									
	6/73		10/75		6/79		11/86		12/87	
	Raw	Treated	Raw	Treated	Raw	Treated	Raw	Treated	Raw	Treated
Alkalinity (HCO)	159	161	171	159	154	155	159	157	159	154
Calcium	90	90	86	90	86	86	78	74	77	72
Magnesium	26	26	22	29	29	28	18	21	18	22
Chloride	90	92	104	108	90	90	53	55	64	64
Fluoride	0.37	0.39	1.00	1.00	0.31	0.28	0.28	0.28	0.29	0.29
Nitrate (N)	1.70	1.70	0.07	0.02	0.48	0.48	0.41	0.44	0.51	0.54
Potassium		3.90	4.30	6.0	6.0	3.7	3.7	4.2	4.2	
Conductivity (μmhos cm¹)	1072	1057	1077	1035	1118	1107	846	857	829	833
Sodium	105	107	109	134	115	115	70	71	63	69
Sulfate	290	290	311	309	301	301	232	216	211	206
pH (units)	8.20	8.20	8.20	7.80	7.95	7.75	7.92	7.84	7.76	7.53

Note: Analysis of Colorado River water in Lake Mead, before and after treatment.
Source: Brothers and Katzer (1).

Angeles. Water to this area is supplied by the Calleguas Municipal Water District, which imports all of its water from the Metropolitan Water District of Southern California. This is imported from northern California through the State Water Project, which is an extensive aquaduct system.

Investigations have shown that the North Las Posas Groundwater Basin has available storage capacity of about 370 Mm^3 (300,000 acre ft), primarily as a result of historic groundwater production. However, surface recharge with seasonally available imported water is unlikely to be feasible and cost effective due to hydrogeologic constraints. ASR investigations were therefore initiated during 1990 to determine the feasibility of using aquifer storage to meet seasonal, emergency, and long-term drought storage needs, while also improving the overall reliability of the Metropolitan Water District's operations.

The basin contains sediments of Pliocene to Recent Age, including two principal freshwater aquifers that are separated by 15 to 30 m (50 to 100 ft) of fine-grained sediments of low permeability. The Fox Canyon aquifer consists of marine and non-marine sands and gravel about 61 to 122 m (200 to 400 ft) thick. This is underlain by the Grimes Canyon aquifer, which consists of marine sands with minor gravel about 30 to 122 m (100 to 400 ft) thick. Both aquifers are confined except in narrow outcrop areas along portions of the hills that border the northern and southern margins of the basin. **Figure 9.17** shows a geologic cross section in the vicinity of the ASR well. Typical well yields are between 5.4 and 8.1 megaliters/day (1000 and 1500 G/min). The basin is believed to be overdrafted; however, there is some uncertainty, since estimates range from zero to the full amount of existing production. A mid-range estimate of the rate of overdraft is about 14 Mm^3/year (11,000 acre ft/year) at current production rates.

Initial testing was conducted at the Ventura County Water Works existing Well 97, which is 283 m (930 ft) deep and penetrates the upper 61 m (200 ft) of the Fox Canyon aquifer, which is 137 m (450 ft) thick at this site. An adjacent observation well was constructed with three sampling intervals above, within, and below the aquifer interval penetrated by the ASR test well. Cores were obtained and analyzed during construction of the observation well in order to gain understanding of the geochemistry issues at this site.

Aquifer hydraulic characteristics at the ASR test site were as follows:

Transmissivity	145,000 G/day/ft
	1800 m^2/day
Storativity	4×10^{-6}
Porosity	23% (from laboratory analysis of cores)

Dispersivity	22 ft (calculated)
Well efficiency	87%
Specific capacity	18 to 20 G/min/ft (initial production)
	10 G/min/ft (injection, 1 hr)
	7 G/min/ft (injection, 1 day)
	4 G/min/ft (injection, 1 week)
	2.5 G/min/ft (injection, 43 days)

Upon completion of testing, production specific capacity was measured at 23 G/min/ft, indicating no residual clogging and no loss in well performance.

Analysis of data from the injection portion of the second test cycle indicated a lower transmissivity value of 546 m²/day (44,000 G/day/ft), reflecting the effects of well plugging, interference from adjacent wells, and the 11°C temperature difference between the recharge water and the native water in the aquifer.

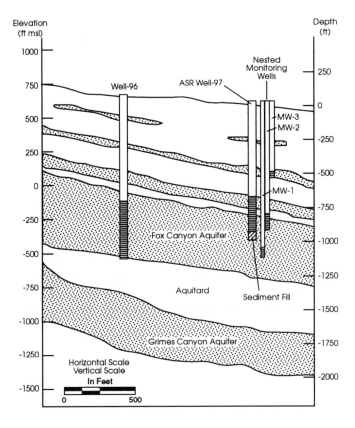

Figure 9.17 Hydrogeologic cross-section, Calleguas Municipal Water District, California.

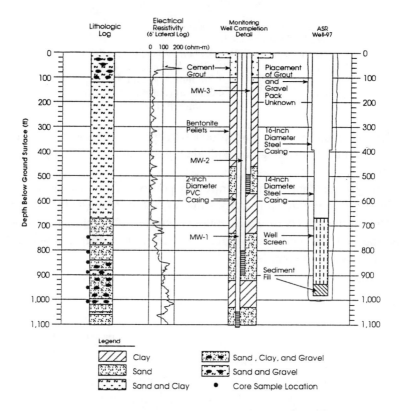

Figure 9.18 Lithology and well construction, Calleguas Municipal Water District, California.

Static water level was about 64 to 67 m (210 to 220 ft) above mean sea level or about 114 m (375 ft) below ground surface, which is at an elevation of about 180 m (590 ft) above mean sea level. The well was constructed in 1947. It has a 350 mm (14-inch) casing perforated from 204 to 283 m (670 to 930 ft) with slots measuring 5 mm (3/16 inches) wide and 200 mm (8 inches) long. **Figure 9.18** shows the well construction and also lithology at this site. The well is equipped with a vertical turbine pump and recharge occurs down the pump column after the pump impellers are lowered against the bowl housing.

Recharge flow rates of up to 3.4 megaliters/day (620 G/min) were achieved during the test program, while recovery rates were up to about 4.3 megaliters/day (800 G/min). Two ASR test cycles were conducted, plus two step-drawdown pumping tests, one constant rate pumping test and an injection test. The first test cycle was brief, to assess initial hydraulic and water quality effects. The second test cycle injected 142 megaliters (115 acre ft) over 43 days. This was then recovered; however, recovery continued until 138% of the recharge water had been produced in order to return water quality to background levels.

Table 9.17 shows native groundwater and also recharge water quality during the test program. Using chloride as a tracer, analysis of the recovered water quality data from Cycle 2 showed that iron and manganese concentrations in the native groundwater (242 and 43 µg/l, respectively) were being reduced, reflecting oxidation. THM concentrations were also reduced substantially. THM formation potential was determined on a sample of the recovered water following chlorination, showing an increase but to a level below the range of THM concentrations in the recharge water. No significant cation exchange took place during ASR testing.

Based upon successful completion of the test program, a second new ASR well was constructed and placed into operation. Five additional ASR wells are to be added in the immediate future. The purpose of these wells is seasonal and long-term storage of water purchased from the Metropolitan Water District of Southern California during off-peak months at discounted rates.

The most important technical issue at this site has been well clogging, which received considerable attention during and after the test program. The test well is located in a portion of the distribution system that has little circulation. Large volumes of solids present in the distribution system, including rust and sand, were swept into the well during initial recharge, causing immediate rapid clogging. Results demonstrated the ability to restore well capacity by pumping. They also suggested the wisdom of flushing the well and pipelines to waste prior to recharge and recovery, and not allowing substantial clogging to occur prior to backflushing the well during recharge.

Existing wells are equipped for power generation during recharge.

9.14 GOLETA, CALIFORNIA*

Goleta Water District, CA, is believed to be the third ASR system to begin operation in the U.S., following Wildwood, and Gordons Corner, NJ. The district is located in Santa Barbara County in southern California and serves about 70,000 people. Average water demand for the district is about 15 MG/day (57 megaliters/day). ASR operations began in 1978.

The original water supply for this area was obtained from wells; however, the quality in some areas was poor, requiring treatment for iron, manganese, and hydrogen sulfide removal. The wells were abandoned or converted to agricultural use when the Cachuma Reservoir became avail-

---

* Goleta Water District, PO Box 788, Goleta, California 93116

**TABLE 9.17   RECHARGE AND NATIVE WATER QUALITY:
CALLEGUAS MUNICIPAL WATER DISTRICT, CALIFORNIA**

Constituent	Units	ML^3	Jensen Treatment Plant 8/1/90	Recharge Water ("Vault-1") 12/11/90	Native Groundwater Well 97 12/11/90
Calcium	mg/L		25	23	68
Magnesium	mg/L		16	15	13
Sodium	mg/L		76	73	36
Potassium	mg/L		4	3	3
Carbonate	mg/L		2	1	0
Bicarbonate	mg/L		99	100	189
Sulfate	mg/L	500	53	46	121
Chloride	mg/L	500	111	105	13
Total dissolved solids	mg/L	1000	355	314	375
Hardness	mg/L	200	128	118	222
Alkalinity	mg/L		85	85	155
Color	TON	15		10	10
Aluminum (unfiltered)	μg/L	1000		96	9
Aluminum (filtered)	μg/L			79	17
Arsenic	μg/L	50		1.80	0.00
Boron	μg/L			158	85
Cadmium	μg/L	10.00		0.00	0.00
Copper	μg/L	1000		0	0
Fluoride (total)	mg/L	1.4–2.4	0.12	0.12	0.25
Iron (unfiltered)	μg/L			0	539
Iron (filtered)	μg/L	300		0	242
Lead	μg/L	50		0.00	0.00
Manganese (unfiltered)	μg/L			0	44
Manganese (filtered)	μg/L	50		0	43
Phosphorous	μg/L			0.04	0.03
Selenium	μg/L	10		0.00	0.00
Zinc	μg/L	5000		7.00	6.00
Ammonia (total): N	mg/L			0.19	0.17
Nitrate (NO_3)	mg/L	45	3.15	1.68	0.18
Silica	mg/L		15.1	15.1	37.5
Total trihalomethane	μg/L			74	0
Chloroform	μg/L			4.2	0.0
Bromodichloromethane	μg/L			15.0	0.0
Dibromochloromethane	μg/L			31.0	0.0
Bromoform	μg/L			24.0	0.0
Total organic carbon	mg/L			2.9	1.1
Total organic halogen as Cl$^-$				170	0
Field parameters					
pH			8.40	8.25	7.52
Eh	mv			419	211
Dissolved oxygen	ppm			8.4	1.0
Temperature	°C		18	16	25
Specific conductance	μmhos			500	600

able as a source of supply during 1956. This source supplies about 10,000 acre ft of water per year (9 MG/day, 34 megaliters/day) to the district. Water from the reservoir is treated at the Corona del Mar water treatment plant, which has a peak capacity of 136 megaliters/day (36 MG/day), and utilizes coagulation, sedimentation, filtration, and chlorination processes to treat the water before it is distributed to the service area.

Since 1963, the yield from the Cachuma reservoir has been supplemented by reactivated local wells that supply up to about 3400 acre ft per year (3 MG/day, 11 megaliters/day) to Goleta Water District and some other water users. In 1972 a "water shortage emergency" was declared due to overdrafting of the aquifer. Some of these wells are equipped for both production and recharge. Well depths range from about 300 to 1300 ft (91 to 396 m).

The ASR storage zone is the same unconfined aquifer that originally supplied the community with water. It is comprised of unconsolidated clayey, silty sand alluvial sediments. Of the 28 wells used to recharge treated drinking water during months when supply exceeds demand, 9 are dual-purpose ASR wells while the remainder are single-purpose injection wells. All were existing irrigation, private or public water supply wells, retrofitted to meet ASR or injection needs. Recharge is typically down the pump column, using both vertical turbine and submersible pumps.

Aquifer hydraulic characteristics are as follows:

Transmissivity	13,000 to 20,000 ft^2/day
	1200 to 1860 m^2/day
Specific capacity	4 to 20 G/min/ft

Depth to static water level varies throughout the service area, reflecting the terrain between the coast and the Santa Ynez mountains, and representing a principal constraint upon ASR system capacity. Minimum groundwater levels are about 60 ft below sea level. However, excessive recharge volume would cause water levels to rise to land surface in portions of this mostly confined system.

Native water quality in the aquifer varies from a TDS of 850 to 1000 mg/L in the wellfield area, deteriorating toward the coast. Other constituents of interest include localized areas of hydrogen sulfide at 0.5 to 1.0 mg/L, and Fe plus Mn at about 1 mg/L. Recharge water quality is treated drinking water from the treatment plant, which has a typical TDS concentration of 550 to 600 mg/L and a chlorine residual of 0.5 mg/L at the wellhead.

Recharge rates in individual wells range from less than 100 G/min (0.5 megaliters/day) to about 2.2 megaliters/day (400 G/min). Low average recharge rates of about 20 acre ft per year (12 G/min, 0.8 L/sec) per well have been experienced in some wells. Recovery rates range from about 200 to 800 G/min (1.1 to 4.3 megaliters/day). The ratio of recharge to recovery specific capacity is typically quite low, averaging about 30%. No backflushing for redevelopment was conducted in any of the ASR wells during the first 5 years of operation, relying upon extended pumping during seasonal recovery to restore well specific capacity.

Since 1978, opportunities for recharge have been quite limited. Between 1978 and 1988, recharge occurred in only 7 years. Of these, all but 2 years had less than 448 acre ft per year (0.4 MG/day, 1.5 megaliters/day) of recharge to the entire system. Maximum annual recharge during this period was 1901 acre ft (1.7 MG/day, 6.4 megaliters/day).

The primary purpose of this system is long-term water storage, to mitigate overdraft. All of the water stored is ultimately recovered, since the basin is almost totally confined against lateral outflow due to the existence of faults.

9.15 PASADENA, CALIFORNIA*

The city of Pasadena is one of 11 water utilities serving customers within the Raymond Groundwater Basin, which is located in southern California near Los Angeles. Groundwater production within this basin has exceeded natural recharge for many decades, as a result of which approximately 493 Mm^3 (400,000 acre ft) of potential storage above the water table is available in the aquifer system, to store imported water flows.

The Metropolitan Water District of Southern California has recently initiated a seasonal storage program, under which water utilities may purchase imported water at discounted rates during winter months when flows are available and demands are reduced. For water utilities with available storage capacity, the savings under this program can be substantial.

During 1989, the city and the district initiated a program to plan and implement recharge facilities, including both surface recharge and ASR wells, in order to meet seasonal peak demands, improve system reliability, meet drought demands, provide for long-term growth, and capitalize on the district's seasonal storage program. For the district, added advantages included more efficient use of existing water conveyance and treatment facilities and storage for surplus imported water supplies when they are available.

The aquifer being recharged is comprised of sand and gravel alluvial deposits with some clays, all under water table conditions. Aquifer hydraulics are variable across the basin, as follows:

---

* City of Pasadena, Water Division, Water and Power Department, 150 South Los Robles Avenue, Suite 200, Pasadena, California 91101

Hydraulic conductivity	2 to 60 ft/day
	1 to 18 m/day
Transmissivity	100 to 25,000 ft²/day or more
	9 to 2,300 m²/day or more
Specific yield	9 to 15%

The transmissivity estimate is based upon the saturated alluvium thickness, which ranges from zero at the base of the San Gabriel mountains to the north, to in excess of 244 m (800 ft) at the southerly end of the basin. Potentiometric surface elevations range from under 152 m (500 ft) to over 427 m (1400 ft) above mean sea level, while land surface elevations in the same area range upward from about 171 m (560 ft) above mean sea level to the south. Both elevations increase northward toward the San Gabriel mountains.

During 1992, two existing wells were retrofitted for ASR operations, commencing recharge in an area that had been subjected to nitrate contamination. From October 1992 to January 1994, these two wells have recharged over 3.7 Mm^3 (3000 acre ft) during months when discounted water was available.

Extensive computer modeling of the Raymond Groundwater Basin was conducted as a part of the recharge feasibility program, using the CFEST model to simulate hydrogeologic response and also movement of the recharge waters throughout the basin. The effect upon location and movement of existing contaminant plumes was also evaluated. The model indicated that, with proposed seasonal operations, the effect upon water levels would be slight. If additional storage is developed to provide reliability in the event of periodic droughts, estimated at 1 year out of 4, the effect upon regional water levels would be greater but still relatively insignificant. In particular, typical water level fluctuations would be in the tens of feet. Further development of available storage capacity could provide the opportunity for regional storage, to provide water to surrounding water utilities during extended droughts. Achieving this benefit would entail institutional changes and water supply agreements that are not now in place.

The Pasadena ASR program has been successful as a key part of a broader regional effort to conserve water and make better use of existing facilities and storage opportunities. It is a particularly good example of a situation that many other utility systems may face. Many utilities have limited local water supplies but excellent local opportunities to store large volumes of water that may become available intermittently. Where hydrogeologic conditions are so favorable, these utilities have the opportunity to store water to meet their own seasonal and drought/long-term storage needs, while also providing a regional reservoir to help supply surrounding utilities that may have less-than-ideal hydrogeologic opportunities for storage. The huge potential savings derived from such regional approaches to developing storage capacity should, in some cases, facilitate enlightened and cost-effective water supply planning and implementation for regional supplies.

Another element of the Pasadena ASR program is the consideration of power generation during recharge. This was discussed under Section 3.2, Design of ASR Wellhead Facilities.

9.16 SEATTLE, WASHINGTON*

The Seattle Water Department operates a surface water supply system capable of meeting peak day demands in excess of 1136 megaliters/day (300 MG/day). To help meet projected increases in demand, a wellfield was developed in the Highline area adjacent to SEATAC Airport during the 1980s. Three wells are currently in operation at 2 sites (Riverton Heights and Boulevard Park), capable of supplying a total yield of about 38 megaliters/day (10 MG/day). Each well is equipped for both recharge and recovery and is provided with facilities for disinfection, fluoridation, and pH adjustment; a vertical turbine pump; a wellhouse; and an observation well.

Early hydrogeologic investigations showed that the Highline area receives insufficient natural recharge to support wellfield production at planned rates, leading to projected water level declines. To offset this decline, an artificial recharge test program was initiated to determine the feasibility of ASR to meet seasonal water needs while maintaining aquifer levels. ASR test operations began in 1991 and the facilities became fully operational in 1993. The source of recharge water is a pipeline conveying treated drinking water from the Cedar River.

The intermediate aquifer utilized for ASR storage is comprised of glacial drift deposits, primarily sands and gravels, silts, and clays. Aquifer hydraulic characteristics are approximately as follows:

Transmissivity	Riverton Heights: 350,000 G/day/ft
	4350 m²/day
	Boulevard Park: 150,000 G/day/ft
	1860 m²/day
Storativity	0.0005 both sites
Specific capacity	27 G/min/ft (2300 G/min, 1 day—Riverton Heights)
	24 G/min/ft (2300 G/min, 1 day—Boulevard Park)

Prior to wellfield operations, static water level was measured at 153.5 ft below ground surface at Riverton Heights, at an elevation of 278.5 ft above mean sea level. At Boulevard Park, static water level was 77 ft below ground surface, at an elevation of 280 ft above mean sea level.

Water quality of the recharge water from the Cedar River pipeline and the groundwater for the Riverton Heights production well are excellent, as indicated by their low mean specific conductance values of 61 and 166 μmhos/cm, respectively. However, diatoms present in the recharge water contribute to well plugging, which can be reversed by pumping.

The objective of the ASR test program was achieved, showing that the Highline wellfield could be operated to meet seasonal peak demands during summer months, while recharging the aquifer during winter months to maintain water levels. Addition of a fourth ASR well to boost summer recovery capacity is under consideration.

The test program included four ASR cycles conducted at an existing unused test well adjacent to the Riverton Heights site. During the test program, some well clogging was apparent. Detailed geochemical and other investigations showed that the source of clogging was diatoms (single-cell algae) present in the recharge water. Periodic backflushing of the well with a frequency of about once every 2 weeks was sufficient to prevent residual clogging and maintain recovery capacity.

To date, areawide water level response to seasonal wellfield operations has been quite small. The opportunity exists at this site to utilize available storage capacity more efficiently. Pumps could perhaps be set at the base of the well screen rather than at the base of the casing, enabling an increase in wellfield production rates and the associated interference between wells during summer months. During the remainder of the year, wells would be recharged to restore aquifer water levels to elevations at or above those occurring prior to wellfield production. An increase in summer peak production rates would be helpful to the Seattle Water Department, and would also achieve operational benefits, since the effort required of operations personnel to operate the Highline wellfield is generally greater per unit of water produced during the summer than that associated with the basic supply of water from the Cedar River. Under pending regulations from the EPA, filtration may be required for water from the Cedar River, in which case the relative effort associated with the two sources would shift.

Another significant potential issue at this site is the pending regulation pertaining to radon concentrations in drinking water. Depending upon allowable radon concentration levels, the recovered water from the Highline wellfield may or may not meet the standards once they are promulgated. Radon is not present in the recharge water but is picked up rapidly during ASR storage. Treatment requirements would probably include aeration and detention time in a ground storage reservoir which already exists at the

---

* Seattle Water Department, Water Management Department, 710 Second Avenue, Seattle, Washington 98104

Riverton site but would need to be added at the Boulevard Park site. Blending of the treated water with that from the Cedar River pipeline should meet radon standards, if this becomes necessary.

9.17 KUWAIT*

During 1989, the Kuwait Institute for Scientific Research (KISR) initiated an investigation of ASR feasibility to establish a strategic water reserve. This work was performed for the Kuwait Ministry of Electricity and Water. The goal was to store a large volume of drinking water in brackish aquifers close to demand centers so that several months' supply would be available, stored safely underground in the event of emergency loss of seawater desalination plants. The water source for recharge would be drinking water, which is primarily water from the desalination plants blended with 5 to 10% brackish groundwater. Once established, this strategic reserve would also be available to help meet summer peak demands within the water supply service area. During the period October 1989 to May 1990, recharge investigations were conducted at 3 existing well sites. The next phase of the program was interrupted by hostilities in 1990 to 1991, but has since continued.

Two sites were tested in the Dammam formation, a limestone artesian aquifer, while one site was tested in the overlying Kuwait Group aquifer, which is comprised of sand intervals between layers of cemented sandstone. Both aquifers are brackish, with TDS levels ranging from 2700 to 5000 mg/L in the Dammam formation and 3900 mg/L in the Kuwait Group aquifer. Aquifer hydraulic characteristics were quite different at the three sites.

Sulaibiya Well SU-10

For Well SU-10 at Sulaibiya, the aquifer was composed of two producing intervals, at the top and bottom of the well, with the following hydraulic characteristics:

Transmissivity	12 m²/day (966 G/day/ft) (Upper Dammam)
	24 m²/day (1933 G/day/ft) (Lower Dammam)
Storativity	2×10^{-5} (Upper Dammam)
	4×10^{-5} (Lower Dammam)
Leakance	1×10^{-4}/day (Upper Dammam)
	5×10^{-4}/day (Lower Dammam)

The well construction included 124 m (406 ft) of 400 mm (16 inch) casing, and open hole to 275 m (902 ft). Of the many wells that have been

---

* Kuwait Institute for Scientific Research, PO Box 24885, Safat, 13109, Kuwait

tested for ASR feasibility, this is believed to be the one with the lowest transmissivity. Static water level at the time of testing was at 37 m (120 ft) below land surface. A head difference of approximately 5 m (16 ft) is believed to exist between the top and bottom producing intervals.

The well is equipped with a vertical turbine pump. A single test cycle was conducted, replacing the pump with an injection tube to beneath the static water level. A recharge volume of 16,416 cm (4.3 MG) was injected during a period of 30 days, at rates that varied from 0.65 megaliters/day (121 G/min) initially to as low as 0.39 megaliters/day (73 G/min) in order to avoid the water level rising above land surface. With redevelopment during the injection period, a sustained injection flow rate of 0.58 megaliters/day (109 G/min) could be maintained. Results indicated substantial plugging, probably due to rust and sand in the distribution system that was swept into the well at the beginning of recharge. When this was removed during redevelopment by backflushing, hydraulic performance improved. There was little evidence of geochemical plugging or air entrainment. Following injection, the well was pumped at variable rates averaging about 0.98 megaliters/day (181 G/min). Pumping continued until backgroundwater quality was reached. The specific capacity ratio between injection and recovery was estimated at about 60%.

Background TDS in the aquifer was about 5000 mg/L. Mixing characteristics were such that over 45% of the recharge volume was recovered before the TDS exceeded 2000 mg/L in the recovered water, and 90% was recovered before the TDS concentration reached 3000 mg/L.

Test results indicated that this site would be suitable for a strategic water reserve, probably using only the lower producing interval of the Dammam formation in order to reduce mixing between stored and native water. New ASR wells would be required, with appropriate design and operation to minimize clogging. Successive ASR cycles were expected to achieve satisfactory recovery efficiency.

Shigaya Well C-105

This well is also in the Dammam formation; however, it is located in the southwestern portion of Kuwait at much higher land surface elevation. Since no freshwater supply is available in this area, water for injection testing was obtained from other nearby wells with similar water quality. Mixing was evaluated using sodium fluorescein dye.

The well has 400 mm (16 inch) casing to 258 m (845 ft), with an open hole to 385 m (1262 ft). Static water level at the time of testing was about 176 m (577 ft) below land surface, dropping at the rate of about 2 m (5 ft) per year. Flowmeter logging indicated no head differences between the top and bottom of the aquifer; however, about 60 to 75% of the flow occurred at a depth of 372 to 375 m (1220 to 1230 ft). The water is brackish, with a TDS concentration of about 2700 mg/L. Testing indicated the following aquifer hydraulic characteristics:

Transmissivity	4000 m²/day (322,000 G/day/ft)
Specific capacity	Injection: 17.4 G/min/ft @ 1080 G/min
Recovery	52 G/min/ft @ 360 G/min

The data is not conclusive; however, it suggests a specific capacity ratio of about 0.40 between injection and recovery, which seems quite low. A single test cycle was conducted in which 46 MG were injected and 50 MG were recovered.

Some plugging occurred during testing; however, this was resolved by pumping the well, without any sign of residual plugging. There was no indication of geochemical reactions that would create problems during long-term operation. Only about 40% of the tracer was recovered, indicating high mixing. This site would probably be suitable for artificial recharge to restore water levels, if water is available, but it may be less suitable as a strategic water reserve.

Sulaibiya Well 135A

This is a screen and gravel pack well with six 400 mm (16 inch) screened intervals at depths ranging from 90 m (300 ft) to 180 m (600 ft). Screen intervals vary from 3 to 12 m (10 to 40 ft) in thickness and total 43 m (140 ft). The depth to static water level was 57.8 m (190 ft) during the test program. The vertical turbine pump was set between the second and third screen intervals at a depth of about 122 m (400 ft).

Hydraulic testing indicated the following aquifer and well characteristics:

Transmissivity	280 m²/day
	22,500 G/day/ft
Specific capacity	9.2 G/min/ft @ 360 G/min (pumping)
	8.0 G/min/ft @ 120 G/min (initial injection)
	2.5 G/min/ft @ 120 G/min (end of injection)

Following this injection period, the specific capacity was substantially restored by pumping the well at various rates, although about 5 to 15% of the capacity had been lost. The well was then acidized to restore capacity and each screen interval was developed with pumping and surging. Specific capacity was restored to levels exceeding those prior to testing, as determined from step drawdown tests. At 360 G/min, the specific capacity was 13.7 G/min/ft. Injection was then resumed but was terminated after 22 hours when the rate of water level rise suggested that rapid plugging was continuing. Total water stored was 4211 cm (1.1 MG), less volumes produced during the step test and pumping of spent acid. The well was then placed back in routine operation and tracer concentrations were monitored until 10,451 m^3 (2.8 MG) had been recovered at a typical production rate of about 1.9 megaliters/day (0.5 MG/day).

Evaluation of potential causative factors suggested that air entrainment and suspended solids in the recharge water probably caused the plugging. Total suspended solids measurements yielded concentrations ranging from 3.2 to 7.4 mg/L during recharge. Substantial quantities of air and turbid water were noted during redevelopment.

Mixing occurred in the well, partly due to differential plugging of the six screen sections. Tracer testing was performed using tritium as a tracer during the latter portion of the injection test following acidization of the well. It was also possible to use sulfate as a natural tracer. TDS concentration of the native water was about 3900 mg/L while sulfate concentration was about 1050 mg/L. Injection water was obtained from the wellfield collection system and had a TDS concentration of about 4400 mg/L but a sulfate concentration of about 2000 mg/L. At 30% recovery, sulfate had declined to 1910 mg/L; at 70%, it had dropped to 1650 mg/L, and at 100% it had dropped to 1390 mg/L.

Although this test was conducted at a very small scale and encountered a variety of complex logistic and technical difficulties, it is notable due to the unique use of tritium as a tracer. Tritium was selected with the assumption that natural tracers would not be sufficient; however, sulfate differential concentrations in the recharge and native water also proved to be useful. Both tritium and sulfate results indicated the same approximate mixing curves during recovery.

Test results were not conclusive; however, they suggested the limited potential value of this aquifer in the Sulaibiya area as a component of a strategic water reserve utilizing both the Dammam and Kuwait Group aquifers. Further effort would be required to design and operate ASR wells in this aquifer in order to meet overall objectives while minimizing plugging and meeting recovery water quality objectives.

References

CHAPTER 1: INTRODUCTION

1. Postel, S., *The Last Oasis: Facing Water Scarcity,* The Worldwatch Environmental Alert Series, Worldwatch Institute, W.W. Norton & Co., 1992, p. 35.
2. Postel, S., *The Last Oasis: Facing Water Scarcity,* The Worldwatch Environmental Alert Series, Worldwatch Institute, W.W. Norton & Co., 1992, p. 30.
3. The World Almanac and Book of Facts, 1988.
4. Shiklomanov, I. A., Global Water Resources, *Nature and Resources,* Vol. 26, No. 3, 1990.
5. L'vovitch, M. I., *World Water Resources and Their Future,* Washington, D.C., American Geophysical Union, 1979, as cited in van der Leeden et al., The Water Encyclopedia.
6. L'vovitch, M. I., *Ground-Water Storage and Artificial Recharge,* Natural Resources/Water Series No. 2, United Nations, New York, 1975.
7. Limaye, S. D., Director, Ground Water Institute, Pune, India, personal communication, 1994.
8. Peters, J. H., *Artificial Recharge and Water Supply in The Netherlands: State of the Art and Future Trends,* A. I. Johnson and R. D. G. Pyne, Eds., Proceedings of American Society of Civil Engineers Second International Symposium on Artificial Recharge of Ground Water, Orlando, FL, July 17-22, 1994.
9. Cederstrom, D. J., Artificial recharge of a brackish water well, *Commonwealth,* Virginia Chamber of Commerce, Richmond, V. 14, No. 12, p. 31, 71–73, 1947.
10. Todd, D. K., *Annotated Bibliography on Artificial Recharge of Ground Water Through 1954,* U.S. Geological Survey Water Supply Paper 1477, 1959.
11. Signor, D.C., Growitz, D.J., and Kam, W., *Annotated Bibliography on Artificial Recharge of Ground Water, 1955–67,* U.S. Geological Survey Water Supply Paper 1990, 1970.
12. Knapp, G. L., *Artificial Recharge of Groundwater. A Bibliography,* Office of Water Resources Research, Washington, D.C., 1973.
13. O'Hare, M. P., Fairchild, D. M., Hajali, P. A., and Canter, L. W., *Artificial Recharge of Ground Water: Status and Potential in the Contiguous United States,* Environmental and Groundwater Institute, University of Oklahoma, Lewis Publishers, Chelsea, MI, 1986.
14. Merritt, M. L., *Subsurface Storage of Freshwater in South Florida: A Digital Model Analysis of Recoverability,* U.S. Geological Survey Water Supply Paper 2261, 1985.
15. Brown, D. L. and Silvey, W. D., *Artificial Recharge to a Freshwater-Sensitive Brackish-Water Sand Aquifer, Norfolk, Virginia,* U.S. Geological Survey Water Supply Paper 939, 1977.

CHAPTER 3: DESIGN OF ASR SYSTEMS

1. Caduto, M. J. and Bruchac, J., *Keepers of the Earth,* Fulcrum Inc., Golden, CO, 1988.
2. Merritt, M. L., *Subsurface Storage of Freshwater in South Florida: A Digital Model Analysis of Recoverability,* U.S. Geological Survey Water Supply Paper 2261, 1985.

CHAPTER 4: SELECTED ASR TECHNICAL ISSUES

1. Rhebun, M. and Schwartz, J., Clogging and Contamination Processes in Recharge Wells, *Water Resources Research,* Vol. 4, No. 6, 1207–1217, 1968.
2. Sneigocki, R. T. and Brown, R. F., *Clogging in Recharge Wells,* Proceedings of Conference on Artificial Recharge, Reading, England, Water Resources Association, 1970.
3. Leenheer, J. A., Malcolm, R. L., and White, W. R., *Physical, Chemical and Biological Aspects of Subsurface Organic Waste Injection Near Wilmington, North Carolina,* U.S. Geological Survey Professional Paper 987, 1976.
4. Huisman, L. and Olsthoorn, T. N., *Artificial Groundwater Recharge,* 1st ed., Pitman Publishing Ltd., Marshfield, MA, 1983.
5. Rancilio, J. A., Los Angeles County Department of Public Works, Hydraulic/Water Conservation Division, personal communication, June 1990.
6. Olsthoorn, T. N., *The Clogging of Recharge Wells,* KIWA-Communications 72, Rijswikn, Netherlands, The Netherlands Waterworks Testing and Research Institute, 1982.
7. American Public Health Association, Standard Methods for the Examination of Water and Wastewater, 17th ed., Part 2540D, *Total Suspended Solids Dried at 103 to 105°C,* Washington, D.C., pp. 2–75 to 2–76, 1989.
8. Schippers, J.C. and Verdouw, J., The Modified Fouling Index, A Method for Determining the Fouling Characteristics of Water, *Desalination,* 32, 137–148, 1980.
9. American Water Works Association Research Foundation, *Aquifer Storage Recovery of Treated Drinking Water,* 1994 (in press).
10. Brown, D. L. and Silvey, W. D., *Artificial Recharge to a Freshwater-Sensitive Brackish-Water Sand Aquifer, Norfolk, Virginia,* U.S. Geological Survey Professional Paper 939, 1977.

CHAPTER 5: GEOCHEMISTRY

1. Rice, G., Brinkman, J., and Muller, D., Reliability of Chemical Analyses of Water Samples — The Experience of the UMTRA Project: Groundwater Monitoring Review, 71–75, 1988.
2. Hem, J. D., *Study and Interpretation of the Chemical Characteristics of Natural Water,* U.S. Geological Survey Water Supply Paper 2254, 1985.
3. Linn, J. C., Schnoor, J. L., and Glass, G. E., Sources and Fates of Aquatic Pollutants, *Advances in Chemistry Series,* No. 216, American Chemical Society, Washington, D.C., 209–228, 1987.

4. Garrels, R. M. and Christ, C. L., *Solutions, Minerals and Equilibria,* Freeman, Cooper Publishing, San Francisco, 1965.

5. Brookins, D. G., *Eh–pH Diagrams for Geochemistry,* Springer-Verlag, New York, 1988.

6. Bichara, A. F., Clogging of Recharge Wells by Suspended Solids, *American Society of Civil Engineers Journal of Irrigation and Drainage Division*, Vol. 112, No. 3, 210–224, 1986.

7. Dragun, J., *Soil Chemistry of Hazardous Materials,* Hazardous Materials Control Research Institute, Silver Springs, MD, 1988.

8. Van Beek, C. G. E. M., Clogging of Discharge Wells in the Netherlands II: Causes and Prevention, in *International Symposium on Biofouled Aquifers: Prevention and Restoration,* American Water Works Association, Denver, 43–56, 1986.

9. Berner, R. A., Rate Control of Mineral Dissolution Under Earth Surface Conditions, *American Journal of Science*, Vol. 278, No. 3, 210–224, 1978.

10. Ball, J. W., Jenne, E. A., and Cantrell, M. W., *WATEQ3 — A Geochemical Model with Uranium Added,* USGS Open File Report No. 81–1183, 1981.

11. Fleming, G. W. and Plummer, L. N., *PHRQINPT — An Interactive Computer Program for Constructing Input Data Sets to the Geochemical Simulation Program PHREEQE,* U.S. Geological Survey Water Resources Investigation Report 83–4236, 1983.

12. Parkhurst, D. L., Thorstenson, D. C., and Plummer, L. N., *PHREEQE — A Computer Program for Geochemical Calculations,* U.S. Geological Survey Water Resources Investigations Report 80–96, 1980.

13. Felmy, A. R., Girvin, D., and Jenne, E. A., *MINTEQ: A Computer Program for Calculating Aqueous Geochemical Equilibria,* U.S. Environmental Protection Agency, Washington, D.C., 1983.

14. Wolery, T. J., *Calculation of Chemical Equilibrium Between Aqueous Phase Solution and Minerals: The EQ3/6 Software Package,* UCRL-52658, Lawrence Livermore National Laboratory, CA, 1979.

CHAPTER 7: ALTERNATIVE ASR APPLICATIONS

1. CH2M HILL Inc., *Environmental Risk and Geochemical Analysis Related to the City of St. Petersburg's Underground Injection and Monitoring System,* Engineering Report for the City of St. Petersburg, FL, March 1993.

CHAPTER 9: SELECTED CASE STUDIES

1. Brothers, K. and Katzer, T., Water Banking Through Artificial Recharge, Las Vegas Valley, Clark County, Nevada, *Journal of Hydrology,* 115, 77–103, Elsevier Science Publishers B.V., 1990.

Appendices

DRINKING WATER QUALITY STANDARDS

United States Environmental Protection Agency
European Community
World Health Organization

CONVERSION FACTORS

UNITED STATES ENVIRONMENTAL PROTECTION AGENCY

REGION IV

345 COURTLAND STREET. N.E.
ATLANTA. GEORGIA 30365

National Primary
Drinking Water Regulations

The Safe Drinking Water Act (SDWA), as amended in 1986, requires EPA to publish Maximum Contaminant Level Goals (MCLGs) for contaminants which, in the judgment of the Administrator, may have any adverse effect on the health of persons and which are known or anticipated to occur in public water systems. MCLGs are to be set at a level which no known or anticipated adverse effects on the health of persons occur and which allow an adequate margin of safety.

At the same time EPA publishes an MCLG, which is a non-enforceable health goal, it must also promulgate a National Primary Drinking Water Regulation (NPDWR) which includes either (1) a Maximum Contaminant Level (MCL), or (2) a required treatment technique. A treatment technique may be set only if it is not economically or technologically feasible to ascertain the level of a contaminant. An MCL must be set as close to the MCLG as feasible. Under the SDWA, "feasible" means feasible with the use of the best technology, treatment techniques, and other means which the Administrator finds are available, after examination for effectiveness under field conditions and not solely under laboratory conditions (taking cost into consideration). NPDWRs also include monitoring, analytical and quality assurance requirements, and specifically, criteria and procedures to assure a supply of drinking water which dependably complies with such MCLs.

DEVELOPMENT OF MCLGs:

MCLGs are set at concentration levels at which no known or anticipated adverse health effects would occur, allowing for an adequate margin of safety. Establishment of a specific MCLG depends on the evidence of carcinogenicity from drinking water exposure or the Agency's noncarcinogenic reference dose (RfD), which is calculated for each specific contaminant.

From the RfD, a drinking water equivalent level (DWEL) is calculated by multiplying the RfD by an assumed adult body weight (generally 70 kg) and then dividing by an average daily water consumption of 2 liters per day. The DWEL assumes the total daily exposure to a substance is from drinking water exposure. The MCLG is determined by multiplying the DWEL by the percentage of the total daily exposure contributed by drinking water, called the relative source contribution (RSC). Generally, EPA assumes that the RSC from drinking water is 20% of the total exposure, unless other exposure data for the chemical are available.

For chemicals suspected as carcinogens, the assessment for nonthreshold toxicants consists of the weight of evidence of carcinogenicity in humans. The objectives of the assessment are (1) to determine the level or strength of evidence that the substance is a human or animal carcinogen and (2) to provide an upperbound estimate of the possible risk of human exposure to the substance in drinking water. A summary of EPA's cancer classification scheme is:

Group A — Known human carcinogen
Group B — Probable human carcinogen
Group C — Possible human carcinogen
Group D — Not classifiable
Group E — No evidence as human carcinogen

Establishing the MCLG for a chemical is generally accomplished in one of three ways depending on its categorization. Each contaminant is analyzed for evidence of carcinogenicity via ingestion. In most cases, the Agency places Group A and B contaminants into Category I, Group C into Category II, and Group D and E into Category III. However, where there is additional information on cancer risks from drinking water ingestion, additional scrutiny is conducted which may result in placing the contaminant into a different category.

EPA's policy is to set MCLGs for Category I contaminants at zero. The MCLG for Category II contaminants is calculated by using the RfD/DWEL/RSC approach with an added margin of safety (usually 10-fold) to account for cancer effects or is based on a cancer risk range of 10^{-5} to 10^{-6} when non-cancer data are inadequate for deriving a RfD. MCLGs for Category III contaminants are calculated using the RfD/DWEL/RSC approach.

DEVELOPMENT OF MCLs:

The SDWA directs EPA to set the MCL as close to the MCLG as is feasible. Based on the statutory directive for setting MCLs, EPA derives MCLs based on an evaluation of (1) the availability and performance of various technologies for removing the contaminant, (2) the costs of applying these technologies, and (3) the ability of laboratories to measure accurately and consistently the level of the contaminant with available analytical methods. Because compliance with the MCL is determined by analysis with approved analytical techniques, the ability to analyze consistently and accurately for a contaminant at the MCL is important to enforce a regulatory standard. This factor is critically important in determining the MCL for contaminants for which EPA sets the MCLG at zero, a number which by definition can be neither measured nor attained. Limits of analytical detection require that the MCL be set at some level greater than the MCLG for these contaminants.

EPA also evaluates the health risks that are associated with various contaminant levels in order to ensure that the MCL adequately protects the public health. For drinking water contaminants, EPA sets as a goal a risk range goal of 10^{-4} to 10^{-6} excess individual risk for carcinogens during a lifetime exposure (arsenic is one exception to this risk range). This policy is consistent with other EPA regulatory programs that generally target this range using conservative models that are not likely to underestimate the risk. Usually the MCLs for noncarcinogenic contaminants are set at the MCL. Since the underlying goal of the SDWA is to protect the public from adverse effects due to drinking water contaminants, EPA seeks to ensure that the health risks associated with MCLs for all contaminants are not significant.

Additional information on this subject matter can be found in Part 141 of the Code of Federal Regulations (CFR) and the January 30, 1991, Federal Register (EPA, National Primary Drinking Water Regulations; Final Rule).

Part 143 of the CFR discusses EPA's National Secondary Drinking Water Regulations (NSDWRs). NSDWRs control contaminants in drinking water that primarily affect the aesthetic qualities relating to the public acceptance of drinking water. At considerably higher concentrations of contaminants, health implications may also exist as well as aesthetic degradation. The regulations are not Federally enforceable but are intended as guidelines for the States.

The NSDWRs represent reasonable goals for drinking water quality. The States may establish higher or lower levels which may be appropriate dependent upon local conditions such as unavailability of alternate source waters or other compelling factors, provided that public health and welfare are not adversely affected.

Provided in the following table is an up-to-date list of all of EPA's MCLs and MCLGs, along with all of EPA's proposed MCLs, MCLGs, and Secondary MCLs. This table will be updated as necessary and the date in the top left hand corner should be checked for time of printing.

A copy of this table will be made available through Region IV's library. If you have any comments or questions about this table please call Glenn Adams at 404/347-3866.

12/22/1992

CURRENT and PROPOSED MCLs, MCLGs, and SMCLs

CHEMICAL	MCL (ppm)	MCLG (ppm)	SMCL (ppm)
INORGANICS			
Aluminum (1/91)			0.05-0.2
Antimony (7/92)	0.006	0.006	
Arsenic (NPDWR)	0.050		
Asbestos (1/91)	7 million fibers/liter (>10 um)		
Barium (7/91)	2	2	
Beryllium (7/92)	0.004	0.004	
Cadmium (1/91)	0.005	0.005	
Chloride (NSDWR)			250
Chromium (1/91)	0.1	0.1	
Color (NSDWR)			15 color units
Copper (7/91)	TT	1.3	
Corrosivity (NSDWR)			Noncorrosive
Cyanide (7/92)	0.2	0.2	
Fluoride (4/86)	4.0		2.0
Foaming Agents (NSDWR)			0.5
Iron (NSDWR)			0.3
Lead (6/91)	TT	0	
(6/90)	0.015 (Action Level)		
Manganese (NSDWR)			0.05
Mercury (1/91)	0.002	0.002	
Nickel (7/92)	0.1	0.1	
Nitrite (as N) (1/91)	1	1	
Nitrate (as N) (1/91)	10	10	
Total (as N) (1/91)	10	10	
Odor (NSDWR)			3 threshold odor #
pH (NSDWR)			6.5 - 8.5
Selenium (1/91)	0.05	0.05	
Silver (1/91)			0.1
Sulfate (NSDWR)			250
Sulfate (7/90)	*400/500	*400/500	
Thallium (7/92)	0.002	0.0005	
Total Dissolved Solids (NSDWR)			500
Zinc (NSDWR)			5
ORGANICS			
Acrylamide (1/91)	TT	0	
Alachlor (1/91)	0.002	0	
Aldicarb (5/92)	Deferred		
Aldicarb sulfone (5/92)	Deferred		
Aldicarb sulfoxide (5/92)	Deferred		
Atrazine (1/91)	0.003	0.003	
Benzene (7/87)	0.005	0	
Benzo(a)pyrene (7/92)	0.0002	0	
Carbofuran (1/91)	0.04	0.04	
Carbon Tetrachloride (7/87)	0.005	0	
Chlordane (1/91)	0.002	0	
2,4-D (1/91)	0.07	0.07	
Dalapon (7/92)	0.2	0.2	
Dibromochloropropane (DBCP) (1/91)	0.0002	0	
o-Dichlorobenzene (1/91,5/89)	0.6	0.6	0.01
p-Dichlorobenzene (7/87)	0.075	0.075	
p-Dichlorobenzene (1/91,5/89)			0.005
1,2-Dichloroethane (7/87)	0.005	0	
cis-1,2-Dichloroethylene (1/91)	0.07	0.07	
trans-1,2-Dichloroethylene (1/91)	0.1	0.1	
1,1-Dichloroethylene (7/87)	0.007	0.007	
Dichloromethane (Methylene chloride) (7/92)	0.005	0	
1,2-Dichloropropane (1/91)	0.005	0	

* - Proposed MCL and MCLG

12/22/1992

CHEMICAL	MCL (ppm)	MCLG (ppm)	SMCL (ppm)
Di(ethylhexyl)adipate (7/92)	0.4	0.4	
Di(ethylhexyl)phthalate (7/92)	0.006	0	
Diquat (7/92)	0.02	0.02	
Dinoseb (7/92)	0.007	0.007	
Dioxin (2,3,7,8-TCDD) (7/92)	3x10E-8	0	
Endothall (7/92)	0.1	0.1	
Endrin (7/92)	0.002	0.002	
Epichlorohydrin (1/91)	TT	0	
Ethylbenzene (1/91,5/89)	0.7	0.7	0.03
Ethylene dibromide (EDB) (1/91)	0.00005	0	
Glyphosate (7/92)	0.7	0.7	
Heptachlor (1/91)	0.0004	0	
Heptachlor epoxide (1/91)	0.0002	0	
Hexachlorobenzene (7/92)	0.001	0	
Hexachlorocyclopentadiene[HEX] (7/92)	0.05	0.05	0.008
Lindane (1/91)	0.0002	0.0002	
Methoxychlor (1/91)	0.04	0.04	
Monochlorobenzene (1/91)	0.1	0.1	
Oxamyl [Vydate] (7/92)	0.2	0.2	
Pentachlorophenol (7/91, 5/89)	0.001	0	0.03
Picloram (7/92)	0.5	0.5	
Polychlorinated biphenyls(PCBs) (1/91)	0.0005	0	
Simazine (7/92)	0.004	0.004	
Styrene (1/91,5/89)	0.1	0.1	0.01
Tetrachloroethylene (1/91)	0.005	0	
Toluene (1/91,5/89)	1	1	0.04
Toxaphene (1/91)	0.003	0	
2,4,5-TP Silvex (1/91)	0.05	0.05	
1,1,2-Trichloroethane (7/92)	0.005	0.003	
1,2,4-Trichlorobenzene (7/92)	0.07	0.07	
1,1,1-Trichloroethane (7/87)	0.20	0.20	
Trichloroethylene (7/87)	0.005	0	
Trihalomethanes (NPDWR) (Bromoform, Dibromochloromethane, Chloroform, Bromodichloromethane)	0.100		
Vinyl Chloride (7/87)	0.002	0	
Xylenes (1/91,5/89)	10.00	10.00	0.02

MICROBIALS

Coliform bacteria (6/89)	< 1/100 ml	0
Giardia lamblia (6/89)	TT	0
Heterotrophic bact. (6/89)	TT	0
Legionella (6/89)	TT	0
Viruses (6/89)	TT	0
Turbidity turbidity)	1 TU (up to 5 TU) (units of	

RADIONUCLIDES

Beta particle and photon radioactivity	4 mrem	0
Gross Alpha particles	15 pCi/l	0
Radon-222 (7/91 *)	*300 pCi/l	0
Radium-226 and Radium-228 (Total)	5 pCi/l	0
Radium-226 (7/91 *)	* 20 pCi/l	0

* - Proposed MCL and MCLG

12/22/1992

CHEMICAL	MCL (ppm)	MCLG (ppm)	SMCL (ppm)
Radium-228 (7/91 *)	* 20 pCi/l	0	
Uranium (7/91 *)	* 20 ug/l or 30 pCi/l	0	

FOOTNOTES

11/85	50 Federal Register (FR), November 13, 1985
4/86	51 FR, April 2, 1986 - Final MCLs and SMCLs
7/87	52 FR, July 8, 1987 - Final MCLs and MCLGs
5/89	54 FR, May 22, 1989 - Proposed SMCLs
6/89	54 FR, June 29, 1989 - Final MCLs and MCLGs
6/90	Action level for lead in drinking water, June 21, 1990, Memorandum from the Office of Emergency and Remedial Response and the Office of Waste Program Enforcement
7/90	55 FR, July 25, 1990 - Proposed MCLs, MCLGs, and SMCLs
1/91	56 FR, January 30, 1991 - Final MCLs, MCLGs, and Proposed SMCLs
6/91	56 FR, June 7, 1991 - MCLGs & NPDWRs for Lead & Copper [Action levels established for lead (0.015 ppm) and copper (1.3 ppm)]
7/91	56 FR, July 1, 1991 - NPDWRs; Final Rule
7/91 *	56 FR, July 18, 1991 - NPDWRs for Radionuclides in Drinking Water
5/92	57 FR, May 27, 1992 - Drinking Water
7/92	57 FR, July 17, 1992 - Final MCLS and MCLGs
MCL	Maximum Contaminant Level
MCLG	Maximum Contaminant Level Goal
NPDWR	National Primary Drinking Water Regulation
NSDWR	National Secondary Drinking Water Regulation
SMCL	Secondary Maximum Contaminant Level
TT	Treatment Technique

* - Proposed MCL and MCLG

ANNEX 1

LIST OF PARAMETERS

A. ORGANOLEPTIC PARAMETERS

	Parameters	Expression of the results (^{1})	Guide level (GL)	Maximum admissible concentration (MAC)	Comments
1	Colour	mg/1 Pt/Co scale	1	20	
2	Turbidity	mg/1 SiO$_2$	1	10	
		Jackson units	0.4	4	— Replaced in certain circumstances by a transparency test, with a Secchi disc reading in meters: GL: 6m MAC: 2m
3	Odour	Dilution number	0	2 at 12°C 3 at 25°C	— To be related to the taste tests.
4	Taste	Dilution number	0	2 at 12°C 3 at 25°C	— To be related to the odour tests.

(^{1}) If, on the basis of Directive 71/354/EEC as last amended, a Member State uses in its national legislation, adopted in accordance with this Directive, units of measurement other than these indicated in this Annex, the values thus indicated must have the same degree of precision.

B. PHYSICO-CHEMICAL PARAMETERS (in relation to the water's natural structure)

	Parameters	Expression of the results (^{1})	Guide level (GL)	Maximum admissible concentration (MAC)	Comments
5	Temperature	°C	12	25	
6	Hydrogen ion concentration	pH unit	6.5<pH<8.5		— The water should not be aggressive. — The pH values do not apply to water in closed containers. — Maximum admissible value: 9.5.
7	Conductivity	µS cm^{-1} at 20°C	400		— Corresponding to the mineralization of the water. — Corresponding relativity values in ohms/cm: 2 500.
8	Chlorides	Cl mg/1	25		— Approximate concentration above which effects might occur: 200 mg/1.
9	Sulphates	SO$_4$ mg/1	25	250	
10	Silica	SiO$_2$ mg/1			— See Article 8.
11	Calcium	Ca mg/1	100		
12	Magnesium	Mg mg/1	30	50	

	Parameters	Expression of the results ([1])	Guide level (GL)	Maximum admissible concentration (MAC)	Comments
13	Sodium	Na mg/l	20	175 (as from 1984 and with a percentile of 90) 150 (as from 1987 and with a percentile of 80) (these percentiles should be calculated over a reference period of three years)	— The values of this parameter take account of the recommendations of a WHO working party (The Hague, May 1978) on the progressive reduction of the current total daily salt intake to 6 g. — As from 1 January 1984 the Commission will submit to the Council reports on trends in the total daily intake of salt per population. — In these reports the Commission will examine to what extent the 120 mg/l MAC suggested by the WHO working party is necessary to achieve a satisfactory total salt intake level, and, if appropriate, will suggest a new salt MAC value to the Council and a deadline for compliance with that value. — Before 1 January 1984 the Commission will submit to the Council a report on whether the reference period of three years for calculating these percentiles is scientifically well founded.
14	Potassium	K mg/l	10	12	
15	Aluminium	Al mg/l	0.05	0.2	
16	Total hardness				— See Table F, page 23.
17	Dry residues	mg/l after driving at 180°C		1500	
18	Dissolved oxygen	% O₂ saturation			— Saturation value >75% except for underground water.
19	Free carbon dioxide	CO₂ mg/l			— The water should not be aggressive.

C. PARAMETERS CONCERNING SUBSTANCES UNDESIRABLE IN EXCESSIVE AMOUNTS ([1])

	Parameters	Expression of the results ([1])	Guide level (GL)	Maximum admissible concentration (MAC)	Comments
20	Nitrates	NO₃ mg/l	25	50	
21	Nitrites	NO₂ mg/l		0·1	
22	Ammonium	NH₄ mg/l	0·05	0·5	

([1]) Certain of these substances may even be toxic when present in very substantial quantities.

	Parameters	Expression of the results	Guide level (GL)	Maximum admissible concentration (MAC)	Comments
23	Kjeldahl Nitrogen (excluding N in NO_2 and NO_3)	N mg/l		1	
24	($K Mn O_4$) Oxidizability	O_2 mg/l	2	5	— Measured when heated in acid medium.
25	Total organic carbon (TOC)	C mg/l			— The reason for any increase in the usual concentration must be investigated.
26	Hydrogen sulphide	S µg/l		undetectable organoleptically	
27	Substances extractable in chloroform	mg/l dry residue	0-1		
28	Dissolved or emulsified hydrocarbons (after extraction by petroleum ether); Mineral oils	µg/l		10	
29	Phenols (phenol index)	C_6H_5OH µg/l		0-5	— Excluding natural phenols which do not react to chlorine.
30	Boron	B µg/l	1 000		
31	Surfactants (reacting with methylene blue)	µg/l (lauryl sulphate)		200	
32	Other organochlorine compounds not covered by parameter No 55	µg/l	1		— Haloform concentrations must be as low as possible.
33	Iron	Fe µg/l	50	200	
34	Manganese	Mn µg/l	20	50	

	Parameters	Expression of the results	Guide level (GL)	Maximum admissible concentration (MAC)	Comments
35	Copper	Cu µg/l	100 — at outlets of pumping and/or treatment works and their sub-stations 3 000 — after the water has been standing for 12 hours in the piping and at the point where the water is made available to the consumer		— Above 3 000 µg/l astringent taste, discolouration + corrosion may occur.
36	Zinc	Zn µg/l	100 — at outlets of pumping and/or treatment works and their sub-stations 5 000 — after the water has been standing for 12 hours in the piping and at the point where the water is made available to the consumer		— Above 5 000 µg/l astringent taste, opalescence and sand-like deposits may occur.
37	Phosphorus	P_2O_5 µg/l	400	5 000	
38	Fluoride	F µg/l 8 — 12°C 25 — 30°C		1 500 700	— MAC varies according to average temperature in geographical area concerned.
39	Cobalt	Co µg/l			
40	Suspended solids		None		
41	Residual Chlorine	Cl µg/l			— See Article 8.
42	Barium	Ba µg/l	100		
43	Silver	Ag µg/l		10	If, exceptionally, silver is used non-systematically to process the water, a MAC value of 80 µg/l may be authorized.

	Parameters	Expression of the results (¹)	Guide level (GL)	Maximum admissible concentration (MAC)	Comments

D. PARAMETERS CONCERNING TOXIC SUBSTANCES

	Parameters	Expression of the results	Guide level (GL)	Maximum admissible concentration (MAC)	Comments
44	Arsenic	As µg/l		50	
45	Beryllium	Be µg/l			
46	Cadmium	Cd µg/l		5	
47	Cyanides	CN µg/l		50	
48	Chromium	Cr µg/l		50	
49	Mercury	Hg µg/l		1	
50	Nickel	Ni µg/l		50	
51	Lead	Pb µg/l		50 (in running water)	Where lead pipes are present, the lead content should not exceed 50 µg/l in a sample taken after flushing. If the sample is taken either directly or after flushing and the lead content either frequently or to an appreciable extent exceeds 100 µg/l, suitable measures must be taken to reduce the exposure to lead on the part of the consumer.
52	Antimony	Sb µg/l		10	
53	Selenium	Se µg/l		10	
54	Vanadium	V µg/l			

55	Pesticides and related products — substances considered separately — total	µg/l		0·1
				
				0·5
					'Pesticides and related products' means:
					— insecticides:
					  — persistent organochlorine compounds
					  — organophosphorous compounds
					  — carbamates
					— herbicides
					— fungicides
					— PCBs and PCTs

56	Polycyclic aromatic hydrocarbons	µg/l		0·2	—reference substances:
					  —fluoranthene
					  —benzo-3, 4-fluoranthene
					  —benzo-11, 12-fluoranthene
					  —benzo-3, 4-pyrene
					  —benzo-1, 12-perylene
					  —indeno (1,2,3-cd) pyrene

E. MICROBIOLOGICAL PARAMETERS

	Parameters	Results: volume of the sample in ml	Guide level (GL)	Maximum admissible concentration (MAC)	
				Membrane filter method	Multiple tube method (MPN)
57	Total coliforms (¹)	100	—	0	MPN<1
58	Fecal coliforms	100	—	0	MPN<1
59	Fecal streptococci	100	—	0	MPN<1
60	Sulphite-reducing Clostridia	20	—	—	MPN≤1

Water intended for human consumption should not contain pathogenic organisms.
If it is necessary to supplement the microbiological analysis intended for human consumption, the samples should be examined not only for the bacteria referred to in Table E but also for pathogens including:
— salmonella,
— pathogenic staphylococci,
— fecal bacteriophages,
— entero-viruses;
nor should such water contain:
— parasites,
— algas,
— other organisms such as animalcules.

(¹) Provided a sufficient number of samples is examined (95% consistent results).

	Parameters	Results: size of sample (in ml)		Guide level (GL)	Maximum admissible concentration (MAC)	Comments
61	Total bacteria counts for water supplied for human consumption	37°C	1	10 (¹)(²)	—	
		22°C	1	100 (¹)(²)	—	
62	Total bacteria counts for water in closed containers	37°C	1	5	20	On their own responsiblity and where parameters 57, 58, 59 and 60 are complied with, and where the pathogen organisms given on page 20 are absent, Member States may process water for their internal use the total bacteria count of which exceeds the MAC values laid down for parameter 62.

MAC values should be measured within 12 hours of being put into closed containers with the sample water being kept at a constant temperature during that 12-hour period. |
| | | 22°C | 1 | 20 | 100 | |

(¹) For disinfected water the corresponding values should be considerably lower at the point where it leaves the processing plant.
(²) If, during succesive sampling, any of these values is consistently exceeded a check should be carried out.

F. MINIMUM REQUIRED CONCENTRATION FOR SOFTENED WATER INTENDED FOR HUMAN COMSUMPTION

	Parameters	Expression of the results	minimum required concen-tration (softened water)	Comments
1	Total hardness	mg/l Ca	60	Calcium or equivalent cations.
2	Hydrogen ion concentration	pH		
3	Alkalinity	mg/l HCO₃	30	The water should not be aggressive.
4	Dissolved oxygen			

NB:— The provisions for hardness, hydrogen ion concentration, dissolved oxygen and calcium also apply to desalinated water.
    — If, owing to its excessive natural hardness, the water is softened in accordance with Table F before being supplied for consumption, its sodium content may, in exceptional cases, be higher than the values given in the 'Maximum admissible concentration' column. However, an effort must be made to keep the sodium content at as low a level as possible and the essential requirements for the protection of public health may not be disregarded.

TABLE OF CORRESPONDENCE BETWEEN THE VARIOUS UNITS OF WATER HARDNESS MEASUREMENT

	French degree	English degree	German degree	Milligrams of Ca	Millimoles of Ca
French degree	1	0·70	0·56	4·008	0·1
English degree	1·43	1	0·80	5.73	0·143
German degree	1·79	1·25	1	7·17	0·179
Milligrams of Ca	0·25	0·175	0·140	1	0·025
Millimoles of Ca	10	7	5·6	40·08	1

ANNEX II

PATTERNS AND FREQUENCY OF STANDARD ANALYSES

A. TABLE OF STANDARD PATTERN ANALYSES (Parameters to be considered in monitoring)

Parameters to be considered	Standard analyses Minimum monitoring (C 1)	Current monitoring (C 2)	Periodic monitoring (C 3)	Occasional monitoring in special situations or in case of accidents (C 4)
A ORGANOLEPTIC PARAMETERS	— Odour (¹) — taste (¹)	— odour — taste — turbidity (appearance)	Current monitoring analyses + other parameters as in footnote 4	The competent national authorities of the Member States will determine the parameters (⁵) according to circumstances, taking account of all factors which might have an adverse effect on the quality of drinking water supplied to consumers.
B PHYSICO-CHEMICAL PARAMETERS	— conductivity or other physico-chemical parameter — residual chlorine (³)	— temperature (²) — conductivity or other physico-chemical parameter — pH — residual chlorine (³)		
C UNDESIRABLE PARAMETERS		— nitrates — nitrites — ammonia		
D TOXIC PARAMETERS				
E MICRO BIOLOGICAL PARAMETERS	— total coliforms or total counts of 22° and 37° — fecal coliforms	— total coliforms — fecal coliforms — total counts of 22° and 37°		

Note: An initial analysis, to be carried out before a source is exploited, should be added. The parameters to be considered would be the current monitoring analyses plus *inter alia* various toxic or undesirable substances presumed present. The list would be drawn up by the competent national authorities.

(¹) Qualitative assessment.

(²) Except for water supplied in containers.

(³) Or other disinfectants and only in the case of treatment.

(⁴) These parameters will be determined by the competent national authority, taking account of all factors which might affect the quality of drinking water supplied to users and which could enable the ionic balance of the constituents to be assessed.

(⁵) The competent national authority may use parameters other than those mentioned in Annex I to this Directive.

Volume of water produced or distributed in m³/day	Population concerned (assuming 200 l/day per person)	Analysis C 1	Analysis C 2	Analysis C 3	Analysis C 4
		Number of samples per year	Number of samples per year	Number of samples per year	
100	500	(¹)	(¹)	(¹)	Frequency to be determined by the competent national authorities as the situation requires
1 000	5 000	(¹)	(¹)	(¹)	
2 000	10 000	12	3	(¹)	
10 000	50 000	60	6	1	
20 000	100 000	120	12	2	
30 000	150 000	180	18	3	
60 000	300 000	360(²)	36	6	
100 000	500 000	360(²)	60	10	
200 000	1 000 000	360(²)	120(²)	20(²)	
1 000 000	5 000 000	360(²)	120(²)	20(²)	

(¹) Frequency left to the discretion of the competent national authorities. However, water intended for the food-manufacturing industries must be monitored at least once a year.

(²) The competent health authorities should endeavour to increase this frequency as far as their resources allow.

(³) (a) In the case of water which must be disinfected, microbiological analysis should be twice as frequent.
(b) Where analyses are very frequent, it is advisable to take samples at the most regular intervals possible.
(c) Where the values of the results obtained from samples taken during the preceding years are constant and significantly better than the limits laid down in Annex I, and where no factor likely to cause a deterioration in the quality of the water has been discovered, the minimum frequencies of the analyses referred to above may be reduced:
— for surface waters, by a factor of 2 with the exception of the frequencies laid down for microbiological analyses;
— for ground waters, by a factor of 4, but without prejudice to the provisions of point (a) above.

Annex 2

Tables of guideline values

The following tables present a summary of guideline values for microorganisms and chemicals in drinking-water. Individual values should not be used directly from the tables. The guideline values must be used and interpreted in conjunction with the information contained in the text and in Volume 2, *Health criteria and other supporting information.*

Table A2.1. Bacteriological quality of drinking-water[a]

Organisms	Guideline value
All water intended for drinking	
E. coli or thermotolerant coliform bacteria[b,c]	Must not be detectable in any 100-ml sample
Treated water entering the distribution system	
E. coli or thermotolerant coliform bacteria[b]	Must not be detectable in any 100-ml sample
Total coliform bacteria	Must not be detectable in any 100-ml sample
Treated water in the distribution system	
E. coli or thermotolerant coliform bacteria[b]	Must not be detectable in any 100-ml sample
Total coliform bacteria	Must not be detectable in any 100-ml sample. In the case of large supplies, where sufficient samples are examined, must not be present in 95% of samples taken throughout any 12-month period

[a] Immediate investigative action must be taken if either E. coli or total coliform bacteria are detected. The minimum action in the case of total coliform bacteria is repeat sampling; if these bacteria are detected in the repeat sample, the cause must be determined by immediate further investigation.

[b] Although E. coli is the more precise indicator of faecal pollution, the count of thermotolerant coliform bacteria is an acceptable alternative. If necessary, proper confirmatory tests must be carried out. Total coliform bacteria are not acceptable indicators of the sanitary quality of rural water supplies, particularly in tropical areas where many bacteria of no sanitary significance occur in almost all untreated supplies.

[c] It is recognized that, in the great majority of rural water supplies in developing countries, faecal contamination is widespread. Under these conditions, the national surveillance agency should set medium-term targets for the progressive improvement of water supplies, as recommended in Volume 3 of Guidelines for drinking-water quality.

Table A2.2. Chemicals of health significance in drinking-water

A. Inorganic constituents

	Guideline value (mg/litre)	Remarks
antimony	0.005 (P)[a]	
arsenic	0.01[b](P)	For excess skin cancer risk of 6×10^{-4}
barium	0.7	
beryllium		NAD[c]
boron	0.3	
cadmium	0.003	
chromium	0.05 (P)	
copper	2 (P)	ATO[d]
cyanide	0.07	
fluoride	1.5	Climatic conditions, volume of water consumed, and intake from other sources should be considered when setting national standards
lead	0.01	It is recognized that not all water will meet the guideline value immediately; meanwhile, all other recommended measures to reduce the total exposure to lead should be implemented
manganese	0.5 (P)	ATO
mercury (total)	0.001	
molybdenum	0.07	
nickel	0.02	
nitrate (as NO_3^-)	50	The sum of the ratio of the concentra-
nitrite (as NO_2^-)	3 (P)	tion of each to its respective guideline value should not exceed 1
selenium	0.01	
uranium		NAD

B. Organic constituents

	Guideline value (μg/litre)	Remarks
Chlorinated alkanes		
carbon tetrachloride	2	
dichloromethane	20	
1,1-dichloroethane		NAD
1,2-dichloroethane	30[b]	for excess risk of 10^{-5}
1,1,1-trichloroethane	2000 (P)	
Chlorinated ethenes		
vinyl chloride	5[b]	for excess risk of 10^{-5}
1,1-dichloroethene	30	
1,2-dichloroethene	50	
trichloroethene	70 (P)	
tetrachloroethene	40	
Aromatic hydrocarbons		
benzene	10[b]	for excess risk of 10^{-5}
toluene	700	ATO
xylenes	500	ATO
ethylbenzene	300	ATO
styrene	20	ATO
benzo[a]pyrene	0.7[b]	for excess risk of 10^{-5}
Chlorinated benzenes		
monochlorobenzene	300	ATO
1,2-dichlorobenzene	1000	ATO
1,3-dichlorobenzene		NAD
1,4-dichlorobenzene	300	ATO
trichlorobenzenes (total)	20	ATO
Miscellaneous		
di(2-ethylhexyl)adipate	80	
di(2-ethylhexyl)phthalate	8	
acrylamide	0.5[b]	for excess risk of 10^{-5}
epichlorohydrin	0.4 (P)	
hexachlorobutadiene	0.6	
edetic acid (EDTA)	200 (P)	
nitrilotriacetic acid	200	
dialkyltins		NAD
tributyltin oxide	2	

C. Pesticides

	Guideline value (μg/litre)	Remarks
alachlor	20^c	for excess risk of 10^{-5}
aldicarb	10	
aldrin/dieldrin	0.03	
atrazine	2	
bentazone	30	
carbofuran	5	
chlordane	0.2	
chlorotoluron	30	
DDT	2	
1,2-dibromo-3-chloropropane	1^c	for excess risk of 10^{-5}
2,4-D	30	
1,2-dichloropropane	20 (P)	
1,3-dichloropropane		NAD
1,3-dichloropropene	20^c	for excess risk of 10^{-5}
ethylene dibromide		NAD
heptachlor and heptachlor epoxide	0.03	
hexachlorobenzene	1^c	for excess risk of 10^{-5}
isoproturon	9	
lindane	2	
MCPA	2	
methoxychlor	20	
metolachlor	10	
molinate	6	
pendimethalin	20	
pentachlorophenol	9 (P)	
permethrin	20	
propanil	20	
pyridate	100	
simazine	2	
trifluralin	20	
chlorophenoxy herbicides other than 2,4-D and MCPA		
2,4-DB	90	
dichlorprop	100	
fenoprop	9	
MCPB		NAD
mecoprop	10	
2,4,5-T	9	

D. Disinfectants and disinfectant by-products

Disinfectants	Guideline value (mg/litre)	Remarks
monochloramine	3	
di- and trichloramine		NAD
chlorine	5	ATO. For effective disinfection there should be a residual concentration of free chlorine of $\geqslant 0.5$ mg/litre after at least 30 minutes contact time at pH < 8.0
chlorine dioxide		A guideline value has not been established because of the rapid breakdown of chlorine dioxide and because the chlorite guideline value is adequately protective for potential toxicity from chlorine dioxide
iodine		NAD

Disinfectant by-products	Guideline value (μg/litre)	Remarks
bromate	25^{b} (P)	for 7×10^{-5} excess risk
chlorate		NAD
chlorite	200 (P)	
chlorophenols		
2-chlorophenol		NAD
2,4-dichlorophenol		NAD
2,4,6-trichlorophenol	200^{c}	for excess risk of 10^{-5}, ATO
formaldehyde	900	
MX		NAD
trihalomethanes		The sum of the ratio of the concentration of each to its respective guideline value should not exceed 1
bromoform	100	
dibromochloromethane	100	
bromodichloromethane	60^{b}	for excess risk of 10^{-5}
chloroform	200^{b}	for excess risk of 10^{-5}
chlorinated acetic acids		
monochloroacetic acid		NAD
dichloroacetic acid	50 (P)	
trichloroacetic acid	100 (P)	
chloral hydrate (trichloroacetaldehyde)	10 (P)	
chloroacetone		NAD

Disinfectant by-products	Guideline value (μg/litre)	Remarks
halogenated acetonitriles		
dichloroacetonitrile	90 (P)	
dibromoacetonitrile	100 (P)	
bromochloroacetonitrile		NAD
trichloroacetonitrile	1 (P)	
cyanogen chloride (as CN)	70	
chloropicrin		NAD

[a] (P) — Provisional guideline value. This term is used for constituents for which there is some evidence of a potential hazard but where the available information on health effects is limited, or where an uncertainty factor greater than 1000 has been used in the derivation of the tolerable daily intake (TDI). Provisional guideline values are also recommended: (1) for substances for which the calculated guideline value would be below the practical quantification level, or below the level that can be achieved through practical treatment methods; or (2) where disinfection is likely to result in the guideline value being exceeded.

[b] For substances that are considered to be carcinogenic, the guideline value is the concentration in drinking-water associated with an excess lifetime cancer risk of 10^{-5} (one additional cancer per 100 000 of the population ingesting drinking-water containing the substance at the guideline value for 70 years). Concentrations associated with estimated excess lifetime cancer risks of 10^{-4} and 10^{-6} can be calculated by multiplying and dividing, respectively, the guideline value by 10.

In cases in which the concentration associated with an excess lifetime cancer risk of 10^{-5} is not feasible as a result of inadequate analytical or treatment technology, a provisional guideline value is recommended at a practicable level and the estimated associated excess lifetime cancer risk presented.

It should be emphasized that the guideline values for carcinogenic substances have been computed from hypothetical mathematical models that cannot be verified experimentally and that the values should be interpreted differently than TDI-based values because of the lack of precision of the models. At best, these values must be regarded as rough estimates of cancer risk. However, the models used are conservative and probably err on the side of caution. Moderate short-term exposure to levels exceeding the guideline value for carcinogens does not significantly affect the risk.

[c] NAD — No adequate data to permit recommendation of a health-based guideline value.

[d] ATO — Concentrations of the substance at or below the health-based guideline value may affect the appearance, taste, or odour of the water.

Table A2.3. Chemicals not of health significance at concentrations normally found in drinking-water

Chemical	Remarks
asbestos	U
silver	U
tin	U

U — It is unnecessary to recommend a health-based guideline value for these compounds because they are not hazardous to human health at concentrations normally found in drinking-water

Table A2.4. Radioactive constituents of drinking-water

	Screening value (Bq/litre)	Remarks
gross alpha activity	0.1	If a screening value is exceeded, more detailed radionuclide analysis is necessary. Higher values do not necessarily imply that the water is unsuitable for human consumption
gross beta activity	1	

Table A2.5. Substances and parameters in drinking-water that may give rise to complaints from consumers

	Levels likely to give rise to consumer complaints[a]	Reasons for consumer complaints
Physical parameters		
colour	15 TCU[b]	appearance
taste and odour	—	should be acceptable
temperature	—	should be acceptable
turbidity	5 NTU[b]	appearance; for effective terminal disinfection, median turbidity ⩽1NTU, single sample ⩽5NTU
Inorganic constituents		
aluminium	0.2 mg/l	depositions, discoloration
ammonia	1.5 mg/l	odour and taste
chloride	250 mg/l	taste, corrosion
copper	1 mg/l	staining of laundry and sanitary ware (health-based provisional guideline value 2 mg/litre)
hardness	—	high hardness: scale deposition, scum formation low hardness: possible corrosion
hydrogen sulfide	0.05 mg/l	odour and taste
iron	0.3 mg/l	staining of laundry and sanitary ware
manganese	0.1 mg/l	staining of laundry and sanitary ware (health-based provisional guideline value 0.5 mg/litre)
dissolved oxygen	—	indirect effects
pH	—	low pH: corrosion high pH: taste, soapy feel preferably <8.0 for effective disinfection with chlorine
sodium	200 mg/l	taste
sulfate	250 mg/l	taste, corrosion
total dissolved solids	1000 mg/l	taste
zinc	3 mg/l	appearance, taste
Organic constituents		
toluene	24–170 µg/l	odour, taste (health-based guideline value 700 µg/l)
xylene	20–1800 µg/l	odour, taste (health-based guideline value 500 µg/l)
ethylbenzene	2–200 µg/l	odour, taste (health-based guideline value 300 µg/l)
styrene	4–2600 µg/l	odour, taste (health-based guideline value 20 µg/l)

	Levels likely to give rise to consumer complaints[a]	Reasons for consumer complaints
monochlorobenzene	10–120 μg/l	odour, taste (health-based guideline value 300 μg/l)
1,2-dichlorobenzene	1–10 μg/l	odour, taste (health-based guideline value 1000 μg/l)
1,4-dichlorobenzene	0.3–30 μg/l	odour, taste (health-based guideline value 300 μg/l)
trichlorobenzenes (total)	5–50 μg/l	odour, taste (health-based guideline value 20 μg/l)
synthetic detergents	–	foaming, taste, odour
Disinfectants and disinfectant by-products		
chlorine	600–1000 μg/l	taste and odour (health-based guideline value 5 mg/l)
chlorophenols		
2-chlorophenol	0.1–10 μg/l	taste, odour
2,4-dichlorophenol	0.3–40 μg/l	taste, odour
2,4,6-trichlorophenol	2–300 μg/l	taste, odour (health-based guideline value 200 μg/l)

[a] The levels indicated are not precise numbers. Problems may occur at lower or higher values according to local circumstances. A range of taste and odour threshold concentrations is given for organic constituents.

[b] TCU, time colour unit.

[c] NTU, nephelometric turbidity unit.

CONVERSION FACTORS

Multiply Metric Unit	By	To obtain English Unit
Length		
kilometre (km)	0.6214	mile
metre (m)	1.0936	yard
centimetre (cm)	0.0328	foot
millimetre (mm)	0.03937	inch
Area		
sq kilometre (km²)	0.3861	square mile
hectare (ha)	2.471	acre
sq metre (m²)	10.764	square foot
sq metre (m²)	1550	square inch
sq centrimetre (cm²)	0.1550	square inch
Volume		
cu centimetre (cm³)	0.061	cubic inch
cu metre (m³)	1.308	cubic yard
litre (L)	61.02	cubic inch
litre (L)	0.001308	cubic yard
litre (L)	0.2642	US gallon
litre (L)	0.22	Imperial gallon
Weight		
metric tonne (t)	0.984	long ton
metric tonne (t)	1.102	short ton
kilogram (kg)	2.205	pound, avdp
gram (g or gr)	0.0353	ounce, avdp
Other		
cu centimetre (cm³)	0.0338	fluid ounce
kilograms/sq cm (kg/cm²)	14.225	pounds/sq in
metric horsepower (CV)	0.9863	hp
kilowatt (KW)	1.341	hp
bar	14.5	psi
Flow		
megaliters/day (ml/d)	0.264	million gallons/day (MGD)

Temperature

$^{\circ}C = 5/9 \ (^{\circ}F - 32)$

Viscosity of Water

Temperature (°F)	32	50	60	70	80	100	120
Temperature (°C)	0	10	15.6	21.1	26.7	37.8	48.9
Kin. Visc. (CS)	1.79	1.31	1.12	.98	.86	.69	.57

CONVERSION FACTORS

Multiply English Unit	By	To obtain Metric Unit
Length		
mile statute (m)	1.609	kilometre
yard (yd)	0.9144	metre
foot (ft)	0.3048	metre
inch (in)	25.4	millimetre
Area		
sq mile (mile²)	2.590	sq kilometre
acre	0.4047	hectare
sq foot (ft²)	0.0929	sq metre
sq inch (in²)	0.000645	sq metre
Volume		
cu yard (yd³)	0.7645	cu metre
cu inch (in³)	16.387	cu centimetre
cu foot (ft³)	0.0283	cu metre
cu inch (in³)	0.0164	litre
cubic yard (yd³)	764.55	litre
US gallon (US Gal)	3.785	litre
US gallon	0.833	Imperial gallon
Weight		
long ton (lg ton)	1.016	metric ton
short ton (sh ton)	0.907	metric ton
pound (lb)	0.4536	kilogram
ounce (oz)	28.35	gram
Other		
fluid oz (fl oz)	29.57	cu centimetre
pounds/sq in	0.0703	kilogram/sq cm
psi	0.0689	bar
horsepower (hp)	1.014	metric horsepower
horsepower (hp)	0.7457	kilowatt
Flow		
million gallons/day	3.785	megaliters/day (Ml/d)

Temperature

°F = 9/5 (°C + 32)

Index